The Recombinant DNA Controversy

A Memoir

Science, Politics, and the Public Interest
1974–1981

The Recombinant DNA Controversy

A Memoir

Science, Politics, and the Public Interest
1974–1981

Donald S. Fredrickson, M.D.
Former Director, National Institutes of Health

The Recombinant DNA Controversy

A Memoir

Science, Politics, and the Public Interest
1974–1981

Donald S. Fredrickson, M.D.

Scholar, National Library of Medicine, Bethesda, Maryland
Former Director, National Institutes of Health, Bethesda, Maryland

Washington, D.C.

American Society for Microbiology
1752 N Street, N.W.
Washington, DC 20036-2904

Library of Congress Cataloging-in-Publication Data

Fredrickson, Donald S.
The recombinant DNA controversy : a memoir science, politics, and the public interest 1974–1981 / Donald S. Fredrickson.
p. cm.
Includes bibliographical references and index.
ISBN 1-55581-222-8
1. Genetic engineering. 2. Recombinant DNA. 3. Recombinant DNA—Government policy—United States—History—20th century. 4. National Institutes of Health (U.S.). Recombinant DNA Advisory Committee. I. Title.

QH442 .F74 2001
660.6′5—dc21

00-067542

Address editorial correspondence to: ASM Press, 1752 N St., N.W., Washington, DC 20036-2804, U.S.A.

Send orders to: ASM Press, P.O. Box 605, Herndon, VA 20172, U.S.A.
Phone: 800-546-2416; 703-661-1593
Fax: 703-661-1501
Email: books@asmusa.org
Online: www.asmpress.org

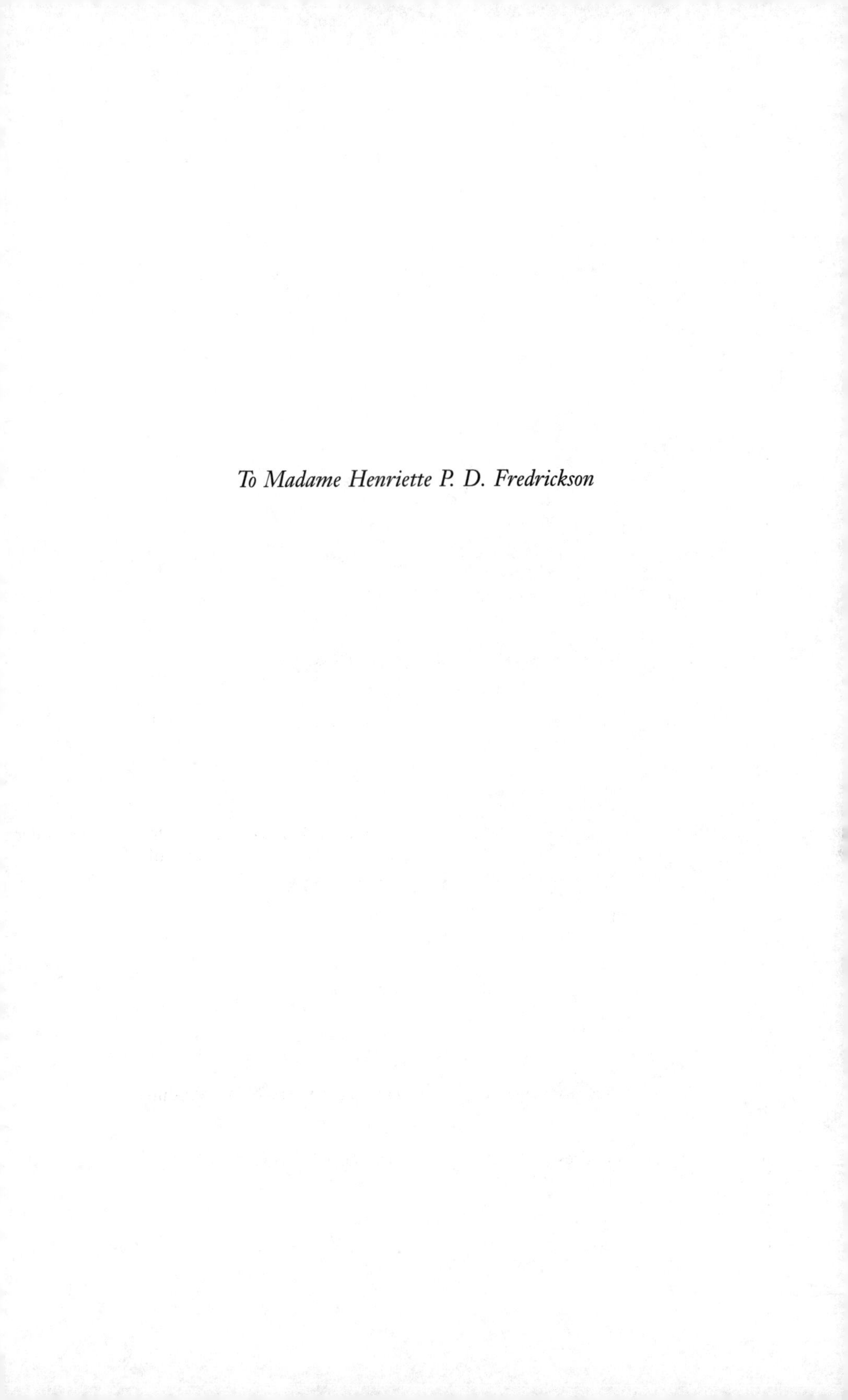

To Madame Henriette P. D. Fredrickson

Contents

Foreword

Donald Fredrickson's memoir provides a unique perspective on the recombinant DNA debate in the United States, especially during the years 1975 through 1981. As director of the National Institutes of Health (NIH), his perspective was that of the leading executive-branch policymaker for recombinant DNA research during this period. In the following pages we see the diverse roles played by the NIH director, other scientists, members of Congress, the Secretary of the U.S. Department of Health, Education and Welfare, multiple federal agencies, business leaders, experts in law and ethics, public-interest advocates, and members of the general public. In addition to mediating among these domestic individuals and groups, the NIH director found it necessary to remain in touch with scientific peers in other nations. The scientific and political leaders of all industrialized countries were wrestling with their responses to this exciting but potentially hazardous new type of research.

The central section of the book (chapters 1 through 10) focuses on one 3½-day meeting, the Asilomar Conference held in February of 1975, and one group that carried forward the work of the Asilomar meeting, the NIH Recombinant DNA Advisory Committee (RAC). This meeting and this group should be considered together because each complemented the other. The conference reached a consensus on reasonable initial rules for conducting recombinant DNA research (and for deferring certain kinds of experiments), while the RAC continued the work of Asilomar, refining the initial guidelines and revising them in the light of further research.

From my perspective as a three-term member of the RAC, the Asilomar meeting was an important and highly constructive endeavor. At the meeting, leading members of the scientific community sought to assess the potential hazards of the research they were planning to conduct. One can quibble about details of the meeting—for example, the narrowness of the invitation list and the meeting's semi-private character. However, in examining the trees one should not lose sight of the forest. An 18-month process of discussion and deliberation, led by Maxine Singer and Paul Berg among others, culminated in an international scientific summit. At the summit a provisional consensus was reached about the most prudent course for the early years of recombinant DNA research. The planners of the summit meeting also proposed that an ongoing advisory committee be created to oversee future research in the field.

For its part, the RAC met for the first time immediately after the Asilomar meeting and devoted its first year to developing an initial set of "Guidelines for Recombinant DNA Molecule Research." After a public review of the draft guidelines, in which I was privileged to take part, they were published in July 1976. During the next 2½ years an increasingly interdisciplinary RAC struggled to revise the 1976 Guidelines in the light of data emerging from the conduct of the research. Meanwhile, legislators in the United States Congress and in some municipalities considered whether the RAC and its NIH-based staff were sufficient to oversee the rapidly expanding field of recombinant DNA research. In the end, a new social compact was implicitly adopted, and, beginning in 1979, an expanded RAC that represented a wider variety of perspectives was permitted to continue as the central oversight body. By about 1982 the task foreseen for the RAC at Asilomar had been essentially completed, and the few vestiges of oversight that remained had been delegated to local institutions.

Did scientists like Maxine Singer and Paul Berg make a mistake in calling attention to the potential hazards of laboratory research with recombinant DNA? With the benefit of hindsight, one can argue that recombinant DNA research turned out not to pose unique dangers. However, the knowledge that was available to the scientific community in 1973 or 1975 gave no firm basis for estimating the harms that this new type of research might cause to laboratory workers or the environment. The potential hazards to laboratory workers of research with infectious microbes and oncogenic viruses were well known, as were the risks of research involving radiation or toxic chemicals. Thus, careful thought and interdisciplinary deliberation before proceeding with at least certain types of experiments seems, even today, to have been a prudent course. In addition, the scientific community's willingness to raise safety questions in the absence of any demonstrated harm enhanced public confidence that in the future scientists also would level with the public in discussing the potential benefits and harms of new lines of research. (This thoughtful warning mode contrasts starkly with the overly optimistic reassurances that the public was given about the purity of the blood supply during the early years of the AIDS epidemic, or about the safety of eating beef despite increasing evidence that mad cow disease was somehow being transmitted from cattle to humans.)

Were the early research guidelines that emerged from Asilomar and the RAC too conservative, too stringent? Again, hindsight increases confidence. By the beginning of 1979 even one of the earliest proponents of caution, Maxine Singer, was expressing concern that the Guidelines were not being relaxed quickly enough and that the RAC was beginning to resemble a "ponderous" regulatory agency.[1] However, the major problem in the years 1976 to 1978 may not have been the initial conservatism of the Guidelines as much as it was the lack of a clearly articulated process for their timely revision. Because of this oversight in the 1976 Guidelines, proposals for revision themselves may have seemed like potential violations of an unchanging (or unchangeable) code. An activist Secretary of Health, Education and Welfare and members of Congress who seemed ready to create a new regulatory agency may also have slowed the revision process. However, the delays of 1978 and 1980 gave way to a rapid relaxation of the Guidelines in the early 1980s, and oversight policies in multiple industrialized countries seem to have been relaxed roughly in parallel. In toto, recombinant DNA research is not likely to have been set back by more than 2 or 3 years at the most by Asilomar and the RAC, and the caution that caused this delay was, at least in part, arguably a prudent response to reasonable concerns about possible biohazards.

Was NIH the most appropriate agency for overseeing laboratory research with recombinant DNA in the late 1970s? While there are theoretical advantages to separating the distinct functions of funding research, on the one hand, and providing oversight for research, on the other, the ad hoc arrangement suggested by the planners of the Asilomar meeting seems to have worked reasonably well for recombinant DNA research. Several factors may have contributed to the apparent success of this arrangement. First, most funding for recombinant DNA research in the 1970s was federal, with the lion's share being contributed by NIH and the National Science Foundation. Thus, a uniform set of federal research guidelines was likely to be adhered to by researchers who knew that failure to act in accordance with the NIH Guidelines could jeopardize their research funding. Second, the NIH Office of Recombinant DNA Activities (ORDA) and the RAC were transparent in all of their early actions, meeting in public and publishing large volumes of material in a public record. Third, the NIH director was strongly supportive of the RAC, meeting with the committee regularly to brief RAC members on recent research developments or his thinking about their role; at the same time, he allowed the committee complete independence in its work. Fourth, in response to gentle prodding by the Secretary of Health, Education and Welfare, the NIH director expanded and diversified the membership of the RAC in late 1978. With this expansion the RAC came to be seen as broadly representative of the spectrum of public opinion on recombinant DNA research. Finally, the staff members at ORDA were dedicated public servants, and most RAC members took their oversight responsibilities very seriously.

The last three chapters of this memoir (chapters 11 through 13) cover scientific and public policy developments in 1980 and early 1981, as well as providing the author's overall perspective on the recombinant DNA research controversy. During this time the RAC took on a quasi-regulatory role vis-à-vis the private sector, formulating guidance on safety standards for large-scale production using recombinant DNA techniques and reviewing proposals voluntarily submitted by companies like Genentech and Eli Lilly. For the first time, RAC members were confronted with decisions about whether to go into executive session to review information that companies wished to maintain as proprietary. As NIH director, Don Fredrickson encouraged RAC members to accept this role for the public good until other agencies—in this case, the Food and Drug Administration (FDA)—could develop their own oversight capabilities. In retrospect, Dr. Fredrickson expresses satisfaction that the RAC helped to prevent delays in the transition from the laboratory to the manufacturing plant.

The year 1980 was critical for the RAC and NIH in another way. Martin Cline of UCLA conducted an unauthorized attempt to use recombinant DNA techniques to treat two patients, one in Israel and one in Italy, who were afflicted with thalassemia.[2] Rather than asking the RAC to perform an investigative and judicial role, the NIH director appointed an ad hoc committee composed of NIH employees. When the investigation verified that both the "Guidelines for Recombinant DNA Research" and federal regulations for the protection of human subjects had been violated by Dr. Cline, Don Fredrickson did not hesitate to impose the rather harsh penalties recommended by the ad hoc committee. Dr. Cline's current NIH grants were affected, as was his ability to secure new NIH grants during the following 3 years.

In 1980 and 1981 the NIH director and the RAC itself also confronted serious proposals to abolish the RAC and its supporting organization, ORDA, and to transform the remaining mandatory Guidelines into a voluntary code of practice. Among the proponents of these changes were some of the organizers of the Asilomar Conference, who 7 years earlier had urged caution in research with recombinant DNA and had strongly supported the creation of the RAC. Don Fredrickson and the new RAC Chair, former Congressman Ray Thornton, advocated and indeed implemented a more moderate approach: further relaxation of the Guidelines and the retention of the RAC as an oversight body that could provide expert advice to the scientific community and that would simultaneously reassure the public and policymakers. In this connection, the NIH director proposed another metamorphosis for the RAC, into a "third-generation" body. The first-generation RAC had been predominantly scientific and had relied on the more broadly constituted NIH Director's Advisory Committee (DAC) for additional perspective. In contrast, the second-generation RAC of late 1978 and 1979 included both scientific and socially oriented viewpoints within the RAC itself. In 1982, shortly after leaving the directorship of NIH, Don Fredrickson called for another transition in the role of the RAC, into an even more inclusive body that would be "better equipped to deal with the emerging problems" while simultaneously being "relieved of some of the detailed burden of reviewing minor administrative concerns."[3]

The next transformation of the RAC took place shortly after this proposal, but the change occurred in response to a report by a presidential advisory commission on bioethics and a new turn in biomedical research—the use of recombinant DNA techniques for human gene transfer (also called "human gene therapy"). In November of 1982 the President's Commission on Bioethics released a report entitled *Splicing Life*.[4] The commission's report sought to de-dramatize the dangers posed by "human genetic engineering" by pointing to similarities between gene transfer for therapeutic reasons, on the one hand, and traditional drugs and biologics, on the other. In the final chapter of its report, the President's Commission also discussed alternative oversight strategies for the emerging field of human gene transfer research. One of three options considered by the commission was the possibility of "revising RAC," adapting its goals and membership to prepare the committee for the new task at hand. In this connection, the commission explicitly referred to Don Fredrickson's notion of creating a "third-generation RAC."[5]

During the next 2 years, under the leadership of a new chair, Robert Mitchell, the RAC considered whether it should accept responsibility for reviewing human gene transfer protocols on a case-by-case basis. In incremental steps RAC members accepted this responsibility in principle, then created a Working Group (later Subcommittee) on Human Gene Therapy to perform the initial review of protocols on the RAC's behalf. In late 1984 and the first half of 1985 this working group developed a set of guidelines called "The Points to Consider"[6] that served as the framework for evaluating human gene transfer protocols. I had the privilege of chairing the working group during the 7 years of its existence.

In the Epilogue to his memoir (chapter 13), Don Fredrickson recounts several important moments in this new phase of the RAC's activity. He briefly reports on the first authorized human gene transfer experiment, which was performed at NIH in 1990 by R. Michael Blaese, W. French Anderson, and their colleagues. He also acknowledges the FDA's acquisition of formidable expertise in cell and gene therapy in the early 1990s but notes that the RAC review process for gene-transfer protocols was a matter of public record, while FDA's reviews of Investigational New Drug (IND) applications occur behind closed doors. With a trace of sadness the author reviews the attempt by NIH director Harold Varmus to abolish the RAC in 1996 and the opposition to the proposed abolition that was expressed by a variety of individuals and groups, including the American Society for Microbiology, "the largest single life science society in the world."[7] With regret Don Fredrickson carries his account forward to 1999 and to the death of Jesse Gelsinger, a relatively healthy participant in a University of Pennsylvania protocol focused on ornithine transcarbamylase (OTC) deficiency. He comments that, in the wake of this subject's death, new and more stringent rules for all research with human subjects are likely to be enacted—especially regarding researchers' financial involvement in the research they are conducting.

Dr. Fredrickson's reference to the financial dimension of biomedical research reminds us quite forcefully of the major changes in the context of this research that occurred between 1975 and 2000, to choose two convenient dates. These breathtaking changes are reflected at several points in this memoir. For example, in 1976, Don Fredrickson notes, there was "the patent"—the Boyer-Cohen patent, held by Stanford University and the University of California on one of the major techniques for combining DNA from different organisms. He notes that this patent was governed by one of 167 Institutional Patent Agreements (IPAs) that had been reached between the Department of Health, Education and Welfare and universities.[8] Under the terms of an IPA, a university could grant an exclusive license to a third party only if it could demonstrate that a nonexclusive license was infeasible. Further, the IPA stipulated that the federal government must be granted a license for use of an invention for research purposes at no cost. While acknowledging the importance of private enterprise for translating researching findings into useful products, Don Fredrickson notes somewhat ruefully that the *Diamond v. Chakrabarty* decision by the U.S. Supreme Court (1980) and the Bayh-Dole Act of 1980 have contributed in major ways to a paradigm shift in biomedical research. Today most innovations in biomedical research, including genes, are aggressively patented by both private companies and academic institutions. Quite clearly the author is concerned that the pendulum may have swung too far in the entrepreneurial direction, to the detriment of basic science.[9]

The sources of funding for biomedical research and development have changed radically since the 1970s, as Don Fredrickson's narrative suggests. According to figures compiled by the NIH, federal expenditures for medical and health-related research and development nearly doubled from $6.8 billion to $13.4 billion between 1986 and 1995.[10] During the same period, industry expenditures for biomedical and health-related research more than tripled, growing from $6.2 billion in 1986 to $18.6 billion.[11] In one part of industry, the segment represented by pharmaceutical companies, the rate of growth in research and development expenditures has been even more dramatic. According to the industry's trade association, PhRMA, pharmaceutical company investments in this sphere have increased from $1.5 billion in 1980—the year of the Cline experiment—to $22.4 billion in 2000.[12]

Is there a role for the RAC and other RAC-like oversight bodies in this new era when private investment in biomedical research and development will clearly outstrip public funding? Or would it be better to have all oversight of this research focused in regulatory agencies, in particular, the FDA? One's answer to these questions will depend largely on one's overall regulatory and political philosophy. If the primary goal of biomedical research and development is to speed useful products to the market and to preserve one nation's competitive position in the global economy, then minimal regulation and a tilt toward approval would seem to be the appropriate regulatory stance. However, if one considers the informing of the public about new biomedical developments and the protection of human subjects in clinical trials to be equally important goals, then a more transparent and inclusive oversight system may be required. From my perspective, the RAC in its early years of overseeing recombinant DNA research and human gene transfer research provides an excellent model for the responsible introduction of a new technology. Again, in my own view, this model is eminently applicable to other emerging fields of biomedical research, for example, xenotransplantation. Where the RAC and parallel advisory committees should be located within the structure of the federal government is, of course, a different question. In the future, there are at least good theoretical arguments for separating the oversight of research from its funding and for stipulating that the RAC and similar bodies should report to a Cabinet secretary or even to an independent agency dedicated to the promotion of ethically responsible research.[13]

One of the most striking statements in Don Fredrickson's memoir appears in chapter 4. There he comments, "Although the directorship of NIH was itself a full-time job, I estimated later that I had to devote at least half of my time to recombinant DNA during 1976–78."[14] This level of commitment to an exciting but potentially hazardous field of research was by no means required of the NIH director. Prior and later directors of NIH would undoubtedly have handled the circumstances of 1976 differently, perhaps deferring to the will of Congress, to the political instincts of the Secretary of Health, Education and Welfare, or to the regulatory authority of the FDA or the EPA. Instead, Don Fredrickson immersed himself in the day-to-day activities required to protect scientific freedom—a freedom that was, from the beginning and almost without exception, exercised in a socially responsible manner. He also nurtured a fledgling advisory committee, the RAC, maintaining regular communication with its members and helping the committee to adapt to changing scientific and social circumstances.

The payoff from this investment of time and energy, both within the United States and in the scientific community around the world, was enormous. Recombinant DNA research techniques were introduced into the world's laboratories in a thoughtful, cautious, and ultimately innocuous way in full view of the public and public policymakers. We owe a great debt of gratitude to the author of this book both for his dedicated public service during those critical months and years and for this vivid account of the major steps in the policy-making process.

LeRoy Walters
Kennedy Institute of Ethics
Georgetown University
Washington, D.C.
December 2000

Preface

I remember the Asilomar Conference as an event both exciting and confusing. Exciting because of the scale of the scientific adventure, the great expanses which had opened to research, and because no one could be indifferent to the debate over the powers and responsibilities of scientists. Confusing because some of the basic questions could only be dealt with in great disorder, or not confronted at all. On the frontiers of the unknown, the analysis of benefits and hazards was locked up in concentric circles of ignorance . . . how could one determine the reality . . . without experimenting . . . without taking a minimum of risk?[1]

PHILIPPE KOURILSKY

The controversy over recombinant DNA broke out suddenly over 30 years ago, when it was discovered that genes from different species of bacteria could be recombined in the laboratory. Fear of the potential hazards quickly grew, and in 1974 a small group of American academic molecular biologists called for a worldwide moratorium on such experiments until the risks could be assessed. In February 1975, 150 scientists from the world's premier laboratories convened for three days at Asilomar Conference Center in Pacific Grove, Calif., to study the problem and formulate an approach to its solution. The conference voted to replace the moratorium with a complicated scheme of rules for containment and restriction of research, which severely limited experimentation and paradoxically hobbled determination of the actual risks. Prior to Asilomar, the conference organizers, led by Paul Berg, had also requested that the National Institutes of Health (NIH) establish guidelines for all research with recombinant DNA.

On the 25th anniversary of Asilomar, in February 2000, I visited the conference site for the first time. In the company of many scientists who had earlier been conferees, I walked along the beach, poked my head into the chapel, and listened to the luncheon bell which had ended that first gathering so long ago. For me there was a special meaning in the occasion, for I had just completed this memoir of how the Asilomar meeting had touched my scientific career, setting the daily calendar and demanding my every attention for six eventful years. I became the director of NIH in July 1975 and unsuspectingly inherited the job of guiding the recombinant DNA controversy through its first exciting, tormented years. The issuance, evolution, and adaptation of the *NIH Guidelines for Recombinant DNA Research* became the focus of more than a decade of suspicion of this audacious new science. This book attempts to describe the actions which NIH and a new Recombinant DNA Advisory Committee (the RAC)—under the careful watch of the scientific community, the government, and skeptical members of the public—undertook to win society's acceptance of this new technology while keeping the science moving cautiously forward.

When the lay public realized the extent of the strange new dangers discussed at Asilomar, the tolerance of many critics suddenly took a turn for the worse. Laymen, scientists, and legislators, on one side or the other, engaged in an angry struggle over the resumption of research and the rules *established by scientists* to control it. Some prominent scientists warned that the new power to join pieces of genes from different sources would create chimeric products that could seed into niches in the environment and possibly spread new diseases beyond control. As expected, and as it should, society reacted. Many hearings, demonstrations, forums, and town meetings were held. In townships, state legislatures, and the U.S. Congress, bills to govern laboratory research were drafted and debated at length. Injunctions to forbid all such experimentation were sought in the courts. More than half a decade of recriminations and anxiety passed before society and biomedical science gingerly patched up the largest rents in their mutually beneficial entente.

Why did this happen? Could it have been avoided? Can we be sure that such a threat to a relationship necessary for the advancement of our civilization will not happen again?

The purpose of this memoir is not to re-tell the story of Asilomar, but to place in context all that subsequently happened. Because I inherited principal responsibility for the *NIH Guidelines for Recombinant DNA Research* that grew from the Asilomar meeting, I became the federal officer answerable for protection of the public welfare as well as the furtherance of the scientific research that had come abruptly to a halt. As such I was the principal spokesman in Congress, and the focal point of attention of the secretary and the hierarchy of the Department of Health, Education and Welfare, on all matters of fear and uncertainty created by recombinant DNA.

Most of what all of us did in that atmosphere of crisis to fulfill our public duties and to preserve the nation's capacity for preeminence in biomedical research has never been published. Thus our successes and our errors have been unavailable for such instruction as they might hold of how best, in the future, to help preserve intact the interface between high science and a powerful government. I attempt here to lay out the roots of that vital relationship as it involved NIH, the nation's single most important biomedical research agency.

Fortunately, great pains were taken to maintain from the beginning most of the vast archives of hearings records, correspondence, and documents relative to our actions.[2] In addition, I preserved a thorough record of my own activity, including extensive diaries covering this period. Across the pages of this memoir move numerous personalities from microbiology, molecular biology, and other scientific disciplines, as well as the leaders among Congress, the administration, and government agencies, environmentalists, and many others who had a role at this time of testing.

At the moment of the Asilomar meeting, the modern world was entering a phase of transition, evolving toward a society in which the once arcane discipline of molecular biology was swiftly becoming a significant force in medicine, commerce, ethics, sociology, politics, and the very nature of science itself. The initial phase of this transition was taken up with determining how dangerous was the new technology and informing the public of every step by a totally open process.

With the booming development of a whole new culture of genomics and medicine, the early fears of physical danger have disappeared, to be succeeded by new controversies—many involving serious moral and ethical issues. The basic scientists, their government sponsorship—joined now by commerce—and the public are striving to preserve a workable social contract. This book describes in detail the earliest of such endeavors, a serious and prolonged struggle that set the stage for more extraordinary times to come.

Donald S. Fredrickson, M.D.

Acknowledgments

I am grateful to Dr. Donald A. Lindberg, director of the National Library of Medicine (NLM), for generous access to the resources and the invaluable assistance of many of the staff of NLM in the preparation of this book. The photos in the insert are from NLM archives. I am also fortunate for the assignment by ASM Press of Mary McKenney as copyeditor, in company with the production oversight of Ellie Tupper. Special acknowledgment is due Dr. Bernard Talbot for both encouragement and redaction during the long period of work on this manuscript, especially its extensive annotation.

1

Asilomar: the End of the Beginning†

1975

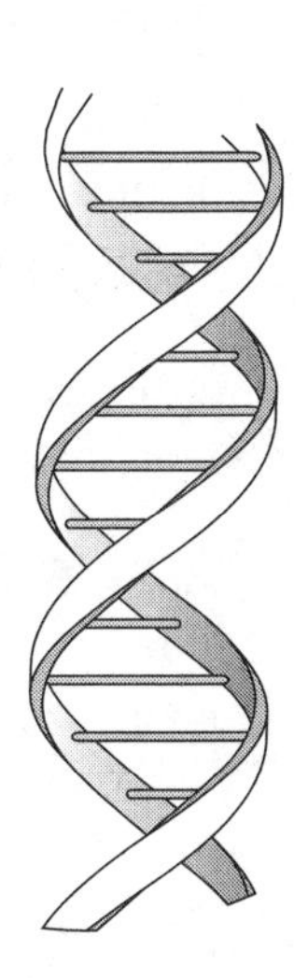

The objective of this first chapter is to reconstruct, from an abundant record,[1] the story of the climactic event of the first act, the 1975 Asilomar Conference on Recombinant DNA Molecules. The subject needs to be viewed in the broadest context. Therefore, we have to zoom in upon it from the past, using a lens of wide angle.

The Coming of Age of Molecular Biology

A landmark in this history is the year 1944. Two noteworthy but unrelated events occurred that precipitated important changes in biomedical research. One was a scientific achievement, the other a political decision.

The Nature of Genes

The scientific achievement was the discovery of the chemical nature of genes. When the first cautious report was absorbed and accepted, it snapped into focus previous research in genetics running back over 80 years, if one counted the careful notes the monk Gregor Mendel put aside in 1865. Following a much earlier trail of research, especially a clue that different strains of pneumococcus were able to exchange certain characteristics like coat appearance and virulence, Oswald Avery, Colin MacLeod, and Maclyn McCarty at the Rockefeller Institute established that the exchanger was a sticky macromolecule or polymer made up of sugar, bases, and phosphoric

† Chapter 1 is adapted from Donald S. Fredrickson, "Asilomar and Recombinant DNA," in *Biomedical Politics*. Copyright 1991 by the National Academy of Sciences. Used with permission of the National Academy Press, Washington, D.C.

acid, known as DNA (for *d*eoxyribo*n*ucleic *a*cid). Needed confirmation that their "transforming principle" was indeed the stuff of the gene came 8 years later with observations that when viruses (bacteriophages) infected bacteria, only the viral DNA entered the host and there led to expression of the complete virus.[2]

Public Support

The symbolic political event in 1944 was a directive from President Franklin Delano Roosevelt to his chief of wartime research, Dr. Vannevar Bush, to find a way to continue federal financing of medical and other scientific research that had proved so successful after the nation's laboratories had been mobilized for war, in what the historian A. Hunter Dupree called the "Great Instauration of 1940."[3] Many members of the Congress, and the heads of at least one government agency, the Public Health Service, were poised to take full advantage of a positive decision to continue this unprecedented effort. The constitutional silence on a federal mandate to support science for its own sake was forgotten. The change also required academic leaders and scientists to overcome a long-held suspicion that taking government money was bound to mean the sale of academic freedom. The details of how this new policy began with the National Institutes of Health (NIH) in 1945, and how this agency became a magician's wand whose touch gave biomedical research an exponential rate of growth for over 10 years thereafter, are major stories in themselves, some details of which are contained in this book. The result was florid expansion of the capacity of America's academic institutions to carry out research and to train young researchers. The greatest growth occurred in basic research, a high-risk activity dependent upon public funds.

The burgeoning scientific community quickly discovered that prewar fears of government interference with scientific freedoms were groundless. From the first, the new resources were primarily distributed to individual scientists on the basis of the judgments of their proposals by scientific peers, managed on a national basis. The briskly expanding network of basic scientists, widely scattered in universities or nonprofit laboratories, was largely self-regulating and became unified in a worldwide profession with the same objectives and intrinsic ethic. Indeed, this shared belief in the autonomy and right of internal regulation of scientific investigation became the central dramatic theme of the recombinant DNA controversy. By restricting themselves voluntarily, the scientists fell into jeopardy of losing forever the freedom absolutely necessary for the vitality and success of their enterprise.

Structure of DNA

In the midst of these boom years in the 1950s, another truly epochal scientific event occurred in England. With dazzling deduction and splendid showmanship, the helical form and base-pairing structure of DNA were unveiled by James Watson and Francis Crick in Cambridge in 1953. The remark, in their report of discovery, "It has not escaped our notice that the specific pairing we have postulated immediately suggests a possible copying mechanism for the genetic material"[4] was a galvanic stimulus for resurgence of the crusade to bring back the answers to fundamental questions of living matter and the evolution of the species.

Such a dramatic expansion of the scientific horizon was perfectly timed to the swelling of the ranks of biomedical researchers. A large fraction of the best and the brightest of the graduate students had begun to move into this pool. Being highly competitive, they shared with budding investment bankers and other entrepreneurs the knack for perceiving where the harvest would someday be most bountiful. As a career, experimental research involves a long apprenticeship to acquire specialized techniques that are applicable to one particular subdiscipline. Thus, the young scientist must select his (or her) special area of interest with care, so that when he embarks—as he eventually must—on a lifetime adventure in independent research, his chosen field will be ripe in opportunities for discovery.

By the early 1950s, an increasing number of aspirants chose to move to the frontier where the outer edges of genetics, biochemistry, and microbiology were merging alongside a flood of new technologies, such as electron microscopy, crystallography, cell culture, and virology, and in parallel with steeply rising capabilities for information storage and analysis. By midcentury, the center of this fluid, expanding field became known as "molecular biology," a term attributed to Warren Weaver, the director of the Rockefeller Foundation support in this period.[5] Already the most interesting molecular forms had become genes, and a limitless series of questions were framed. What was the fuller nature of genes? How were they organized in the chromosomes? Were they conserved in evolution? Could they be interchangeable among species? What were the mysterious codes they carried? How were the codes translated? How was expression regulated with such exquisite timing of differentiation throughout the growth and decline of such a complex machine as the human? What was the nature and origin of abnormal genes that failed their assignments or caused disease?

It should not be overlooked in passing that the birth and early growth of the discipline now centering upon genetics were hastened and greatly enlivened by the participation of scientists, many of them British or European, who were attracted to biology from disciplines such as mathematics, physics, and chemistry. Their presence among the leaders on the new frontier helped lend élan and an elite nature to the cadre of young scientists calling themselves molecular biologists.[6]

Fruit Flies, Corn, Molds

Available techniques to get at the gene, however, were crude and cumbersome, and the field took time to mature. In early studies of recombination of genes, which is an important process for reproduction, pioneers like Thomas Hunt Morgan had profitably used the fruit fly (*Drosophila*), a creature that is still invaluable for this purpose today. Others, like Barbara McClintock, turned to corn or other plants to learn about the organization of genes in the chromosomes and their mobility or susceptibility to rearrangement. In their classical work in the 1930s and 1940s, George W. Beadle and Edward L. Tatum used bread molds (*Neurospora*), easy to culture and rapidly reproducing by genetic crosses. Simple as they were, the molds taught these pioneer geneticists the fundamental tenet of the central dogma: one gene controls the structure of one protein.[7]

The Need for Germs

Those who were primarily interested in studying growth, differentiation, and genetics in mammalian, including human, tissues, however, now turned by necessity to the microbiological world for answers. The inhabitants of this ancient kingdom of living things had been the most instructive tutors of biologists since the promulgation of the germ theory of disease by Pasteur and Koch in the 19th century. The bacteria were readily available, had short generation times, were cheap and simple to culture, and were generally predictable and reliable in behavior. Up to 1950, a large share of the growth in understanding of biochemistry, nutrition, and the great maturation of enzymology was attributable to studies of bacteria.

For genetic studies, there are fundamental differences between the bacteria and most other living things. The former are termed prokaryotes because they have no cellular nucleus and the chromosomes are free in the cell juice, or cytoplasm. In bacteria, some of these genes are in circular DNA molecules, or plasmids, which are often exceptionally mobile and can transfer genes to other bacteria. All the other cellular forms are called eukaryotes and their cell nuclei hold all but a few of their genes arranged in a certain number of pairs of chromosomes. All the genes of either a prokaryote or a eukaryote are known collectively as the genome. The major processes of exchange of genetic characters between organisms, so-called transductions or transformations, could only be observed in a few strains of microorganisms in 1950. One of these was the intestinal bacterium *Escherichia coli*, a laboratory partner in many invaluable studies. Of particular importance was, and still is, a stable strain of *E. coli* designated K-12, which was cultured from a patient at Stanford University Hospital in the 1920s and eventually used in laboratories around the world. It was in this strain that a precocious Joshua Lederberg, while studying with Tatum at Yale, observed a third method of the transfer of genetic characters, called conjugation. In this process—the first intimation of sexuality in bacteria—a "male" and a "female" *E. coli* bacterium link up side by side, and an end of the male chromosome enters the female. The entering DNA recombines with the host genome and, after replication and cell division, the new recombinant cell has genetic features of the two parental DNAs.[8]

Viruses

Viruses also began making invaluable contributions to molecular biology after techniques for cultivation of cells in culture were devised in the 1940s. Viruses are packets of genes and proteins so small they can pass through filters that capture bacteria. In the simplest sense, the virus is a basically a "transportable genome," stealing entry into the host cell, where the viral genes proceed to replicate and sometimes combine with the host genome but invariably must direct the cell machinery to synthesize their products, called virions, which in turn enable the viral genes to be transferred to other cells. Certain viruses are the only organisms in the biosphere that utilize a genome containing not DNA but RNA (*r*ibo*n*ucleic *a*cid). These are molecules that are complementary to DNA in structure and that have essential functions in the cell in translation of the DNA code to proteins. The RNA viral genome of one class of RNA viruses, the retroviruses, contains the code for the enzyme reverse transcriptase, which transcribes the RNA to DNA.[9] The DNA from such retroviruses may also recombine with DNA in the host genome. By such "natural" recombinations, retroviruses and mammalian cells may exchange and activate cellular genes called oncogenes, because their presence and expression may underlie cancerous transformation in the host.[10]

Viruses have long been known to be capable of causing tumors in animals. As long ago as 1906, Peyton Rous found a retrovirus that causes sarcoma in chickens. Since then, many other RNA and DNA viruses have been identified that are tumorigenic in animals, particularly rodents. The Epstein-Barr virus, a DNA virus isolated from a rare human tumor called

Burkitt's lymphoma, is one of the very few viruses suspected of being tumorigenic in humans.

The potential hazards of infections from some species of bacteria and viruses did not retard early molecular biology. By the second decade after the discovery of the "transforming principle," the laboratories of the virologists and microbiologists were thoroughly invaded by biochemists, geneticists, and cell biologists and molecular biologists. The whir of the Sharples centrifuge, surrounded by its misty aerosol of *Escherichiae* in harvest, was commonplace in the most advanced laboratories and a sign that higher science was in progress. Nonpathogenic viruses were handled on open laboratory tables, and—there being yet no better methods—cultures were transferred by mouths separated from the contents of the pipette by a cotton plug. The microbiologists had learned, in their apprenticeships, respectful behavior toward organisms known to cause disease (pathogens) and compulsively washed down the lab tops and their hands if a drop of viral culture was spilled. However, a general belief could be said to have prevailed that humans and nonpathogenic microbes had reached a state of equilibrium that was not likely to be easily upset by human manipulation.

The interests of most of the molecular biologists did not lie in classical bacteriology, and it is safe to say that many had received only rudimentary instruction in the handling of pathogens or in the ecology of microorganisms. Any anxieties they harbored were directed more toward maintaining a competitive edge in the hunt for the new paradigms that might arise any day, and their laboratory technique with respect to germs often reflected these priorities.

The possibility of using the insights and methods of molecular biology to better the lot of humanity was already being discussed by the mid-1960s.[11] It would only be a little longer before the discovery of restriction enzymes, tools capable of cutting DNA selectively and with precision at points along the chain.[12] And just a few years later, a particularly useful such enzyme would be the precipitating cause of the recombinant DNA controversy.

An International Frontier

The ever-expanding territory of molecular biology spread across two continents and occupied floors in the top universities and research centers in a number of countries. A half dozen British laboratories, including those at Cambridge, London, and Edinburgh, largely supported by the Medical Research Council, were highly productive. Europe was developing the European Molecular Biology Organization (EMBO), with a major communal laboratory in Heidelberg, Germany. In the 1950s and 1960s, France had its centers, particularly in Paris, at both the university and the Institut Pasteur. At the latter, there were many workers, such as André Lwoff and François Gros, whose specialty was bacteriophage, viruses that can live parasitically within bacteria, sometimes fatally turning upon their host. Also at the Pasteur Institute, the laboratories of François Jacob and Jacques Monod were a particularly popular center of intellectual ferment that attracted many Americans for training and later collaborative work. Here an elegant conception of how the expression of (bacterial) genes is regulated was being shaped. Bacteria, prominently including *E. coli*, were exposed to mutagens, including ultraviolet light, and then their capacity to adapt to stringent change in growth medium was tested. From these experiments gradually emerged the concept of the operon, a cluster of genes controlled by a single promoter. This idea led to an understanding of repression and induction of gene expression.[13]

By far the largest number of molecular biologists were working in the United States, in laboratories extending from Boston and Cold Spring Harbor, N.Y., to southern California. The NIH was a major source of support, its grants also going to European laboratories, including those of Jacob and Monod. The NIH intramural laboratories had many commitments to this field in the 1960s, with the heaviest concentration being in virology. The National Cancer Institute (NCI) would soon erect one of the very few maximum security laboratories in the world to search for the elusive viruses some thought were at the root of human cancers.

Important financial support to nonmedical scientists was also being provided by the National Science Foundation (NSF). During the 1960s, Herman Lewis, the head of its Section on Cellular Biology, organized an informal Human Cell Biology Steering Committee. Its stated purpose was to advise on large-scale cell cultures at different sites to foster a scale-up of studies in molecular biology, but it also was a clearinghouse for ideas among some of the leaders in the field. The steering committee met fairly regularly and usually in Washington, D.C. Its membership included several faculty members from Stanford University.[14] It was at Stanford in the early 1970s that experimentation in molecular biology would first lead to serious controversy.

The Experiment

In the late 1960s, Paul Berg, professor of biochemistry at Stanford, took sabbatical leave in the laboratory of the virologist Renato Dulbecco at the Salk Institute. Berg had worked on molecular aspects of protein synthesis and was no stranger to the use of *E. coli* mutants. Like many others, he had become interested in the molecular genetics of viruses. His curiosity about whether a virus might be used to transfer a foreign gene into eukaryotic cell cultures led him to become familiar with the virus called simian virus 40 (SV40). Berg considered the relationship between phage and bacteria to be closely analogous to that of SV40 and eukaryotic cells, and he wondered if the virus might work more efficiently as a vector for a bacterial gene. Berg enlisted two coworkers, David Jackson and R. H. Symons, to determine if they could insert a bacterial gene held by a modified lambda phage into the SV40 genome. Janet Mertz, a graduate student newly arrived from the Massachusetts Institute of Technology (MIT), was intrigued by the possibility that SV40 chromosomes could be reproduced in bacteria. Sheldon Krimsky described the Stanford laboratory activity at that time, including Mertz's growing ambivalence about such an experiment.[15]

SV40 was first isolated from monkeys in 1960 and was carried in cultured monkey kidney cells. Within a short time, it was discovered that the virus-infected cells caused tumors in hamsters.[16] This finding was of exceptional interest to the makers of poliomyelitis vaccine, for monkey kidney cells had also been indispensable for cultivating the poliovirus for the first vaccine. When it was looked for, the contamination of rhesus monkey kidney cells with SV40 was indeed high. It was by then no surprise, yet still a most unpleasant revelation, that some lots of the polio vaccine also contained SV40. A survey in 1961–62 revealed that many of the recipients of the vaccine had antibodies to both the poliomyelitis virus and SV40.

Using the fairly cumbersome techniques then available at the time, Berg and his coworkers were able to delete portions of the circular, helical coils of the SV40 genome. In the spring of 1971, they began to make preparations to couple SV40 genes to bacterial genes for insertion into eukaryotic cell cultures.

Critique

In June, Mertz attended a workshop at Cold Spring Harbor Laboratory, and there discussed with others the proposed experiments at Stanford. John Lear opens his book with a full-stop rendition of the outcome of her revelation.

> On the afternoon of Monday, June 28, 1971, Robert Pollack, a 31-year-old microbiologist on the research staff of the Cold Spring Harbor Laboratory, Long Island, made a telephone call that would fundamentally change the relationship of American science to the society that sheltered it.[17]

Pollack's call was to Paul Berg. It did not catch Berg completely unawares, for Mertz had already relayed to him some of Pollack's criticisms. Pollack told Berg in effect that he should "put genes into a phage that doesn't grow in a bug that grows in your gut," and reminded him that SV40 is a small animal tumor virus and transforms human cells in culture, making them look like tumor cells. Pollack later described the idea as a "pre-Hiroshima condition—it would be a real disaster if one of the agents now being handled in research should in fact be a real human cancer agent."[18] At the end of the course, Pollack is said to have complained in his final lecture that "No one should be permitted to do the first, most messy experiments in secret and present us all with a reprehensible and/or dangerous fait accompli at a press conference."[19]

After Pollack's call, Berg undertook further sampling of the opinions of his peers about the proposed experiment and renewed discussions he had had much earlier about the general nature of such research. In 1970, Berg dined at the home of Maxine Singer, a molecular biologist at NIH, and her husband, a lawyer and trustee of the Hastings Institute.[20] Another guest was Leon Kass, who in 1971 was to publish a widely read article on the social consequences of the new biology.[21] Kass and Berg later exchanged correspondence on the subject of their dinner conversation. On a later trip to Bethesda, Md., in 1971, Berg paid a visit to scientists in the so-called "Memorial Laboratories" of Building 7 on the NIH campus, which had been dedicated to several scientists who had contracted fatal diseases during laboratory or field studies. There he talked to virologists who were working on SV40. One of these NIH scientists was Andrew Lewis, who still remembers Berg's admission that some of the scientists he had talked to felt that there was some line out there that shouldn't be crossed in manipulating the genome until more was known.[22]

The Encounter

Shortly before Berg's visit, Lewis had had a reminder of the rising tensions in the competitive field of molecular biology. In August 1971, he had gone to Cold Spring Harbor to make a presentation of his work on hybrids of adeno-SV40 viruses. Adenoviruses are large viruses that can cause respiratory infections. Experiments in which these viruses had been grown in monkey kidney cells for purposes of preparing vaccines had led to the discovery of hybrid viruses, in which the genomes of adenoviruses also were contaminated with the genes of SV40. Most of these hybrids were defective, i.e., unable to reproduce. For a decade they had attracted little attention. Lewis's hybrids, however, were nondefective and therefore capable of independent growth. Because these hybrids were much more likely to lead to information about the tumorigenic property of SV40, interest in them was rising sharply at laboratories like Cold Spring Harbor. Berg's proposed experiment was now well known at this institution, and Dan Nathans, who was at the same meeting, described headway in dissecting the circular SV40 genome with one of the first restriction enzymes.

After his talk, Lewis had an unexpected encounter—shocking for a young scientist—with one of the Wunderkinder of molecular biology. Lewis had never met and didn't recognize James Watson, who recently had become director of Cold Spring Harbor Laboratory. Without introduction, Watson expressed his displeasure that Lewis had failed to share samples of the viruses with his laboratories and proceeded to enumerate ways by which he could force Lewis to provide them. Lewis responded by relating his concern that he felt the hybrid DNA in these nondefective viruses might be hazardous and that he was reluctant to share samples without agreement by the recipients to acknowledge the possible hazards and to follow certain precautions. The next month, he supplied samples of this type of hybrid to the Cold Spring Harbor Laboratory, and he stepped up efforts to convince his NIH superiors that they should endorse a policy requiring a "memorandum of understanding" to accompany the sharing of the nondefective hybrids and other potentially hazardous viruses. Lewis's friends and coworkers at NIH did not share his serious concerns about the hazards of his experimental material, but NIH eventually organized to undertake such precautions.[23]

When Berg returned to his laboratory in the fall of 1971, he informed his coworkers that they should postpone that part of their proposed experiment that would transfect the SV40-lambda hybrid into *E. coli*. He called Pollack and told him this and asked Pollack to help him organize a conference on the hazards of tumor viruses the following year. The departure of Berg's coworker David Jackson to start a new laboratory at Ann Arbor, Mich., in the spring of 1972 made the postponement of the original experiment indefinite.

The "First" Asilomar Conference

In 1972, the controversy over recombinant DNA was still well contained within the community of molecular biologists, and there had been no organized attempt to deal with the major single source of anxiety—the DNA of cancer viruses. Paul Berg, however, had refused to let the matter drop. In August 1972, the Human Cell Biology Steering Committee informed its members that a meeting on "containment" would be held in Asilomar on January 22–24, 1973. As Herman Lewis remembers it, the idea for the conference had come from Berg. NSF was willing to pay for the conference, but Alfred Hellman of the NCI wanted his institute to participate, and NCI therefore shared the costs. Pollack, Hellman, and Michael Oxman of the Children's Medical Center in Boston assisted mainly by proposing names of participants to Berg. Berg selected the final list and handled most or all of the preliminaries. It is certain that he picked the location, for the conference center in Pacific Grove, Calif., had long been a favorite place of campus scientists at Stanford.

Sometimes dubbed "Asilomar I," the January 1973 meeting involved about 100 biomedical scientists, all but one or two of whom were Americans. There was no effort to invite the press, but the proceedings were edited by Pollack, Oxman, and Hellman and later published. The up-to-date information on many viruses was summarized by experts. The evidence or lack of evidence that the known viruses caused human cancer was thoroughly vetted. There was no evidence to support a single case. In the case of the polio vaccine that had been contaminated with SV40 in the late 1950s, the available knowledge about the several million recipients of the vaccine did not suggest any alteration in cancer rate or type. It was obvious, however, that a fuller epidemiological search would someday be required to raise the level of certainty. Finally, safety precautions, especially for those engaged in the search for any virus causing human cancer, were outlined. In closing, Berg stated that "prudence demands caution and some serious efforts to define the limits of whatever potential hazards exist." Recombinant DNA experiments were not mentioned in the proceedings. Asked for his impression of the effect of the exercise, Andy Lewis answered, "After the conference we felt less concerned about the hazards of [laboratory] viruses causing cancer." Some of the recorded comments or exchanges from the conference floor indicated that other anxieties were causing tempers to fray. There was concern that fear was being spread unnecessarily. It was also evident that many scientists were becoming alarmed that research money would not be adequate to cover potential escalating costs of new containment, epidemiological studies, or other safety requirements.[24]

*Eco*RI

In the spring of 1972, R. N. Yoshimori, working on his doctoral thesis in the University of California at San Francisco laboratory of Herbert Boyer, isolated from *E. coli* a new restriction enzyme designated *Eco*RI. The enzyme was quickly shared with other laboratories. At Paul Berg's suggestion, John Morrow examined its action on the SV40 genome. He found that the SV40 DNA was cleaved at a unique site. This finding provided a reference site for mapping the SV40 genome. To her great excitement Janet Mertz discovered that when *Eco*RI cleaved the circular DNA, it produced a linear segment with "sticky" ends that adhered to other ends similarly cleaved. Electron microscopist Ronald Davis quickly confirmed her impression, and Boyer came quickly to see. Within a short time, his associate Howard Goodman showed how *Eco*RI cuts left complementarity of the bases in DNA, so that a perfect splice to other DNA similarly cut would result.[25]

Scientific Exchange and Scrutiny

At the end of September 1972, about 50 molecular biologists from 12 countries, including nearly a score from the United States, attended an EMBO workshop in Basel, Switzerland, on DNA restriction and modification. An evening was devoted to "an open discussion of the use of restriction endonuclease to construct genetic hybrids between DNA molecules . . . the implications this may have as a useful tool in genetic engineering and the potential biohazards." A few weeks later, Honolulu, Hawaii, was the site of a 3-day U.S.-Japan conference devoted to plasmids, including recombination and genome transformation.[26] This latter meeting also gave Boyer and Stanford's Stanley Cohen opportunity to discuss close collaboration in experiments to probe the enzyme's utility in plasmid manipulation. Within a few months, the partnership had established that *Eco*RI uniquely cleaved a local plasmid (named pSC101, for Stanley Cohen), and they had combined two antibiotic-resistant plasmids, inserting the hybrid genes into *E. coli*.[27]

These international scientific meetings in the autumn of 1972 were but two examples of the constant worldwide exchange among scientists, interactions that sometimes foster long-range collaborations but are also vital to maintain communication among scattered workers in fast-moving fields of research. Experimental science is an open process, having an existential quality that is the antithesis of secrecy. A scientist who has made a discovery can usually be counted upon to leak it. Proof of the existence of a claim, a full report of the evidence, and its submission to confirmation and validation are required to ensure the precious priority of discovery that is still the paramount personal reward of scientific discovery. The worldwide scientific community, including the corps of peer-reviewed publications that serve the different fields, judges and protects these priorities as international properties.

Key judgments about the worth and priority for support of a scientist's work are largely national decisions, however. Judgments of the ethics or morality of individual scientists or their experiments likewise remain within national boundaries. The major reason for this is the national or regional character of public support for scientific research. Cultural expectations are a major force in the maintenance of fiscal support of science. The continuing public approval of generous appropriations through agencies like the NIH is based on expectations of improved public health and conquest of diseases. The basic research, which laypersons cannot always identify as keyed to their aspirations, is nevertheless tolerated and tacitly understood as necessary to maintain the tide upon which practical benefits eventually arrive. The currency of these transactions is the continued credibility of the scientists and the ultimate satisfaction of the consumer public, including the public's pride of sponsorship of a worthwhile and popular enterprise. In the early 1970s, the biomedical community began to experience concern about increasing tension on the vital public-science connection.[28]

The Gordon Conference on Nucleic Acids of 1973

The most effective and continuous self-monitoring of the scientific tribes is derived from regular gatherings of the warriors and the elders to examine in depth recent performance and progress. Among the favorite kinds of such meetings are the Gordon research conferences, which have played a formative role in the careers of nearly all biomedical researchers. For a week in the summer, members of a subdiscipline take over the furnishings abandoned by vacationing students at a number of New England schools and engage in highly informal, intensive review of their particular field. On June 13, 1973, the Gordon Conference on Nucleic Acids began at New Hampton, N.H. The first 3 days were dedicated to the synthesis of DNA, the structure of RNA, and the interaction of proteins and DNA, themselves topics in which movement was rapid and fascinating. The fourth day was given over to bacterial restriction enzymes in the analysis of DNA. In a session chaired by Daniel Nathans, Herbert Boyer was scheduled to speak. According to John Lear, Stanley Cohen had obtained Boyer's agreement to say nothing about the current work of their partnership at the Gordon conference. Sheldon Krimsky, however, cites chairperson Maxine Singer recalling how Boyer had shared with the conferees information about the capabilities of the restriction enzyme *Eco*RI to splice DNAs of different origin, and how two plasmids bearing genes specifying resistance to two different antibiotics had been joined.[29] It was after Boyer's comments that someone loudly sounded the excited comment that "Now we can combine any DNA."

Other reaction to this hint that biology was approaching proximity to something akin to the nuclear physicists' chilling arrival at "critical mass" was delayed until late afternoon, when two researchers at the Cambridge Molecular Biology Laboratory, Ed Ziff and Paul Sedat, sought out the two chairpersons of the conference. These scientists, Maxine Singer of NIH and Dieter Söll of Yale, listened to the urging of Ziff and Sedat to have the conference discuss the potential hazards of the experiments disclosed in the afternoon's session. Left with only a half day's conference to go, Singer and Söll nevertheless agreed to arrange for the matter to be taken up at the beginning of the Friday morning session. Within about a half hour, the approximately three-quarters of the enrollees at the conference who were still on hand reached a consensus that a communication should be sent to the presidents of the National Academy of Sciences (NAS) and the Institute of Medicine.[30] In a separate vote, it was also agreed that the letter should be published.

As later drafted by Singer and Söll and sent to the conferees for approval before dispatch, this letter began:

> We are writing to you, on behalf of a number of scientists, to communicate a matter of deep concern. Several of the scientific reports at this year's Gordon Research Conference on Nucleic Acids . . . indicated that we presently have the technical ability to join together, covalently, DNA molecules from diverse sources. . . . This technique could be used, for example, to combine DNA from animal viruses with bacterial DNA, or DNAs of different viral origin might be so joined. In this way new kinds of hybrid plasmids or viruses, with biological activity of unpredictable nature, may eventually be created. . . .

The letter further noted that the experiments had the potential for advancing fundamental knowledge and alleviating human health problems but that some hybrid DNA molecules might prove hazardous to laboratory workers and the public. It suggested that the academies should establish a committee to consider this problem and to recommend specific actions or guidelines.[31]

The die was cast. The Gordon conference reaction was unprecedented. Its expression of deep concern could not go unheeded, and the train of events was set in motion to inexorably roll on to the principal subject of this narrative.[32]

The Academy's Turn

Receipt of the Singer-Söll letter, dated July 17, 1973, was acknowledged by NAS president Philip Handler a few days later.[33] The conference letter appeared in the September 13 issue of *Science*. Quite coincidentally, an editorial in the same issue by Amitai Etzioni dealt with a recent poll of public attitudes toward institutions and concluded that friends of science had no grounds for "hysterical alarm."

Having consulted with the council of the NAS in late August, Handler informed the executive committee of the brand-new Assembly of Life Sciences (ALS) that he was referring the Singer-Söll letter to it. Paul Marks, the chairman of ALS's Division of Medical Sciences, replied to Handler that he agreed that the ALS should establish a study committee and indicated that he was "as concerned with the potential hazards of certain of the hybrid molecules being studied as I am with the potential of unreasonably gloomy predictions [of] these hazards."[34] The ALS executive committee heard directly from Maxine Singer in September and, when asked for a suggestion as to who might head the study committee, she suggested Paul Berg. Berg was requested by Handler to take charge, and early in January, Berg informed the ALS that he would organize a small group of individuals (fewer than 10) for a 1-day meeting to consider mechanisms for reviewing potential dangers (as well as benefits) stemming from the ability to generate a wide variety of hybrid DNA molecules.[35]

The Meeting of the Planning Committee

Berg convened the meeting that he had in mind at MIT on April 17, 1974. The six other participants selected by Berg independently were David Baltimore, James Watson, Dan Nathans, Sherman Weissman, Norton Zinder, and Richard Roblin. Herman Lewis of the NSF was also there as observer. Maxine Singer was unable to attend. Much has been written about this historic 1-day meeting—that Jim Watson had been keen on an international meeting, that Berg had recalled how Norton Zinder had said, "If we had any guts at all, we'd tell people not to do these experiments," and how Roblin came to participate.[36] The importance of the details of this historic event is submerged by the essence of its conclusions contained in the report released 3 months later in a press conference at the NAS on July 18, 1974.[37]

The report began with a summation of recombinant DNA achievements since the July Gordon conference, i.e., the creation of new bacterial plasmids carrying antibiotic resistance markers, the insertion of toad ribosomal

DNA into *E. coli*, where it synthesized RNA that was complementary to the inserted DNA, and unpublished experiments involving incorporation of *Drosophila* DNA into DNA from plasmids, and phage ready to be inserted into *E. coli*.[38] There then followed the committee's conclusion that this type of unrestricted activity could create artificial recombinant DNA molecules that might prove biologically hazardous. As an example of this, the report cited the possibility that *E. coli* exchanging such new DNA elements with other intestinal organisms could have unpredictable effects.

The committee therefore made four recommendations, which are summarized as follows.

First, a *moratorium* on certain experiments was proposed:

> . . . most important, that until the potential hazards of such recombinant DNA molecules have been better evaluated or until adequate methods are developed for preventing their spread, scientists throughout the world join with the members of this committee in voluntarily deferring the following . . . experiments.

Those experiments to be deferred were broken into two types. Type I had to do with the creation of new, autonomously replicating plasmids that could carry antibiotic resistance to strains not now having such genes or that could enable toxin formation in now innocent strains. Experiments of type II included the linkage of DNA from oncogenic or other animal viruses to plasmid or other viral DNAs.

Second, to "carefully weigh" experiments to link animal DNA to plasmid or phage DNA.

Third, the director of NIH was requested to give immediate consideration to establishing an advisory committee to evaluate hazards of recombinant DNA, develop procedures to minimize those risks, and devise guidelines for work with recombinant DNA.

Fourth, a call was issued for an international meeting of involved scientists from all over the world early in the coming year to discuss appropriate ways to deal with the potential hazards of recombinant DNA molecules.

The relationship of Berg's committee to the NAS, the endorsement of its recommendations by the ALS-National Research Council, as well as the stress on the international nature of the proposed conference were very important touches added at the final stages of report preparation and review. They were a credit to both the NAS and the committee and materially helped to buffer a possible inference that a gang of seven (or was it 11?)

American scientists had impulsively doused the boiler of what arguably would become the most powerful scientific engine of the century.[39]

The Asilomar Organizing Committee

On September 10, 1974, the committee to organize the February 1975 Asilomar meeting gathered in room E17 at the MIT Center for Cancer Research. The committee, consisting of chairman Paul Berg, David Baltimore, Richard Roblin, Maxine Singer, Sherman Weissman, and Norton Zinder, were joined by others. Donald Brown, Richard Novick, and Aaron Shatkin were summoned because they had agreed to play key roles as chairmen of three working parties that would be formed to issue formal reports. Herman Lewis attended in his familiar role as patron and rapporteur for the Human Cell Biology Steering Committee (a good share of its members being directly involved). William Gartland had been hastily sent as an observer for NIH, the principal underwriter of the forthcoming conference.

A first order of business was a discussion of foreign participation, ending with the additions to the organizing committee of Sydney Brenner, of the Molecular Biology Laboratory at Cambridge, and Niels Jerne, chairman of the EMBO Council. (Jerne was unable to participate in the committee or the conference.) Brenner, a highly articulate and gifted molecular biologist, was also a member of the Ashby working party in Britain that had quickly been set up by the Board for the Research Councils to determine how British science should react to "the Berg letter."[40]

A nearly complete format for a three-and-a-half-day conference was produced by the time the meeting ended.[41] Berg solicited many suggestions for possible participants, but the final invitation list was of his selection. The slate was in keeping with the intent expressly stated in the July 1974 report: an international meeting of involved scientists. About 90 of the scientists invited were American, and another 60 came from 12 different countries; all were among the elite in one segment of molecular biology or another. No organizations were represented per se. Sixteen members of the press were invited, all accepting the condition that no copy would be filed until the three-and-a-half-day conference had ended.[42]

The three discussion panels met in November to become acquainted and to work on the substance of their reports. Richard Novick's plasmid panel began to work on an extensive analysis that finally would cover most of the potential areas of hazard. When Aaron Shatkin's panel arrived in Asilomar on the night before the conference, they were unaware that their draft report was to have already been completed. The virologists gathered after dinner to draw one up.

The Asilomar Conference, February 24–27, 1975

Monday—Opening Day

The organizing committee had decided at their September meeting that there would be no publication of the conference because of the manpower and time required. It was eventually decided that a tape recording run throughout the sessions would be maintained as an archive. Any conferee could ask that the recording be stopped during his or her discourse, but none so requested. When the participants then noticed the small forest of microphones set up by members of the press, the ensuing discussion ended by permitting the press to use their recorders for preparation of their stories. Allowing any part of the tapes to be broadcast was declared to be against the rules.

David Baltimore opened the conference on Monday morning. He gave a short history of how this meeting had come about, its auspices, and who had been the organizers. People had been invited to the meeting on the basis of their expertise or involvement in the science. He explained that the meeting had been convened to lay out the existing technology and what had been done to find answers to questions of what (experiments) should be done or not done, and what should be done before an experiment was undertaken. Baltimore emphasized that the balance of risks and benefits would be considered, but he stated that discussion of the hazards was a subject more important than either the benefits or molecular biology per se. His summary of the program ended with the observation that on the last morning, the organizing committee expected to present a summary statement including general guidelines for discussion and consensus. Baltimore reminded the audience that if it couldn't reach consensus there was no one else to turn to.

Paul Berg next stepped to the podium to review the basics of recombinant DNA technology. This set the tone of much of the first 3 days of the meeting, the format and content tending toward the highly technical, with presentations in the traditional style of experts talking to experts. It reflected scientists doing what they do best, talking about their own work. There was another requirement to be satisfied by such intercourse, however, and that was the need of the participants to be exposed to the different techniques, personalities, and scientific jargon peculiar to each of the three or four major subcultures assembled: the nonphage virologists, the "plasmid engineers," the specialists in phage ("lambda people"), and the eukaryotic cell biologists. The insularity of these narrow subspecialities predictably bred suspicions that one's own area of research could emerge from such a meeting unfavorably restricted by strangers.

The expertise on hand at Asilomar was impressive (see Appendix 1.1). Speaking after Berg was Stanley Falkow, who combined a medical background with an encyclopedic knowledge of bacteria. After him came Ephraim Anderson from the Public Health Laboratory Service in Britain, who also had medical training and who had dealt with epidemics of intestinal infections before concentrating on plasmids. Anderson had taken umbrage at the type I recommendations in "the Berg letter," partly because in the version printed in *Nature*, a dropped word led to the interpretation that his long-time research had been banned. As soon as he read it, Anderson shot off a note to the editor for the next issue expressing the wish that the "NAS statement had been presented less pompously." At Asilomar, Anderson and a British colleague, William Smith, were requested to present their experimental evidence that *E. coli* K-12 had a low risk for transferring plasmids to other enteric bacteria. After it was all over, however, Anderson's criticism of the conference was unmollified.[43]

Another speaker on the first day, Roy Curtiss III from the University of Alabama, had displayed a very different reaction to the Berg letter. A month after it appeared, he sent off a 16-page memorandum to the signatories and distributed hundreds of copies to the world community of molecular biologists, in which he stated:

> I heartily endorse the aims, but not necessarily the scope of your recommendations. . . . I personally pledge to cease Type I experiments (to construct bacterial plasmids that are not now known to exist) that I was currently engaged in . . . and not to initiate Type II experiments. . . .[44]

Curtiss, moreover, argued for specific heightening of the restrictions and spelled out conditions under which he believed *E. coli* might be hazardous. Berg and many others responded to the Curtiss letter, and the reiteration of prior arguments now enriched the debate.

The last speaker in the postdinner session that evening—after presentations by Herbert Boyer and Stanley Cohen—was Ken Murray of the team of molecular biologists in Edinburgh. He described phages as cloning vehicles. In a far more conciliatory tone than that used by Ephraim Anderson, Murray had written a companion note in the July issue of *Nature*, which he closed with a line borrowed from the *Manchester Guardian*'s earlier comment on the Berg letter:

> While welcoming the NAS initiative . . . if we follow the moderate tone set by the NAS we shall be careful not to oversell the social benefits devolving from the recent experiments.[45]

When Berg had begun his session on the morning of the first day, he mused aloud that the writers of the original letter had not anticipated how it would affect the scientific community and that the organizing group was not prepared or experienced in how one should arrive at a decision. He said therefore that a panel of lawyers arranged by Dan Singer would present views on law and public policy issues on the third day. Harold Green, another Washington lawyer and a trustee of the Hastings Institute, spoke after lunch on the first day.[46]

Green told the scientists that the conference and its unique moratorium on research—for which he gave them high praise—would serve as a moral precedent and a model of how science should proceed to deal with such issues. He was asked several questions about how the responsibilities for risk, or the framework for proceeding with experimentation, would be determined, and he offered his opinion that the government would ultimately determine the public policy. Green then held out astringent balm to any injured by this forecast: "It has been said that all institutions in society are imperfect and of these the government is the most imperfect."

Tuesday—Getting Down to Guidelines

The second day began with the presentation by Richard Novick of the report of his Plasmid-Phage Working Group on "Potential Biohazards Associated with Experimentation Involving Genetically Altered Microorganisms, with Special Reference to Bacterial Plasmids and Phages." Anyone doubting the seriousness or scientific literacy of the experts at Asilomar—some of whom by now were beginning to feel the tightening coils of a growing dilemma—should read through the 35-page, single-spaced document completed by this working group just before the start of the Asilomar meeting.[47]

The conclusions of this first of the working group reports were most conservative. The document contained extensive recommendations for classifying, monitoring, and designing many classes of experiments, and it would provide a template for the future recommendations of the NIH Recombinant DNA Molecule Program Advisory Committee. The mass of information it contained, however, seemed to overwhelm the capacity of the participants to absorb it so quickly. A day later, the organizing committee would find it necessary to amend its construction for purposes of proposing a framework for consensus. Long after Asilomar, it would often be said that the conference had failed to consider the high improbability that *E. coli* strain K-12 could be converted to a dangerous enteric pathogen or engage in harmful genetic transmission to other organisms in vivo under normal circumstances. The working group report did not neglect such calculations, but the pace of the Asilomar debate outstripped the time for adequate reflection upon them. It would not be until 1977 (the Falmouth conference) that similar deductions led to the dismissal of the much-used *E. coli* strain K-12 as a serious hazard.

Tensions Build. The mesh of protection proposed by the plasmid-phage panel grated upon the nerves of some of the listeners. Michael Rogers, the correspondent from *Rolling Stone*, later reported some sample reactions. There was Josh Lederberg's rising to express grave concern about the danger of the panel's recommendations "crystallizing into legislation." Ephraim Anderson then demanded the panel indicate, by a show of hands, which of its members "have had experience with the handling and disposal of pathogenetic organisms capable of causing epidemic disease." When all rather sheepishly admitted they had probably had too little, their tormentor added insult to injury by nipping away at the grammar and syntax of the report. Suddenly, Jim Watson uttered a call for an end to the moratorium, and moreover, "without the kind of categorical restrictions called for in the plasmid report." Rogers recalled that Maxine Singer was on her feet immediately to ask what had changed in the last 6 months to cause Watson to jump the movement he had helped to launch.[48]

In line with the assessment of a number of other subsequent commentators, Rogers much admired Sydney Brenner ("the single most forceful presence at Asilomar") and describes him as rising shortly thereafter to ask waverers in the crowd, "Does anyone in the audience believe that this work—prokaryotes, at least—can be done with absolutely no hazard?" After a dramatically long pause, Brenner continued, "This is not a conference to decide what's to be done in America next week; if anyone thinks so, this conference has not served its purpose." During the afternoon, Brenner led a session on the desirability of "biological containment," the designing of plasmids, phage, or other vectors that could not survive in a new ecological niche outside of the laboratory and thus would do no mischief if they escaped the "physical containment" that had been thus far discussed. It was not a completely new idea, but Brenner's enthusiasm stimulated much discussion and encouraged thinking about other ways to open up the blocked channels of research.[49] That night a group of "lambda people," concerned that the plasmid group had overly emphasized crippled plasmids in their proposals for biological containment, worked late and by morning had a design on paper of a potentially safer phage vector.[50]

A heavy barrage of virology was laid down in the late afternoon and evening session of the second day. Undoubtedly, in the minds of some of the scientists—especially those to whom viruses lay in unfamiliar territory—any anxieties over *E. coli*-triggered epidemics paled in contrast to a concern about human cancer being caused by some devilish recombination of DNA from tumorigenic viruses. Among the presentations was that of Andrew Lewis, who described his work on the adeno-SV40 hybrids, accompanied by the precautions he considered desirable for the use and sharing of the nondefective forms of these organisms. But after Aaron Shatkin came forward with the recommendations of the virus working group, the panel appeared to disappoint some who considered viruses to be the greater menace. The report consisted of two pages, the first signed by seven of the eight members of the panel. It began with a reaffirmation of benefits ahead, a theme the organizers at Asilomar had requested be muted. The preamble to the report read:

> The construction and study of hybrid DNA molecules offer many potential scientific and social benefits. Because the possible biohazards associated with the work are difficult to assess and may be real, it is essential that investigations be re-initiated only under conditions designed to reduce the possible risks. Although the need for the development of new and safer vectors is clear, we believe that the study of these recombinant DNAs can proceed with the application of existing National Cancer Institute guidelines for work involving oncogenic viruses . . . with the exceptions noted below [highly pathogenic viruses] we recommend that self-replicating recombinant DNA molecules be handled according to guidelines for moderate risk oncogenic viruses. . . . the vast majority of experiments will fall into the moderate risk category.

The second page of the report was Andrew Lewis's minority report of one, which called for experiments on recombination of DNA from animal viruses to take place in moderate-risk facilities as defined by NCI, and only when theoretically safe vectors had been developed.

A day later, an amended report was issued by the viral group that endorsed the desirability of both physical and biological containment for experiments in which viral or eukaryotic cell DNA would be inserted into prokaryotic hosts. The number of signatories of this unanimous statement had increased by one.[51]

It should be noted that while there were other participants at Asilomar, such as Curtiss, Falkow, and Robert Sinsheimer, who expressed conservative views, Andrew Lewis was the one who most clearly and steadfastly demonstrated his convictions against a popular tide.

Wednesday—Dissonance, Lessons in the Law

On the morning of the third day, copies of several communications were passed out. One was an open letter to the conference from Science for the

People. The principal message in this three-page document was that "decisions at this crossroad of biological research must not be made without public participation" and that the signers did "not believe that the molecular biology community . . . is capable of wisely regulating this development alone." It called for a continuation of the moratorium until several proposals for widening public input were put into effect. The authors were bacterial geneticists and molecular biologists, among whom was Jonathan Beckwith. In 1969, a Beckwith team had become the first to isolate a gene, the *lac* operon. There was no formal discussion of the letter at the conference, and scientific presentations filled the morning.[52]

After lunch, the report of the Working Group on Eukaryotic Recombinant DNA was presented by Donald Brown. This group envisioned that recombinant eukaryotic DNA could be hazardous in the following three ways:

> 1) a gene could function in the bacteria in which it is cloned and produce a toxic product; 2) a DNA component could in some way enhance the virulence or change the ecological range of the bacterium in which it is cloned; or 3) a DNA component could infect some plant or animal, integrate into its genome, or replicate, or by its expression could produce a modification of the cells of the organism.[53]

As they were painfully aware, the scientists here were grappling with questions for which existing knowledge was woefully inadequate. The very experiments proscribed as potentially hazardous were the ones from which the answers would ultimately have to come. Already there was skepticism that *E. coli* might simply replicate animal genes and never translate them into proteins, but the fundamental difference between translation of DNA by prokaryotes and eukaryotes was yet to be discovered. The frustration engendered by tireless invention of scenarios invited baroque and temporary constructions. The recommendations of Brown's working group included a classification of three major levels of hazard, with additional subclasses, for which a complicated hierarchy of containment conditions was arbitrarily assigned. "Shotgun" experiments in which a vector might be exposed to pieces of the total genome of a eukaryotic cell were all consigned to the highest hazard class, with mammalian DNA being particularly suspect, because it "more likely contained pathogens for humans." Such rulings caused dismay among those who would be forced to carry out their particular experiments inside scarce high containment resources. Disagreements over classification of hazards quickly cropped up and reappeared until the final hour of the conference.

Legal Opinions. After dinner, at the Wednesday evening session, Daniel Singer was introduced. He would preside over a small panel of lawyers he had selected in hopes of strengthening the framework for the final discussion on the following morning. Singer began by complimenting the scientists on the exercise of public responsibility which he perceived in their undertaking. He also wished to remind them that he saw the issues as being not only scientific, but that benefits and risks were social issues, and the public which was paying for the research would have to have its say.

Alexander Capron, a law professor at the University of Pennsylvania, began with his impressions of Asilomar so far, likening it to typical scientific meetings, highly technical in content—"like Cold Spring Harbor." "In other words," Capron snapped, "counter-phobic behavior." He too believed that the public would have to have its inning, and "the public" meant government and the law. Capron then coursed across the terrain of regulation, rule making, and legislation and concluded that he hoped that he had led the scientists to accept three things: some regulation is necessary, it may lead to restrictions, and public and governmental bodies would insist on having a say.

The third speaker, Roger Dworkin, a law professor at Indiana University, led the scientists out into the chilling atmosphere of legal liability. He described dangerous crevasses with names like proximate cause, negligence, and strict liability, and created courtroom scenes of litigation. Dworkin hit a particularly sensitive nerve as he discussed workers' compensation and regulatory agency involvement, including the roles of the Occupational Safety and Health Administration. Here he offhandedly suggested that even the secretary of labor might have the final authority over the rules for recombinant DNA research. Like Banquo's ghost, this specter reappeared several times before the discussion died down late that evening. Listening to the lawyers predict what might happen to the decisions made on the morrow, the scientists stiffened their resolve to close up ranks so that the world would see that the scientific community was able to finish what it had begun. More than all the others, the members of the organizing committee now realized that the product of their long last-evening's work had to be definitive.

The Final Hours

The final session, on Thursday, February 27, opened at 8:30 a.m. Paul Berg, who was keenly conscious that his deadline was noon, began by recapitulating the three responsibilities that the organizing committee had accepted.

> 1) to organize this conference to bring experts together for discussion of risks; 2) to attempt to determine what consensus exists and embody this in a statement; and 3) to prepare a statement to the NAS concerning the outcome of these deliberations.

Each member had that morning received a copy of the provisional statement that the organizing committee had spent the night preparing. There were six sections in this statement, and he opened discussion on the first. It was a statement of scientific accomplishments and an intimation that the situation was somewhat clearer than in the previous July. Several participants immediately raised procedural questions about how everyone might have input into the wording. Others inquired if all chance for modification ended with the close of this last session. A member of the organizing committee reminded the conferees that their report was not "written in a vacuum, but reflected the Committee's views of what seemed to have been agreed upon thus far." "Will we get to vote on each paragraph?" someone asked, and the chairman replied that he would prefer trying to detect informally what the consensus was.

Notwithstanding his reluctance to begin a series of time-consuming ballots, Berg quickly found that a vote was being forced by Brenner's suggestion that reaction to the following simple statement be tested: "Work should proceed, but with appropriate safeguards; the pause is over." Hands flurried up, and the chairman said that he perceived an overwhelming consensus on this statement. There was a palpable sense of relief at this forward movement, and discussion turned to the second point. It too was greeted by suggestions for improvement of grammar and form. Encouraged to concentrate on substance, the participants allowed the chair to decree a general agreement that "with reservations, some form of experiments should proceed; some, however, should not."

Upon entrance into the issues of actual levels of containment for experiments, the discussion deteriorated, but the chair gamely kept order. He listened to great differences of opinion on details and permitted polls whenever it appeared they might be useful. Feelings ran high. There were numerous attempts to amend some definitions of hazard from the floor. A voice cried out to protest that the carefully prepared statement of his working party had "been prostituted."

As the first lunch bell sounded, the discussion was jostling back and forth between the ultimate and the penultimate paragraphs. The moment for the final question could no longer be delayed. Berg, making himself heard over the commotion, flatly said, "All those in favor of this as a provisional statement, please raise your hands." Stanley Cohen protested loudly that he could not support something if he could not see the wording of it. "All those opposed to the statement," Berg now demanded. Richard Roberts counted "somewhere about four hands." Two of these belonged to Lederberg and Watson. A third was Cohen's. Waclaw Szybalski recalls, "I was strongly opposed, vocally objected, and raised my hand when negatives were requested." Philippe Kourilsky, agreeing with the count of "four or five," says his was also a negative vote. Thus, the statement with which they had begun the morning—although frayed and variously patched along the way—had made it through, still holding to the framework fashioned by the organizing committee in their last night's vigil.[54]

Someone had asked the Russian delegates to remain to the end. A spokesman for the group rose to indicate, in a few words, that a world partitioned politically could nevertheless hold an undivided scientific community.[55]

By 12:15 on February 27, Asilomar II had ended.

A press conference was held the following day. The members of the press who had attended throughout were now freed from their bonds; and having earned honorary degrees in molecular biology, they released generally laudatory and respectful commentary. On that same day, the new NIH Recombinant DNA Molecule Program Advisory Committee met for the first time and adopted the provisional statement of the conference as interim rules for federally supported laboratories in the United States.

Conclusion

Berg, Baltimore, Brenner, Roblin, and Singer submitted the final report of the Asilomar Conference on Recombinant DNA Molecules to the ALS under a cover letter from Berg, dated April 29. In keeping with the NAS tradition for the reports of its numerous committees, it was reviewed on this occasion by members of the ALS executive committee, who also received some comments from academy president Handler. It was approved on May 20, and appeared as a "summary statement" in the June 6, 1975, issue of *Science* and the academy *Proceedings*.[56]

Read today, this statement still stands as a lucid and fair description of the conference consensus as it has just been described. It does not seek to go beyond the facts as they were considered by the participants, neither in prediction of benefits nor in dismissal of any of the putative biohazards considered possible at the time. As Handler had commented in passing the report to the ALS, it was written "only to the cognoscenti in the field" and did not seek to deal with the question of being sure other audiences comprehended the conclusions.[57]

Judging Asilomar

Twenty-five years have now passed since the participants in the Asilomar conference went home to explain to anxious coworkers and laboratory staff what the new restrictions meant. Many also went to university leaders and institute administrators to argue for the new security facilities now required for their work. A few soon found themselves "at the barricades" in their own communities, such as Ann Arbor, Cambridge, and the Institut Pasteur, where tensions were rising. As fears diffused into the general population, not only laypersons, but dissident scientists as well, turned militant, and—as the lawyers had predicted—representatives of the U.S. government and those of several other countries rose to play their different roles.

2

A Federal Case

1975

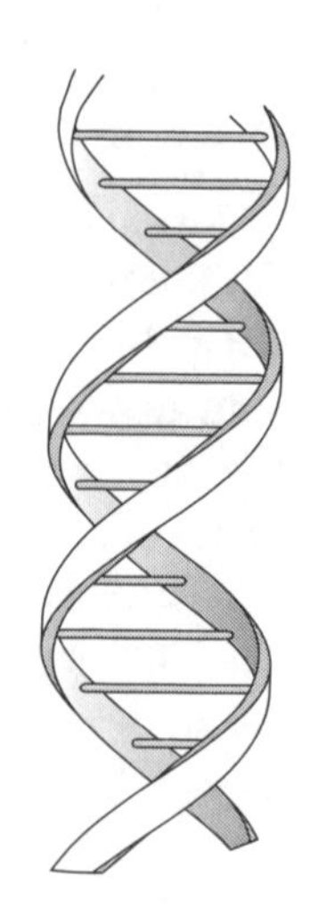

. . . We cannot hope to anticipate all possible lines of imaginative research that are possible with this powerful new methodology. Nevertheless, the Committee has received a considerable volume of written and verbal contributions from scientists in a variety of disciplines. In many instances the views presented to us were contradictory. At present the hazards may be guessed at, speculated about, or voted upon, but they cannot be known absolutely in the absence of firm experimental data—and unfortunately, the needed data were, more often than not, unavailable. Our problem then has been to construct guidelines that allow the promise of the methodology to be realized while advocating the considerable caution that is demanded by what we and others view as potential hazards.[1]

America's Move toward the Federal Solution

February 26, 1975. As the conference participants milled around the conference center following the last night's session, David Baltimore bumped into William Gartland, the "official" NIH observer. "Well," Baltimore said, "Tomorrow it will be all yours."[2] There were doubtless some of the molecular biologists who believed that a safety net built by the specialists would be enough of a response for both the scientific community and the public. But for most who had listened to the lawyers' dire warnings on the last evening, it now seemed highly possible that the outcome of the Asilomar conference could become a federal case—a public policy question, with the government misunderstanding the earnest search for the proper way to proceed with the solution to a problem that only they were poised to understand.

Almost instinctively, the scientists at the fateful Gordon conference in 1973 had turned neither to the NIH nor to any other government agency, but first to the National Academy of Sciences (NAS), preferring to share their dilemma "in the family," seeking an "in-house" solution. Clifford Grobstein later faulted the academy for not alerting the scientific community to the issues at the time it had the initiative,[3] but as the lawyer, Harold Green, observed years later, "the most important fact about the recombinant DNA controversy is that the 'organizing group' who called for the moratorium and who steered the Asilomar Conference chose to 'go public' with the problem whether or not they consciously intended or desired to engage the public [fully in the controversy]."[4] Indeed, the academy-blessed organizing committee that later met at MIT to develop plans for the international meeting had realized that at least the major funding agencies would have to participate in the solutions. Thus, the first "Berg letter" had invited government to play a role but one that was carefully circumscribed:

> Third, the Director of the National Institutes of Health is requested to give immediate consideration to establishing an advisory committee charged with (i) overseeing an experimental program to evaluate the potential biological and ecological hazards of the above types of recombinant DNA molecules, (ii) developing procedures which will minimize the spread of such molecules within human and other populations, and (iii) devising guidelines to be followed by investigators working with potentially hazardous recombinant DNA molecules.[5]

NAS vice president James Ebert observed at the time that he found this division of labor between NAS and NIH to be attractive. "The Academy has played a critical role in bringing the question to public attention . . . and the follow-up conference clearly falls within our mission. However the operating body to accomplish the further steps appears quite properly an NIH responsibility."[6] To the scientists, NIH was a known, trustworthy component of government, and few felt that any more government involvement should be necessary.

As far as the academy was concerned, the assessment of major, circumscribed roles for the premier scientific institution and federal biomedical research agency was reasonably on target. It is noteworthy and inescapable, however, that if the dangers of recombinant DNA experimentation inherent in the Asilomar discussions were to be taken seriously, the philosophical construct of a truly international scientific community was soon to be put to a practical test. An environment in which an unpleasant chimera can obtain even a remote niche is a global threat. Theoretically, a manmade cancer virus created in Prague could be in Philadelphia a week later if uniform worldwide precautions were not taken. At the time, the transcendent unity of a world scientific community, was, for all practical purposes, an idea shared by most scientists but never tested and quite unprepared for the consensus derived at Asilomar.

For the hard sciences, including biology, the means to judge the proofs are established and bond the community of scientists as no other convention might. However, science is carried out and supported by nations or states that are sovereign in their control of the profession and the institutions housing the laboratories. Very different customs, traditions, and governmental mechanisms are involved, and science is a source of pride and social and cultural esteem that have political reference to national boundaries within which the government, industrial, and scientific communities maintain an informed but unorganized relationship. Public governance of academic science in the mid-1970s mainly consisted of decisions about how much government funding would be allocated to science agencies. In the last two decades, the health sciences have been directed to emphasize research relating to particular diseases of population sectors. Fortunately, when the Congress has selected new targets, it has usually provided additional funding and not demanded that agencies redirect present activities.

By 1975, the amount of research in molecular biology in the United States was overwhelmingly greater than that of any other country and mainly funded by the NIH and National Science Foundation (NSF). The level of sophistication in Great Britain, France, and one or two other European countries was comparable to that of North America.

Parallel Actions in the United Kingdom

The British had examined the proposition of the first Berg letter from the day it had been received in 1974. Hans Kornberg, a professor of biochemistry at Cambridge, first took his copy of the Berg letter to Sir George Godber, then in charge of a committee organized by the new Health and Safety Executive to cope with a fatal laboratory spill of smallpox virus. The Medical Research Council (MRC), which was responsible for funding all the recombinant DNA research in Britain, quickly became aware of the Berg letter as well, and in July 1974—months before Asilomar—called it to the attention of the government committee which oversaw civil science. They in turn saw to the appointment of a committee, under the leadership of Lord Ashby, to examine the issues and Britain's position on the moratorium so abruptly declared by a small group of Americans. Ashby's committee, a distinguished collection of scientists and health officials,[7] rendered its report on January 15, 1975, shortly before the Asilomar meeting. It advocated immediate resumption of research under appropriate voluntary controls.[8] Partly because of the reaction of the strong union position of British laboratory workers, another committee under Sir Robert Williams, director of the Public Health Service and a member of Ashby's committee, was given the mandate by the Department of Education and Science to develop the genetic engineering controls advocated by Ashby. All but one of the members of the Williams committee were scientists. A single member represented the Health and Safety Executive. The latter agency, established in 1974, had authorities similar to those of the Occupational Safety and Health Administration (OSHA) in the United States but had more extensive rights of entry and means of control of practices over all laboratories. There would be considerable jockeying between Health and Safety and the Department of Education and Science before the equilibrium of executive control was achieved in Britain. The situation was not dissimilar to the American scientists' preference for the NIH to be in control of recombinant activities as opposed to any of the regulatory agencies, including the Environmental Protection Agency (EPA), the Center for Disease Control (CDC), the Food and Drug Administration (FDA), OSHA, and its cousin, the National Institute for Occupational Safety and Health (NIOSH), the latter being located in the Department of Health, Education and Welfare (HEW). The story of NIH in regard to these other agencies, just awakened to the recombinant DNA controversy, is about to unfold (see chapter 4).

Thus, during 1975, two committees would be working across the Atlantic to construct two sets of guidelines for recombinant DNA research, one British[9] and one American. This is the story of the American one.

The NIH Recombinant DNA Molecule Program Advisory Committee

The NIH Recombinant DNA Molecule Program Advisory Committee, popularly known as the RAC, was formed by NIH in response to the Berg letter of July 1974. The original name was a misnomer in one important sense. The NIH never had a "program" of recombinant DNA molecule research.

The NIH director in 1974 was Robert S. Stone. Paul Berg wrote a letter to NAS president Handler in June 1974, urging him to be sure of a swift response from NIH.[10] Director Stone in turn wrote Handler in July 1974 that NIH had been apprised of the report "prepared by the committee of the Assembly of Life Sciences, chaired by Dr. Paul Berg," and was "developing guidelines, soon to be published, for assuring containment of such agents."[11]

The director appears to have been referring to new guidelines for viral pathogens that had been in progress at NIH before the RAC was established. Nevertheless, *Nature* lauded him for his "swift and positive response" in indicating he would form a committee to define the possible hazards.[12] In a box on the same page, Michael Stoker[13] of the Imperial Cancer Research Fund noted that most of the technology involved in all this had been developed in the United States and "It is encouraging that the very leaders in the field have taken the initiative and been supported by the academy. . . . for many it will be a test of self denial and social responsibility in the face of strong intellectual temptation."

In addition to the protracted self-denial, the molecular biologists had unwittingly matriculated in a tutorial in federal procedure. When he wishes to form a national advisory group, the director of a federal agency is not only bound by the 1972 Federal Advisory Committee Act but is also at the mercy of the rules and procedures of the executive branch. The draft charter of the RAC had to make its way progressively to the desks of the surgeon general, the assistant secretary for health, the HEW general counsel, and the executive secretary of the HEW secretary. The time required for the eventual arrival in the HEW secretary's work basket of such a document was the inverse of the diligence of the HEW staff in pushing it along. Once it was returned to the agency with the seignorial stamp, the nominations for the members would often have to retrace the same route for approvals.

If the White House advisors or the president became involved, the pace of a charter could be glacial, and, as I was later to discover, charters might be withdrawn by executive whimsy.[14]

The charter of the RAC was eventually signed by HEW secretary Caspar Weinberger on October 7, 1974. It carried a sunset clause, requiring renewal after 2 years.[15] Twenty-five years later, the RAC is still in operation, and many biologists are unaware of its origins. In 1993, a reporter for the *Washington Post* incorrectly described the RAC as having been "created [recently] to review experiments for transplanting genes in man."[16]

Recruiting the Members of the RAC

In anticipation of the successful passage of the charter during the summer of 1974, Stephen Schiaffino, the veteran NIH associate director for scientific review, began a procedure he had performed scores of times before for other committees. He solicited suggestions for the RAC membership from the executive secretaries of the genetics, virology, molecular biology, and microbial biochemistry study sections, who in turn canvassed the members for their suggestions. By early September, he had a list of over 40 experts from which he and Leon Jacobs would select 11 members. Jacobs was a microbiologist of the old school, with the title of NIH associate director for collaborative research. He would serve as cochairman of the RAC and was a vocal participant in the Asilomar conference. Neither Schiaffino nor Jacobs could have dreamed that 3 years later, a prominent senator would move legislation to do away with the RAC, replacing it with a commission, each member of which was to be appointed by the president, and the chair confirmed by the Senate. The commission, furthermore, would have a minority of members who "have been professionally engaged in biological research." This part-time body was to be responsible for both the writing and the policing of the guidelines and, in its spare time, was to undertake "a study of the basic ethical and scientific principles which shall underlie the conduct of recombinant DNA research."[17]

NIH director Stone appointed DeWitt Stetten, Jr., to be the first chair of the RAC. DeWitt (Hans) Stetten, then NIH deputy director for science, had a distinguished career in both research and administration. An expert in biological chemistry, he had been a faculty member at Columbia University and at Harvard before he became a division chief at the Public Health Research Institute of the city of New York. Stetten had then come to NIH in 1954 as scientific director of the then National Institute of Arthritis and Metabolic Diseases.[18] For several years also director of the National Institute of General Medical Sciences, Stetten moved in 1974 to the position of deputy director for science, the head of the entire intramural program of NIH. Given his scholarly temperament, Stetten was an excellent chairman of the sometimes tempestuous RAC.[19] Stetten's high morality and idealism left him vulnerable as the RAC's course headed into the turbulent waters of political and social conflict. When he resigned from the chairmanship of the RAC, his farewell remarks were tinged with exasperation and disillusion. Particularly, he disagreed with the earliest action on the NIH Guidelines that I took as Dr. Stone's successor.[20]

By December 1974, Paul Berg was very concerned with the time it had taken to gather up the RAC in time for the approaching Asilomar conference.[21] To mollify Berg, Philip Handler suggested they could let NIH stall for awhile and route the outcome of the Asilomar conference through the Assembly of Life Sciences.[22] Just before Christmas, however, Ronald Lamont-Havers, now acting director of NIH after Stone's departure in

February, wrote Berg a long letter to explain the apparent foot-dragging of the NIH. It had been easy to comply immediately with the request for support of the forthcoming conference, "an issue solely within the control of NIH," he wrote, before going into some detail about the chartering of a committee in the federal milieu. However, he could now state that the nominees had all just been approved and would be at Asilomar.[23]

The first members of the RAC were Edward Adelberg, molecular geneticist from Yale; Ernest Chu, molecular geneticist from the University of Michigan; Roy Curtiss III, microbiologist and constructor of vectors from the University of Alabama; James Darnell, cell biologist from Rockefeller University; Stanley Falkow, microbiologist from the University of Washington; Donald Helinski, molecular biologist from the University of California at La Jolla; David Hogness, molecular biologist from Stanford; John Littlefield, molecular biologist from Johns Hopkins; Jane Setlow, molecular biologist from Brookhaven National Laboratory; Waclaw Szybalski, phage expert from the University of Wisconsin; and Charles Thomas, molecular biologist from Harvard.

Three members of the RAC (Adelberg, Hogness and Thomas)[24] were actively conducting recombinant DNA experiments. The majority were not, but every member was an expert in the technologies immediately concerned with recombinant DNA experimentation.

Additional Members

From the first, the RAC also felt it was deficient in certain types of expertise and urged that its membership be expanded to include persons experienced in epidemiology of infections, virologists, and scientists with broad views of biology. Within a few months, several others had joined the RAC: Elizabeth Kutter, biologist from Evergreen State College, Washington; John Spizizen, microbiologist from the Scripps Research Foundation in La Jolla; and Wallace Rowe, chief of viral diseases, National Institute of Allergy and Infectious Diseases, initially the only member from the NIH intramural staff. One layperson was added to the RAC late in 1975: Emmette S. Redford, professor of government and public affairs from the University of Texas. LeRoy Walters, an ethicist from Georgetown University, became the second layperson added to the RAC, early in 1976.

The Tasks of the RAC

The all-points appeal to scientists in the Berg letter of July 1974 to temporarily halt all attempts to open Pandora's box was as effective as it was unique and presumptuous. As we have seen, 7 months later at Asilomar, the historical precedents were extended. An assemblage of the world's foremost experts had concluded that the attempts on the lock must be resumed but very carefully and by working together. Experimentation was to be resumed only under conditions of containment, and effectiveness of containment was to match the assumed risk. The spokespersons added an almost plaintive codicil crucial to this agreement: "Hopefully, through both formal and informal channels of information within and between the laboratories of the world," the match of containment to risk should be consistent.[25]

These statements on the conclusions of the Asilomar conference, first drafted by the organizing committee on the last evening of the conference, finally appeared in print in June 1975. The summary was a scientific paper to be read by experts and as such was a fair statement of the consensus reached on how to resume potentially dangerous laboratory experiments, the risks of which were frankly unknown. Scenarios of potential dangers were omitted, as were defensive estimates of the rich rewards in new knowledge anticipated.

Containment was to be physical and biological. Biological containment meant creation of genetically crippled "fastidious" bacterial hosts and vectors that could not survive outside the laboratory. Many scientists realized the urgency of this directive as they heard the stringency of physical containment proposed for their particular experiments. Laboratories with negative air pressure locks and means of containing fluid leaks were found in very few academic laboratories in 1975.

The report scaled the experiments from *minimum risk* to *maximum risk* in four steps. Minimum meant the use of prokaryocytes, phage, and plasmids with known limits of exchangeability. The maximum risk lurked in the employment of dangerous pathogens as hosts or vectors of new recombinant genes. In between lay precautions built, for lack of knowledge, on the most conservative views. A fear that cryptic viral genes might be unwittingly attached to bacterial genomes and spread potential carcinogens led to scaling of risks of using genes from humans and other warm-blooded creatures as among the most dangerous. The shotgun experiments in which uncharacterized eukaryotic DNA would be randomly attached to the sticky ends of restriction enzyme-treated prokaryotic genes also provided great anxiety. That prokaryotes could not translate the full message of eukaryotic genes was but one of the myriad unknown facts ultimately to be learned from recombinant technologies. The paradox was that the evidence needed for its liberation lay inaccessible without experiments using the recombinant technology. This is part of the story of all treks by humans into the unknown.

The First Steps of the RAC

The state of the art was to remain relatively stationary through the remaining months of 1975 during which the RAC labored to convert the Asilomar statements into formal guidelines. The committee met the day after the Asilomar conference in the no-frills government-class setting of the Capri Room of the Hotel Bellevue in San Francisco. The meeting was open to the public, as were all sessions of the committee, a requirement of the Federal Advisory Committee Act.[26] A few of the reporters from Asilomar attended.

The RAC was a cross section of the participants at Asilomar. Chairman Stetten observed that "a wide diversity of viewpoints was apparent among [them], ranging all the way from those who felt that recombinant DNA experiments posed little or no increased risk to those who were deeply concerned about the possibility of massive adverse impact on our environment and our eco-system."[27] A bit uncertain about whether they were a distinct national body or an extension of Asilomar, the RAC members were read their charge by Leon Jacobs. In addition to formulating guidelines, they were to advise the NIH director on (i) the conditions which the NIH should impose on its grantees and contractors working with recombinant DNA molecules, (ii) the level of effort the NIH should make to provide high containment facilities, and (iii) steps NIH should take to stimulate research to reduce the biohazards.

For this meeting and its second in Bethesda the following May, the RAC adopted the Asilomar report as interim guidelines, but it quickly recommended other conditions regarding support of recombinant DNA experiments. One of the first was that grants were to be awarded only after a review of the risks involved by a "biohazards" review committee of the applicant's home institution, accompanied by a certification by the institution that provisions were adequate for risk containment. The adequacy of the certification would be judged, along with the quality of the research proposal, by the NIH study section reviewing the applicant's proposal. Because the Asilomar conference report was not yet a public document and could not be officially cited, it was suggested that the NIH send applicants copies of the 1974 Berg letter and the British Ashby report, which had appeared just before Asilomar. An appendix to the grant application would include data on experiments proposed, the vector to be used, the investigator's estimate of the potential biohazard, and a statement of understanding of the Asilomar and Ashby reports.

The requirement of review by an institutional committee was an adaptation of a Public Health Service rule for clinical research. This rule required that research on human patients carried out in any grant-supported institution had to be overseen by a local body, usually designated an institutional review board (IRB).[28] Comparable committees would be set up for oversight of recombinant DNA research. At first, these committees were called institutional biohazards committees (IBCs). In its earliest prescription, the RAC was adamant that the local committees were to make no judgment of the *quality* of the research proposal. This was reserved for the NIH-run study sections. The preference of American scientists for judgments by their peers serving in national study sections remains one of the greatest strengths of America's scientific system. Submission to the decisions of the RAC reflected a faith in the RAC as being, at the least, truly a peer group. This was an act of assurance that the traditional canons were operative.

The RAC further urged NIH to encourage grant applications for the development of bacterial strains incapable of living outside special laboratory-produced environments and to promote their free exchange. Also, a newsletter should be established to keep all scientists in the world au courant. Grantees of the NSF, then the only other significant source of support for this research, were requested to inform the NSF that they were aware of the Asilomar principles and agreed to follow good laboratory practices. Initially, there was conjecture about how uniformity of adherence to a single set of guidelines would eventually be obtained in the United States, let alone the rest of the world.[29]

The Hogness Draft

Having been distracted with organizational matters up to now, the RAC realized they had to bear down at their third meeting on the grinding task of working out the guidelines. To try to get this under way, David Hogness agreed to draw up a model set with the assistance of a small committee consisting of Chu, Helinski, and Szybalski and a few outside consultants in advance of the next meeting.

The "Hogness draft" was taken up by the RAC at their third meeting in the comfortable summer headquarters of the NAS at Houston House in Woods Hole, Mass. Only eight members of the RAC were there. Observer participants also present were Peter Day, an agricultural expert who was a constant consultant; Louis Siminovitch, the foremost recombinant expert in Canada; and Elizabeth Kutter, a biologist from Evergreen State College who was soon made an RAC member. At the end of the 2-day session, the latest revision of the guidelines was informally circulated.

A week later, DeWitt Stetten was abruptly reminded of the potential divisiveness of the recombinant DNA dilemma and the attentiveness of the scientific community to the work of the RAC. He received a letter from RAC member Stanley Falkow, who apologized for his inability to get to Woods Hole, adding that he "strongly dissents from the Committee majority's approval of the revisions" in the Hogness draft. The RAC ought to "re-read its Charter" and "if these guidelines are adopted as NIH policy, it would seem hypocritical for me to continue to serve." Falkow's objections included his noting that the prohibited experiments did not include a single example of an experiment with a eukaryotic gene product or an animal virus (except viruses in classes 3, 4, 5). Furthermore, the data he had provided the committee at its first meeting, showing *E. coli* K-12's ability to carry some transferred gene outside into the local disposal system, had been ignored, and the idea that all cold-blooded animals and all other "lower" eukaryotes could be handled in the P2+EK1 host-vector system seemed "built on a foundation of quicksand."[30]

Then arrived a letter bearing nearly 50 signatures proclaiming that the Woods Hole guidelines had been read at a Cold Spring Harbor bacteriophage meeting and there were objections to the "lowered safety standards . . . failure to prohibit at this time the most hazardous experiments . . . that no mammalian DNA should be cloned under less than P3 containment and that the RAC should consider limiting shotgun experiments of mammalian DNA to P4 containment until proven safer vectors are available. . . . the committee should have more scientists who are animal virologists, plant pathologists, geneticists, and epidemiologists. . . . why not some scientists who are not directly involved in cloning experiments? . . . representation of the public at large would be advisable. . . ."[31]

Helinski, who had also missed the Woods Hole meeting, wrote to say that he found the downgrading of containment for shotgun experiments and cloning of animal viruses an error. Stetten replied that Curtiss (who had been present) and Falkow felt exactly as Helinski did but that Adelberg (who had also attended) felt the revisions were "entirely appropriate."[32] Marshall Edgell complained that the revision was too strict. "I do not believe the right to free inquiry is absolute. However, it is so precious that we should be exceedingly cautious in its abridgement."[33] Paul Berg's detailed critique was thorough and did not hide his displeasure:

> The Hogness draft recommended P3+EK3 (their most stringent requirement) and even for so-called safe cell lines they suggested P3+EK2. The new version suggests P3+EK2 for all mammals and P3+EK1 or P2 (open labs)+EK2 for warm-blooded animals other than mammals.[34]

The letter is illustrative of Paul Berg's unrelenting concern for the course of the NIH Guidelines, as well as the opalescence of the detail into which they were descending.

The "Final" Draft

The Woods Hole version of the guidelines was placed in the hands of Elizabeth Kutter, who worked with another subcommittee to bring another draft to the next meeting of the RAC on December 4 in La Jolla, Calif.

At this meeting, DeWitt Stetten's recall of Furness's Variarum Variation edition of Shakespeare provided the necessary rack from which the NIH Guidelines were finally synthesized from the winners of innumerable parliamentary votes on text and syntax of each of the differing sections of the Hogness, Woods Hole, and Kutter versions.[35] The "tightness of the final version" was attributed by Nicholas Wade to the presence at the La Jolla meeting of Paul Berg, Maxine Singer, and Sydney Brenner, three of the spiritual leaders at Asilomar.[36] Spirits were high, too, because Roy Curtiss had only a few weeks earlier announced that he had been able to produce a disarmed strain of *E. coli*, permitting the stringent biological containment prescribed for some of the experiments.

The Guidelines Compared with the Asilomar Consensus

The skeleton of the Guidelines emerging from this 10 months' work by the RAC was faithful to the Asilomar consensus. Generally, the RAC had selected the more conservative positions on physical and biological containment for some classes of experiments. It also established as official the new shorthand for describing such containment with its P1 through P4 terminology for physical containment and EK1, 2, and 3 for biological containment, the latter being more idealistic than actual at the time. The Ps and EKs quickly became the basic jargon for intercommunication of molecular biologists worldwide. The NIH benefited enormously from the services of Emmett Barkley, then head of the safety office of the National Cancer Institute, whose documents describing the containment systems and recommending standards for safety procedures would serve as information sources for scientists and institutions in much of the world.

Prohibitions

The list of experiments that were prohibited under the Guidelines was explicitly presented in the final draft.

1. Cloning of recombinant DNAs derived from the pathogenic organisms in classes 3, 4, and 5 of the "Classification of Etiologic Agents on the Basis of Hazard" by the Center for Disease Control
2. Deliberate formation of recombinant DNAs containing genes for the biosynthesis of toxins of very high toxicity (e.g., botulinum or diphtheria toxins)
3. Deliberate creation from plant pathogens of recombinant DNAs that are likely to increase virulence and host range
4. Widespread or uncontrollable release into the environment of any organism containing a recombinant DNA molecule unless a series of controlled tests leave no reasonable doubt of safety
5. Transfer of drug-resistant traits to organisms that are not known to acquire them naturally should be deferred if they could compromise the use of a drug to control disease agents in medicine or agriculture
6. In addition, we recommend that at this time large-scale experiments (e.g., more than 10 liters of culture) with recombinant DNAs known to make harmful products not be carried out.

The last restriction immediately bothered industrial concerns that were eager to capture the possibilities of this new technology. The resolution of these problems is taken up in chapter 11.

Before the Guidelines were promulgated, it was decided to remove the qualification for any release of organisms containing recombinant DNA into the environment. This was an important rule to reduce anxiety in the first several years of use of the technology. Later, when recombinant DNA technology began to be used to create altered plant and animal products that were designed to survive when released into the environment, the RAC would become enmeshed in a number of interagency constructions attempting to provide broad expertise (and perhaps spread the responsibility for any grave error). The Agriculture Department and the EPA eventually became more appropriate centers for this responsibility.

Napoleonic Code versus Common Law

As the American Guidelines were coursing toward completion, the U.K. working party under Robert Williams was proceeding to develop some ground rules for British scientists. The Department of Education and Science released the Williams report in August 1976, a month after the NIH Guidelines had been accepted and promulgated. As expected, the Health & Safety Executive released draft regulations for compulsory notification of experiments involving genetic engineering. The Williams party categorization of experiments was not too far from that of the NIH Guidelines. There was no mention of biological containment, and the British P1, P2, and P3 requirements were a bit stricter than those of NIH. There was no restriction on *E. coli* K-12 as a host. The Williams report also recommended a national-level central advisory service named the Genetic Manipulation Advisory Group (GMAG).

Struggle over whether the GMAG should be a creature of the Health & Safety Executive or have some more "academically minded" chaperon went on for several months. By the time the report was released, GMAG was the responsibility of the Department of Education and Science, with the Medical Research Council providing its secretariat. At first, the draft regulations requiring notification of experiments to Health & Safety fell to objections of the scientists.

Susan Wright has suggested that "the convergence of the British and American systems" was primarily the product of political interests, and did not represent an international scientific consensus.[37] This is a debatable opinion. What is beyond question are the marked procedural differences between the British and American modes of drafting and administering guidelines.

The proceedings of GMAG were to go on behind closed doors, in accordance with the British Official Secrets Act. Moreover, the decisions would be made case by case under the guidance of the simpler rules laid down by the Williams report. The version of the NIH Guidelines approved at the La Jolla meeting was massive, covering in great detail the precise means of selecting containment and an ample start at methods of implementation that would later undergo further growth. In contrast to our British cousins and the common law tradition they had bequeathed to America, the *NIH Guidelines for Recombinant DNA Research* were a voluminous Napoleonic Code.

A partial explanation of this difference rests in the insistence of the Americans, who had dominated Asilomar and its preliminaries, that there be a level playing field, as reiterated in the summary report of Asilomar submitted to the NAS. The American proposals would have to be made in the open, according to the fully detailed rules. The detail of the NIH Guidelines was also compatible with the maximum consistency of decisions by local institutional committees and by study sections.

The greater number of American laboratories, supported by multiple agencies and institutions, including commercial activities, justified an expectation that one set of guidelines would have to apply to all the laboratories in the United States. The RAC did not foresee how this would be done, beyond the belief that few other American participants would take the trouble to write their own rules if the coverage in the NIH Guidelines was overwhelming. By 1976, many countries were examining the U.S. and U.K. guidelines and preparing to select one or the other or a mixture of the two. The extension of the Guidelines to all recombinant DNA experimentation in the United States was considered very important from the first, but given the limitation of NIH authority to only its own employees or grantees, further extension to other laboratories remained an intermediate problem to be overcome. The story of the eventual solutions to this problem offers lessons in what is likely to occur in future dilemmas of science policy and practice.

The Chairman's Disillusionment

Hans Stetten had passed a grueling examination for pilot rank in bringing the RAC into the harbor with its catch of rules designed to open exploration of the most promising of all conquests of biology. Moreover, he had performed with no instruction from Bethesda, for the NIH director's chair had been vacant for half a year, and I, as the successor since mid-1975, had offered no intervention.

In his memoirs, Stetten describes his feelings in making final comments at the La Jolla meeting of the RAC: "It must be stressed at once that our charge was to write guidelines, which we construed as advice to those concerned. There was no suggestion that we should write regulations, much less laws."

Stetten records his impression of the terminal event of that otherwise triumphant day.

> The committee had come to a document upon which there was a high level of agreement. Tired but encouraged, I picked up the telephone and placed a long-distance call to the Director of the NIH, Donald Fredrickson. When I reported what had been accomplished, his initial response was, "We will have to submit the guidelines to the NIH Director's Advisory Committee."
>
> I blush to state that my response was "Oh, my God!" Knowing full well what the bureaucracy would do to our fine document. . . ."[38]

Later, those who saw him come back from our brief telephone conversation told me that all his energy seemed to have disappeared, his shoulders sagged, and his face was drained of all color as he relayed to the others, "He wants to have a public hearing on them!"

I understood the emotions expressed by the chairman of the RAC. Hans Stetten clearly considered that I was opening the gates of the campus to an unprecedented Philistine invasion. He was a scientist who rejected any reference to politics and to whom the idea of having laymen determine scientific prerogatives was simply unacceptable. This had always been the quintessential attitude of scientists working in independent institutions. The scientists accept any lay support on their behalf as long as it exclusively leaves to them the selection of *specific research* projects and the actual *doing* of the science.

Before I can explain my decision conveyed to Hans Stetten and describe the beginning of my guardianship of the covenant at Asilomar, I owe the reader a brief transitional explanation.

Transition

Coincidence

February 26, 1975. On the very day of David Baltimore's proffer of "the federal case" to William Gartland in Pacific Grove, Calif., I was in Washington considering a message from the White House. President Gerald R. Ford had sent word to the Senate asking for my confirmation to fill the empty seat of the NIH director. On July 1, 1975, I took the oath from HEW secretary Caspar W. Weinberger, in the presence of the president.[39]

Scientist

Twenty-two years earlier, in 1953, I had begun my career as a scientist at the National Heart Institute as a member of the first class of the institute's clinical associates.[40] I was assigned to a group headed by the protein chemist Christian B. Anfinsen, whom James A. Shannon, then scientific director of the institute, had hired to lead the team setting out to understand lipid metabolism and its relationship to arteriosclerosis. With Anfinsen, and later with many colleagues and postdoctoral students, I had a glorious 21 years of clinical and laboratory research in the new field of plasma lipoproteins, where we made some useful discoveries, characterized a couple of new diseases, and allowed me to lecture around the world.[41]

Leader of the Institute of Medicine

I left NIH in 1974 to become the president of the Institute of Medicine of the NAS. I had not completed a full year before I was asked to return to NIH as director.[42]

3

The Gathering Storm

1975

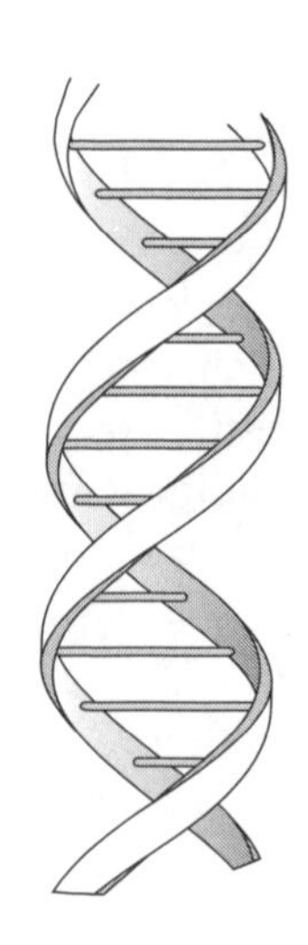

Scientists understandably balk at the notion of regulations of science and technology. Not to regulate is as much a public decision as the opposite and only postpones the time of regulation. Controls will eventually be called for. Concepts of "risk". . . need to be broadened to include some of the social and ethical consequences.[1]

LEON KASS

July 1975. I took up my new post innocent of what was ahead. The hearing in the Senate, in a small room on May 2, had been conducted by Senator Harrison Williams (D-NJ).[2] Ronald Lamont-Havers, who had been caretaker of NIH during much of the preceding 3 years, came to my side and handed me a note to tell me that "everything was fine on the campus." Nothing said about a place called Asilomar or what that new committee called RAC was working on off-site. The lawmakers I met during the hearing or on my preliminary visits were not anticipatory of anything out of the usual, no questions about genetic perils. The reports from Asilomar by the press attending had been muted and in no way alarming, and news of other manifestations had not penetrated Washington.

I can no longer remember when during the summer, or by whom, I was first told that the Recombinant DNA Molecule Program Advisory Committee would soon be finishing the *NIH Guidelines for Recombinant DNA Research* and that when they came I would be directly responsible for them. The NIH virologist Andrew Lewis later told me that no one had early approached me to discuss the recombinant DNA issue, because "you were a cardiologist and nobody knew what your reaction would be."[3] I believe it was Wallace Rowe, the lone intramural scientist member of the RAC, who came into my office to alert me to what was coming. A large man with bright blue eyes under disheveled hair, who was said to work through most of the night, Rowe was a world-class virologist whose ethical views I came much to admire in the years before he later succumbed to cancer while I was director. After Rowe's descriptions I asked him who else—on the campus and elsewhere—could give me further information about Asilomar and what they felt about the coming guidelines. I'm sure that he, or Hans Stetten, brought me in contact with Maxine Singer, Malcolm Martin, and other informants. I don't recall early visits from any other members of the planning committee, with the possible exception of Norton Zinder.

As I look back now upon the actions we took and the decisions that we made, I am emboldened to do justice in relating this story to cover all sides of the controversy that followed. I am also reminded of Stephen Toulmin's remarks at a National Academy of Sciences forum later in 1977:

> I think a disinterested outsider is justified in saying that Paul Berg, Don Fredrickson and the high command at NIH did a very respectable and conscientious job of working on those safeguards in the absence of any proper institutional set-up for dealing with the societal aspects of science policy.[4]

He was condemning, as did many others after Asilomar, the "lack of more representative and responsible institutional machinery to assure that the interests of science and the larger community are served. . . ."

We had no time to supply this missing societal machinery, and certainly we would gain nothing by reentering the complete moratorium that had preceded Asilomar. We had to move rationally within the frame that had been imposed upon us.

Obviously, I knew less than the experts at Asilomar about the detailed risks of the experiments proposed. My first task was to discern who were the experts, a judgmental process that is standard equipment for anyone for whom science is a profession. Fortunately, NIH was located in the center of the science, including many who had participated not only in Asilomar but also in the events leading to it. I listened carefully to the opinions that this novel class of experiments might produce new pathogens causing human disease or create new organisms that would find niches in the environment and perhaps alter it irrevocably. I also comprehended that within the technology lay the route to an understanding of human genetics and development that would be revolutionary. It seemed obvious that only further careful experimentation could determine whether the assumed hazards of recombinant DNA (rDNA) were real. I was made aware of the impatience of the molecular biologists to resume the experimentation they had voluntarily stopped. I also harbored confidence in the sincerity of the precipitous actions taken by the organizing committee and was grateful that they had at least obliquely dealt NIH into the decision process. It was apparent that, whatever flaws lay in the proceedings at Asilomar, the overwhelming majority of the collective expertise believed it desirable to proceed under universal cautions. Only a handful had favored a continued moratorium, but many were obviously uncertain whether the precautions were great enough. It was without question that we were caught up in a move that would come to be regarded as extraordinary in the history of science. My judgment of the quality of the scientific leadership involved in starting and successfully concluding the meeting at Asilomar was high enough to inspire my confidence.

Nevertheless, I had to ask, were they certain enough at the time that the level of risk to the public did not outweigh the potential benefits? Was I? Recently, I reread an essay on the DNA controversy and NIH's dilemma with regard to a regulatory role. It defines the limitations of a scientist in a position of responsibility for an institution like NIH in coping with a major decision such as that presented by the subject of Asilomar. It was written by Peter Hutt, who frequently played a useful role as a critic of my actions as the NIH conductor during this affair.

> We must recognize that this type of issue presents fundamental differences in philosophical principles, not simply a narrow dispute on technical details. It raises the most basic questions of personal beliefs and human values—the degree of risk or uncertainty that any individual is willing to accept in his daily life. Attempts to resolve it on the basis of rigorous scientific testing or analytical discourse, therefore, simply miss the point. . . . *Nor, indeed does a scientific background equip one with any greater insight into this type of policy issue or any more impressive credentials or greater authority to act as arbiter in resolving these matters* [italics mine]. As long as we remain a free society, these basic philosophical principles will, and properly should, remain subject to intense public scrutiny and debate.[5]

Firm Convictions

I did not pretend to have articulated any set of well-defined philosophical principles. I had formed, however, three firm convictions about the next steps in dealing with the rules coming from the RAC. The first was that the public must somehow be consulted before the Guidelines were promulgated and consistently during their subsequent evolution. The second was that the rules must remain in a flexible format admitting their constant revision as the evidential base changed. The third was that NIH should be the venue where these issues could best be handled. I shall deal primarily with the last issue, the uniqueness of NIH in potentially handling the rDNA controversy.

Uniqueness of the Agency

Positives

As an agency of HEW, and part of the Public Health Service, NIH was under the authority of the executive branch and subject to congressional oversight. It had the mass of scientific expertise, both in its intramural scientific staff and in its extramural grantees and their institutions, to cope with the technical details involved. Moreover, as the world's largest supporter of nonprofit biomedical laboratories and institutions, it had the fiscal power to ensure compliance among its grantees and contractors. I also regarded it as an advantage that NIH had no formal regulatory authorities, even though we would be scorned as amateurs as we sought to avoid the regulatory thicket. None of our companion government agencies, such as FDA, CDC, OSHA, NIOSH, or EPA[6] then had the scientific expertise to match their potential regulatory authority over this area of science. And none had the confidence of the molecular biologists. Although we should have to take steps later to be sure, NIH thus seemed in the strongest position among all federal agencies to achieve the objectives of appropriate protection of the interests of both science and the public.

Negatives

NIH, however, also lacked certain strengths. Its grasp extended only to its employees, grantees and contractors, and their institutions. It had no control over other branches of the government and indeed had not ascertained if any of the regulatory authorities of the other agencies might supersede NIH authority. NIH did not have control over the many expert molecular biologists who were employees or grantees of the National Science Foundation or the Department of Agriculture, and NIH had no control over activities in the private sector, including commercial use of molecular biology. It was essential that these gaps be closed if the guidelines were to work and survive. In decisions dealing with public safety as well as that of laboratories, NIH would also "come with its own baggage," the readily admissible and unavoidable appearance of conflict of interest (to some, the stigma of "regulation by experts"). We would have to proceed most judiciously and with maximum consultation with all parties of interest in order to achieve and retain acceptance of NIH dominance.

All in all, the NIH was nearly a unique institution, and no other body had yet called out or been summoned to take its place. To be sure, I also knew NIH very well[7] and had inherited from James Shannon that sense of relative independence from the Public Health Service and the greater mobility of the institution that would be useful in maintaining control of this (hypothetically) potentially hazardous new period of scientific exploration.

Mobilizing the Troops

By this time, word came of several manifestations, in Paris and in Michigan, against acceptance of the hazards of recombinant DNA research. Therefore, we were urged to get prepared. To cope with immediate objectives and for pursuit of a long campaign, and in accepting responsibility by delegation of the secretary of HEW for decisions on which the public safety was dependent, I would need all the forces at my command. During my first months as director, I made an important appointment. I convinced Joseph Perpich, whose background had caught my attention at the Institute of Medicine during my brief presidency, to join me at NIH as associate director for planning and program evaluation. The title of the position was now an anomaly,[8] and I converted it to an office similar to executive secretariat and general counsel combined. Perpich had invaluable training and experience for such duties. An M.D. with specialty training in psychiatry and a degree in law, he had been clerk to Judge David Bazelon of the U.S. Third Circuit Court of Appeals. He had also spent time on the staff of Senator Edward Kennedy and had an invaluable sense of the etiquette and choreography of "the Hill" as well as its sensitivity to the public. Throughout the rDNA controversy, he served as my closest advisor and assistant, and his previous juridical and political experience was reflected in the style imposed on the format of many documents and decisions. It also kept us ready to turn a flank to repel predicted or unexpected assaults.

The Kitchen RAC

Perpich and I, with valuable help from Bernard Talbot, who was special assistant to Hans Stetten and later my "special assistant for recombinant DNA," set out to enlist the support of an informal, unchartered team of highly informed, independent-minded advisors from the NIH staff who would help the NIH director to understand, and approve or decline, all recommendations of the RAC and to defend those actions from challenge. The immediate availability of numerous expert NIH intramural scientists gave the NIH director a unique and enviable advantage over any other federal science agency. Maxine Singer, chief of the Nucleic Acid Enzymology Section of the National Cancer Institute (NCI) and one of the original organizing committee that had created the Asilomar meeting, played a continuous, unique role in the events moving toward the resolution of its consequences. The virologist Wally Rowe, an ingenious researcher and a complex person, whose conscience would often make him alternately conservative and daring, served me as a very instructive model of the "molecular biologist's mind," a world in which I occasionally found myself wandering. His compatriot Malcolm Martin was an expert in molecular biology, microbiology, and virology who possessed a Wellsian (H. G., not Orson) view of the future of gene translation and who, with Rowe, was a solver of special problems, including the polyoma experiment and the Falmouth and Ascot meetings. An invaluable scientist was Susan Gottesman, Laboratory of Molecular Biology, NCI, who came to be the ranking authority in the world on the interpretation of the Guidelines. Sue was more, for she had a talent for leading work groups through parliamentary resolutions of reconstruction of the Guidelines. Her votes on the amending of RAC motions were a barometer that frequently alerted me to issues unperceived and meriting special care in fulfilling my responsibilities.

Other talents were needed and readily came to help, like Emmett Barkley, NCI's young safety expert, who was conversant with containment, departmental procedures, and relevant laws and regulations. (". . . the Spencerian literacy, stubborn streak of honesty . . . indispensable people skills; could anyone else—except Diderot perhaps—have put together Barkley's Compendium of Containment?"[9]). There was also Joe Hernandez, Division of Legislative Analysis, NIH, to whom I wrote a farewell: "What are we to make of your choosing to leave us now? . . . Who is to carry the mysterious drafts through the musty corridors, pen the little notes in the margins of defective bills? . . . Recite the gaffes of lawmakers as they stumble on the stones in Gregor Mendel's Garden?" Joe's path was followed by Michael Goldberg. I was beholden, too, to the others among the legislative intelligence forces who routinely kept us on top of such activities.

William Gartland, at first loaned to us from the genetics grants division of the National Institute of General Medical Science, became an emissary (he and Leon Jacobs were the only NIHers at preparations for Asilomar) and then took the beleaguered post of director of the newly established Office of Recombinant DNA Activities. Woefully understaffed, yet able to carry most of the enormous files of the Guidelines under his hat, Gartland performed an enormous service to molecular biology. We had volunteers from many of the institutes, in addition to the NIH director's staff. It was the accessibility and performance of this "Kitchen RAC" which facilitated and much influenced the evolution of the NIH Guidelines.

A peculiar feature of NIH, then and now, was that its staff was infiltrated by some members whose fealty was sworn to other divisions of the department. One of those uneasily sitting in the bosom of the Kitchen RAC was NIH legal advisor Richard Riseberg, who officially was on the staff of the general counsel of HEW. Toward the end of our long campaign, I wrote Riseberg that "your role has not been easy . . . a double agent with cover blown from the start, yet fair and honest in representing both servant and master . . . both enlightenment and the legal dark."[9] William Carrigan, who sat in on every session and compiled the voluminous archives represented by the big yellow volumes of *Recombinant DNA Research*, merited the accolade, "the 'DNA crises' have revealed a splendid editor in residence. . . . You've not had Maxwell Perkins' class of authors to save from error . . . but you've loyally pretended our fiction was no less immortal. . . ."[9] Backing up the Kitchen RAC was the unstinting and invaluable support of Belia Ceja, Elizabeth Shelton, Florence Hassell, and Richard Curtin.

Guidelines, Not Regulations

In the Berg letter of 1974, the invitation for NIH involvement had requested guidelines. One cannot imagine the term regulations having appeared in any earlier drafts. Clearly, the scientists wanted a translation of their safety prescriptions into a form that would permit the flow of new knowledge to continue and the rules to change right along with it. It was entirely my second condition as well. Some of the American participants were aware that the NIH directors often issued guidelines for the conduct of grantees. These issuances usually emanated from Bethesda without consultation higher than the Public Health Service. They were not published prior to promulgation, and public comment was not requested. Regulations were something else. As the lingua franca of the FDA, OSHA, sometimes the CDC, and the other regulatory agencies, they were associated—in the minds of the scientists and NIH administrators alike—with a halting rule-making process subject to endless and sometimes inexplicable delays as each version was submitted for approvals by a bureaucratic chain of command and prescribed period of public comment, with reiterations and explanation of revisions. Delay—something intolerable to experimental scientists—and language insertions unwittingly creating restrictions and unintentionally threatening scientific freedom presented a fearsome view of what might lie on the road to the first real understanding of the nature of genes.

No, I too, was convinced that guidelines it must be, if the process could be controlled by the scientific community. I would subsequently take refuge in defending our stubborn position by admitting that we had no regulatory authority or expertise. Later, it would also be obvious that we could have used some regulatory experience during the deliberations of the RAC. The final product of the RAC delivered to NIH in January 1976 failed to provide administrative provisions for either amending the Guidelines or making exemptions, exceptions, and other small changes which we would sorely need.

The Crucial Differences

The doubts of the general counsel's office that were faithfully sent us through Riseberg were initially daunting: guidelines would not be permissible. In response, we organized a tutorial, with Susan Feldman, the NIH regulations officer, summoned before the Kitchen RAC. We listened once more to the fine-print analysis outlining the differences between guidelines, regulations, and rule making. And could you explain the relevant laws once again? Here are her notes:[10]

> **Guidelines.** Simply a statement of rules or procedures that people are expected to follow . . . does not have the force of law . . . NIH has implied authority to issue guidelines without higher level clearance.
>
> **Regulations.** As used in government circles has a precise technical meaning: Refers to substantive rules of general applicability adopted as authorized by law . . . that have been published in the Federal Register for the guidance of the public . . . Usually subject to long delay and iterative process for revision. Note: The Director, NIH does not have authority to sign or publish a regulation. They must be signed by the Assistant Secretary for Health and approved by the Secretary.
>
> **Notice of Proposed Rule Making (NPRM).** It is a requirement of the Administrative Procedure Act (APA) that all rules of general applicability governing Federal programs be published in the Federal Register. APA also requires that the public be given an opportunity (a minimum of 30 days) to comment before final adoption and implementation. Note: Comments of the public have to be dealt with, and their disposition explained in a preamble. After the approval of this preamble of explanations by the HEW Secretary, the entirety must be published once again. The rules continue to read "upon final publication in the Federal Register regulation is codified by the Administrative Committee ...and published in the Code

of Federal Regulations (CFR)." *The Director, NIH does not have authority to sign or publish a Notice of Proposed Rule Making (NPRM)* [italics mine].

FACA and APA

Two vigilant guards located at the gate to the pass had to be reckoned with. The Federal Advisory Committee Act (FACA) is a younger brother of the Administrative Procedure Act (APA). Both rose out of legislation or rules established in the 1950s. FACA, particularly, was a by-product of a perennial concern of different administrations and Congresses that the government had too many advisory groups, which failed to report, wasted government funds, and rarely expired. A 1972 bill in the Senate slightly preceded FACA, and its definitions[11] would be altered when it was joined by the conference report on the companion House bill.[12] The Senate bill rather loosely defined "advisory committee" as one "established or organized" by statute, the president, or an executive agency, and noted that the phrase "established or organized" was to be understood in its "most liberal sense." The many groups covered by the provisions of the bill specifically included advisory councils to the NIH.

The House bill in its amended form became FACA.[13] Adopting wholesale many of the provisions of an older executive order no. 11007, issued by President John F. Kennedy, FACA stipulates rather severe requirements on agency heads in using advisory groups. Each such group must be given life by a charter; it must keep detailed minutes of its meetings; these must be chaired or attended by an officer or employee of the federal government, who may adjourn them if he or she determines it is in the public interest to do so; must provide advance notice of meetings and open them to the public, unless the agency head to which the advisory committee reports determines that it may be closed to the public in accordance with the Government in the Sunshine Act; minutes of meetings must be available to the public unless they fall within one of the exemptions of the Freedom of Information Act; and insofar as the agency can arrange, the advisors must represent a balanced group. The existence of the group is limited to 2 years, unless the agency requests otherwise.

The APA's more explicit definition of the "administrative agency" over which it has comparable or often joint jurisdiction has more teeth, i.e., "a part of government which is generally independent of the exercises of its functions" and which "by law has authority to take final and binding action affecting the rights and obligations of individuals, particularly by the characteristic procedures of *rule-making and adjudication*" [italics mine].[14] The General Services Administration is the agency responsible for administering APA and FACA.

The heart of any scientist is chilled at the thought of commencing to submit his or her experimental plans to rule making and embedding them in the Code of Federal Regulations to be administered by the General Services Administration, a pragmatic organization which administers public buildings and services. It conjures up recurring visions of a frozen river under which was entrapped an enormously energetic giant demanding release.

Suddenly our tutorial ended with a final note. Someone remembered that all the rules contained one exception almost forgotten. "*NPRMs carried as 'General Notices' can be published without being codified in the Code of Federal Regulations and without having the force of law as regulations*" [italics mine].

Could we not meet most or of all the requirements of APA-FACA, while somehow greatly accelerating the standard process of handling public viewing and comments and treating Guidelines as "general notices"? Could this be our Northwest Passage between guidelines and regulations?

I declared firmly that we would go for guidelines on "public policy grounds." The somewhat gossamer quality of such pseudo-regulations were accepted for the moment by Perpich and all the rest of the Kitchen RAC—with the exception of our dour legal advisor from the HEW. We would have to convince our administrative superiors of this virtue, however, and enlist their official support. But first we had another major consultation to perform. It was time for a public appraisal of the Guidelines, a landmark exercise described in chapter 4.

Warnings of What Lay Ahead

Parisian Distress (February 1975)

The first outbursts of popular anxiety immediately after the Asilomar conference were intense, relatively brief, and occurred several thousands of miles away in Paris. When they returned home from the conference, French scientists lost little time in describing to their compatriots the proceedings and the forthcoming rules for containment that were agreed upon. Shortly, a recommendation was made by the Délégation Générale à La Recherche Scientifique et Technique that 300,000 francs be allotted for a containment laboratory at the Institut Pasteur. The staff held meetings to inform coworkers of what experiments would be carried out in the new laboratory designed to be placed in the Duclaux Building, where the laboratory and office of Director Jacques Monod was located. A news item about the Parisian incident in *Nature* (July 1975) reported that "Nobody expected the explosion that occurred." One group of research workers and technicians undertook to fight both the decision to place the special laboratory within the department of molecular biology and the experiments themselves. An atmosphere of panic briefly overran the institute. Manifestos and demands for public meetings arose, and movements questioned the present reliance on the way science is run, and for whom science is done—the scientists or the people? The intended location of the new containment laboratory was shifted to a rear building, and any excitement died down fairly quickly. Undoubtedly, there was an upset in Paris, but some who were there find the reporting highly inflated in tone.[15]

Deceptive Calm at Home

Reactions to Asilomar required a longer incubation period in the United States. The first muscle flexing occurred before I returned to the NIH campus in July 1975 and while the RAC was struggling to prepare the Guidelines for delivery in the winter. In April, Senator Kennedy's Health Subcommittee held a 1-day hearing on the theme of the relationship of science to a free society. Witnesses included Stanley Cohen and Donald Brown, who argued for retaining control of rDNA technology by the biomedical research community. Willard Gaylin, an ethicist, and Halstead Holman, a professor of medicine at Stanford, spoke in favor of public participation in decisions on control. This brief obbligato was orchestrated by Kennedy staffer Lawrence Horowitz, an M.D. from Stanford.[16] Horowitz and I were to have frequent encounters in the years ahead. Shortly after the hearing, in a talk at the Harvard School of Public Health, those in attendance reported that Senator Kennedy had warned that regulation by scientists alone was "elitist and parochial and that the public must give informed consent." Several papers relating to the Asilomar meeting appeared, one by Josh Lederberg and another by Robert Sinsheimer.[17] A few more conferences dealing with the social responsibilities of science would be held during 1975, at the New York Academy of Sciences and the University of Chicago. The local, intense, and protracted debate on such experimentation was virtually ignored by the national press.

The University of Michigan

As in Paris, the first significant American reaction was initiated by a decision of a major research institution to provide a P3 laboratory for faculty molecular biologists. Beginning in early 1975, a series of events began on the campus of the University of Michigan[18] and soon involved the people of Ann Arbor and its surrounding Washtenaw County. Asilomar was barely over when faculty scientist David Jackson, who had left Berg's laboratory when the latter decided to indefinitely postpone his crucial experiment with SV40, sought permission to erect a high-containment laboratory (defined as P3 in the emerging guidelines). In April, the university's microbiological hazards committee conveyed both its support of recombinant DNA research as of importance to the university and a recommendation for the building of such a facility to the vice president for research, Charles Overberger. He in turn forwarded the request to the Biomedical Research Council, and in November the regents stamped their approval of the request.

The institutional biohazards committee had been set up in the vanguard of the recommendation of the early draft guidelines of the RAC. By this time, however, diverse opinions on the subject led to formation of several other campus committees to join a debate. One group (Committee A), largely a technical group of molecular biologists, emerged, and then another group (Committee B) was formed of faculty members, chaired by Al Zander, the associate vice president for research. Committee B set out to develop a recommendation concerning whether or not this new technology should be housed on the campus. As described by Jane Goodfield, several faculty members of the Department of Humanities in the College of Engineering, including Henryk Skolimowski, a philosopher, and Susan Wright, a lecturer in history and a member of Science for the People, were opposition leaders. They addressed an appeal to Committee B to face the ethical and moral implications of the research over and above the technical questions of estimated risks. Their intent was to delay approval of the rDNA research "until the university and surrounding community had been informed on all sides by the best technical, ethical, and legal opinions."[19] The Michigan test of citizen reactions was not to be resolved for another 13 months. By February, the Senate Assembly unanimously decided to set up public forums, including outside experts. A team from NIH (comprising Maxine Singer and Joe Perpich) visited in the first week of January to discuss the proposals for renovation of three sites.

Others among the outside participants were Asilomar luminaries such as Paul Berg, Singer, and David Baltimore. Jonathan King, the molecular biologist from MIT, presented arguments for Science for the People. This organization began its activities in Boston in the late 1960s and subsequently spun off into chapters in Berkeley, San Francisco, Chicago, New York, and Ann Arbor. A public debate between Baltimore and King ended in a shouting match. Three months later, Committee B issued a report supporting the recombinant DNA research on campus and concluding that the NIH

Guidelines, delivered to NIH in January—and by now about to be promulgated—provided an acceptable basis for its regulation. In May 1976, the regents voted 6 to 1 to allow the research.

In later reflections, Jane Goodfield, a most perceptive writer and observer of science, whose book is a sensitive account of the thinking of scientists and others in these events, found that the debate within Committee B in Ann Arbor framed a crucial philosophical question: "The ethical relationship that exists between society and [other] professions which marks the nature of their social contract is supported by the twin pillars of mutual responsibility and accountability. . . . Such an accountability is missing from the social contract between the scientific profession and society, however, and this is the focus of the present dilemma."[19] More such questions revolving about epithets condemning "regulation by expertise"[20] and the "politics of expertise"[21] were going to be our fare for months (or years) to come.

The forum had been aired on the campus news station and had accrued bountiful coverage in the University of Michigan newspaper, but no national attention was focused on this important series of events,[22] and few American citizens living elsewhere were aware of it.

Planning a Public Airing

Lifting a self-prescribed moratorium on the advance of a scientific revolution, surely the first of any scale in the history of biomedical science in America, had to be done properly. It seemed to us that extraordinary conditions to ensure the safety of the workers and of the public had been set up. It appeared that most of the scientists, understanding the origin of safety rules, would go along. Would the public, however, understand them and accept them as sufficient? We had to explain them to persons who could represent a sampling of public opinion. Obviously, this offering also had to be done in a manner acceptable to the majority of the people, particularly the authorities and opinion leaders who would undertake to decide for them. They would also be deciding on whether we, the NIH, were to be entrusted to continue as the agent for both science and the people.

Prior to Hans Stetten's negative receipt of my opinion of what we had to do, we had not notified the administration and representatives of the Congress. We set out to pass the word that we would forthwith have a public meeting to obtain reactions to the new *NIH Guidelines for Recombinant DNA Research* and that we would then return to the government with a better plan to cope with the questions and problems that would arise. We already had the elements of that plan in mind. The first was convincing our audiences to accept our belief that further cautious experimentation must proceed. Second, the Guidelines must be maintained so that timely changes could be made according to the results of the careful forward probing. We would not stress any potential benefits over risks until these were calculable. We did not stress that we were developing a plan for ensuring that all federal agencies would use the same guidelines for similar science they supported or conducted and for determining whether our moves impinged on their real or presumptive regulatory authorities to intervene.

We were aware of concern among scientists that the rules had taken almost a year to be put in place. The moratorium was incomplete, and international competitors were already conducting rDNA research. The French had instituted controls comparable to those agreed upon at Asilomar as they reviewed and awarded grants for continued research. The Williams committee in Britain was still preparing its rules, but the Ashby report in January 1975 had not recommended cessation of all such research. Reports filtered in that experiments were also proceeding in a prominent molecular biology laboratory in Zurich, Switzerland.

The NIH Mind-Set on Exposing the Guidelines to Public Critique

Some critics would accuse us of having concluded from the start that there was no danger in rDNA experiments and that our only concern was to prevent panicky reactions from upsetting the resumption of the stalled procession. Quite the contrary, we accepted the conclusions of the experts at Asilomar and the subsequent deliberations of the RAC that we were at ground zero of an enormous revolution in biology and that there could be dangerous consequences. There could be harm to laboratory workers and conceivably to innocent populations and the ecology.

Certainly one element in any decision making was lacking so far. We had not had an opportunity to learn what a properly informed, reasonable group of citizens without vested interest in the technology would think of the putative benefits and hazards and of the manner in which the Guidelines had been constructed. There was evidence that it might soon become a legislative initiative to place the judgment about such a recondite area of technology into the hands of others more concerned about the ethics and eventual effects of such power. We could not predict how far this resentment might spread.

A Changing Community of Scientific Research

I fought back any hesitations at the moment to remind myself that biomedical research had changed dramatically since I had first come to NIH. When we came to the laboratories in the early 1950s, NIH was like the

Elysian Fields, where one had unhampered liberty to romp about satisfying the curiosity of the experimentalist. Everyone could count on the ultimate reward, recognition for adding a few tiles to the gigantic mosaic of knowledge of life which had so many empty spaces to be filled. If there were constraints in this idyll, they were the unyielding standards of James Shannon and his generation, who insisted that if NIH intramural scientists were to have total scientific freedom, they had reciprocal obligations to work at it full-time and eschew commercialization. Academic scientists in the hard sciences such as chemistry and physics had long worked out arrangements for consulting with industry without apparent harm to their objectivity or their institutions. But biomedical research in the 1950s was in an age of innocence. Both inside NIH and without, this was the era when serendipity was considered sacred, and targeted research was anathema in most subdisciplines. Frankly, few of the great basic discoveries were ready for practical applications. Looking back, it is easy to see why a separation of not-for-profit and commercially supported medical research was easy to maintain, and Shannon's rules were rarely challenged.

Now the scene was rapidly changing. Biomedical science was undergoing unbelievable expansion. An increasing number of discoveries were leading to opportunities for development, and commercial interest was beginning to penetrate the isolation of the ivory towers. With the inception of the "biological revolution" (the emergence of genetic engineering), the race to commercial application would lead to practices of "technology transfer" that made obsolete many of the rigid ideals. The gateway of serendipity as the route to discovery was being bypassed by a newer generation who now possessed instruments that gave them vision of specific targets of opportunity and, increasingly, anticipation of rewards more fungible than prizes and citations.

Great as was my public responsibility, it included a mandate to protect this burst of energy and creativity that might (and I believed that it could) have incalculable benefit to humanity.

Preparing for Administration of the Guidelines

The RAC was continuing to tune up their new set of Guidelines. The use of such implements required laying new duties on scientific institutions already seeking to put up new physical barriers and costly equipment in anticipation of the release of the rules. Memoranda of understanding were going to be required of every university or not-for-profit research institution which intended to house recombinant DNA research. It was a pledge to faithfully follow the Guidelines. Each such place must also set up an institutional biohazards committee. Investment was increasing, and Asilomar would soon be 10 months behind us. This was now the time to inform the public.

4

Extending the Seed of Power to the Public

1976

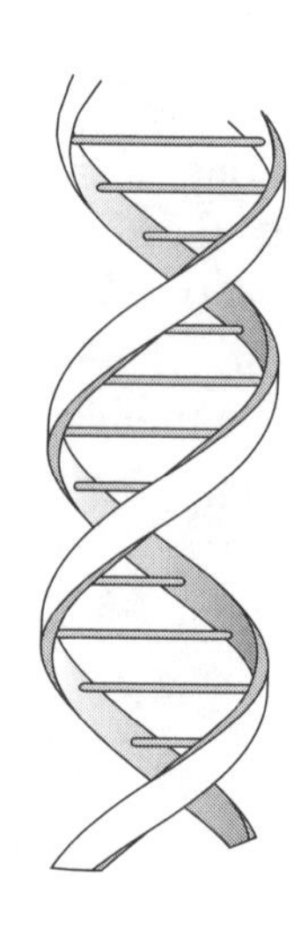

Because the works of the intellect . . . become the sources of strength and riches one must consider each scientific development, each morsel of new knowledge, each new idea, as though it were a seed of power placed in the reach of the people.[1]

ALEXIS DE TOCQUEVILLE

The Director's Advisory Committee

A Viable Venue

Our hand would be partly forced by the circumstances under which we would have to make the arrangements. In early January 1976, we did not have a suitable venue for a public forum where we might air the NIH Guidelines. No one wanted to risk requesting a departmental charter for a new high-level review group. Not only would the course likely be subject to delay, but it would be full of hidden bays and estuaries from which the vessel might emerge deprived of NIH as the pilot. Fortunately, someone remembered that there had been an NIH Director's Advisory Committee (DAC) established sometime toward the end of Shannon's directorship. Was the charter still alive?

The DAC had been established in 1966, following the recommendation in 1965 of a presidentially appointed committee to make a study of NIH. Its purpose was to advise the NIH director and the secretary of HEW on NIH policy matters and to make recommendations regarding program development, resource allocations, policies, and regulations. The committee was chaired by the secretary or the secretary's designee, who was invariably the NIH director. Its charter called for 16 members: 11 researchers from the private sector and academic research communities and 5 representatives of the public. Each director used the DAC in different ways and to various degrees.[2] Of the authorized membership of the DAC, the terms of 5 of the 16 were still active. These included Joseph Dodds, a physician from Chattanooga, Tenn.; Roy Hudson, president of the Hampton Institute; James Kelley, former HEW comptroller and now executive vice chancellor of the State University of New York; Robert Petersdorf, chairman of the Department of Medicine at the University of Washington; and Charles Sprague, cardiologist and president of the Health Science Center of the University of Texas in Dallas. Petersdorf's specialty was bacteriology and infectious diseases but not molecular biology.

Filling the Complement

The depletion of the DAC had been due partly to the lack of an NIH director for 8 months and partly to a tendency of HEW or the White House to exchange proffered nominees for political favorites. Without departmental consultation, we quietly appointed fourteen ad hoc replacements to fill the vacancies.

Four were scientists with working knowledge of microbiology or molecular biology. They were Marion Koshland, professor of bacteriology and immunology at the University of California; Joseph Melnick, professor of virology at Baylor University; Marjorie Shaw, both a geneticist and a lawyer at the University of Texas; and Robert Sinsheimer, chairman of the Division of Biology at California Institute of Technology.

Two of the ad hoc appointees were professional ethicists: Daniel Callahan, director of the Hastings Institute of Society, Ethics and the Life Sciences, and Leroy Walters, director of the Center of Bioethics of the Kennedy Institute at Georgetown University. The law was represented by Marjorie Shaw; Peter Barton Hutt, former general counsel of the FDA and now in private practice; and—through the efforts of the judge's previous clerk, Joe Perpich—David Bazelon, chief judge of the United States Court of Appeals for the District of Columbia Circuit, was persuaded to come, taking the risk that the subject matter might some day course through his jurisdiction. Also added were Walter Rosenblith, a physical scientist and provost of MIT, and Philip Handler, a biochemist and president of the National Academy of Sciences. Two young persons, Margo Haygood, a laboratory technician, and Alan Ladwig, officer of a student association, agreed to serve. Finally, seeking a respected layperson known to the public, we asked Esther Peterson, who had been President Lyndon Johnson's consumer adviser.

Additional Participants

Five speakers were invited to explain the elements of research and the history of the Asilomar conference, along with the possible risks and benefits of the technology and the nature of the NIH Guidelines. All attending had received a copy of the proposed Guidelines in advance. The chair of the Recombinant DNA Advisory Committee (RAC) would describe how the RAC had constructed them, and explanations of physical and biological containment would be offered. David Hogness from Stanford and Roy Curtiss from the University of Alabama were invited as individuals who had made signal contributions to the technologies but appeared to have different views of the science—one had initially renounced the experiments and only later returned to his laboratory activities.

A half dozen other speakers had requested to be heard, and all were accommodated, each to be initially allotted about 15 minutes and requested to submit summaries of their positions. These individuals included David Baltimore and Donald Brown, already stellar members of the scientific cast at Asilomar; Richard Goldstein, molecular biologist at Harvard and organizer of the multiauthored letter to the RAC protesting the Woods Hole version of the Guidelines; Charles Madansky, graduate student in the Department of Microbiology and Molecular Genetics at Harvard Medical School and a member of Science for the People; John Sedat, molecular biologist from Yale; and Susan Wright, historian at the University of Michigan, one of the leaders of the opposition at the Ann Arbor demonstrations.[3]

The announcement of the meeting as open to the public was publicized. In addition, the leaders of a number of organizations interested in environmental protection and other public interests were invited to attend.[4]

Agenda

The meeting of the DAC took place all day February 9, 1976, and adjourned late in the afternoon of the next day. The members of the committee had seats around the great table in NIH's conference room 10 in Building 31. Several rows of chairs surrounded them, filled with a dense circle of participants, including the invited experts, other speakers, and listeners from the public. All participants had received much background material, which few could have had much time to study.[5]

As director of NIH, I presided. Hans Stetten, who was to explain the role of the RAC, was at my side throughout, despite his initial rejection of the idea of such a meeting.[6]

The meeting began promptly at 9:00 a.m., when I introduced the committee members to each other, explaining that I would also introduce the speakers, keep the time, and otherwise ensure that all the participants who desired to speak would be heard. It was announced that the entire meeting would be taped, and it was obvious that television cameras were recording the meeting as well. Unless otherwise noted, excerpts of the extensive record of this meeting are identified in brackets by page numbers in volume 1 of the NIH publication *Recombinant DNA Research* (*RDR*).[7] Individual testimony is identified by page numbers in brackets.

I began with the charge to the committee:

> The purpose of this meeting is to seek your advice on proposed guidelines setting conditions for the conduct of certain experiments with recombinant DNA molecules. . . . Your responses to the guidelines will assist in the task of defining scientific and public interests in the research. [p. 149–151]

A Tutorial in Recombinant DNA Research

Appropriately, the meeting began with presentations of the leading organizers of the Asilomar conference.

Paul Berg

The first to speak was Paul Berg, who described at length the events leading to the Asilomar conference and the proceedings that resulted in this meeting in Bethesda. He did not let it be overlooked that the moratorium was an act originating within the circle of scientists who, while wanting most to race down this new avenue to discovery, had made the decision to hold back. He summarized his own feelings briefly.

> . . . How did the scientific community respond to that unorthodox call for a pause in what was clearly a particularly exciting line of research? To my knowledge, during the time the appeal was in force, there was no explicit violation. . . . I believe the most important outcome of [the NAS letter in 1974] was the debate and response among the scientific community itself. . . without any power, the only thing we had available was the reasonableness of our arguments, and moral suasion . . .
>
> My own personal view is that the guidelines are stricter than they need be on the basis of the scientific evidence alone . . . but I and many of the scientific community are prepared to accept them as the price for exploiting this powerful new tool of scientific research. . . . [p. 148–169]

Hans Stetten

DeWitt (Hans) Stetten, the chair of the RAC, followed. He summarized the manner in which his committee had gone about its task of constructing guidelines. His remarks were a brief summary of the events told in chapter 2. He concluded with this observation:

> It is my belief and . . . [others] would agree that there is no final guideline. . . . new experimental data will continue to come which will cause us, from time to time, to modify our estimate of hazard and the constraints imposed. Therefore this [the RAC] is a standing committee. Its next scheduled meeting is April 1st and 2nd. [p. 168–174]

Maxine Singer

Maxine Singer now summarized the Guidelines. She employed a number of charts during her explanation. She provided much illumination on the technology but wisely chose to move early in her long and lucid discourse to the apex of the rules, displaying the list of "experiments not to be initiated at present":

1. Those involving DNA from pathogenic organisms (classes 3, 4, 5 [CDC])
2. Deliberate formation of recombinants when genes for dangerous toxins are present
3. Deliberate formation of recombinants from plant pathogens if likely to increase virulence or host range
4. Widespread release into environment of any organism containing recombinant DNA except when there is no reasonable doubt of safety
5. Transfer of drug resistance traits to new microorganisms if transfer would compromise drug use
6. Large scale experiments if harmful products made. (Exceptions allowed only with the permission of the RAC.)[8]

Singer concluded emphatically that the inclusion of experiments involving organisms with genomes that had evolved over millennia, allowing them to express substances lethal for humans, was not to be sanctioned. And she noted that, at this stage, no one dared state for sure that the new technology could not leapfrog the processes of mutation and selection that maintained the tenuous equilibria among the living creatures on the earth.[8]

Singer wrote us after that meeting that she had considered it not appropriate for her as an expert witness to convey several of her own views on the latest version of the Guidelines, which she had critically perused. She had proposed avoiding use of SV40 when viral particles are produced, as well as modest strengthening of the roles of the institutional committees. It was possible for us to work these suggestions into recommendations for the RAC to consider prior to final issuance. None of the experts felt that the Guidelines could not be improved.[9]

W. Emmett Barkley

The chief of the safety division of the National Cancer Institute (and an invaluable member of the Kitchen RAC) was already emerging as one of the world's premier authorities on the details of physical containment prescribed in the Guidelines. His presentation was introduced, and he answered multiple questions, but his demonstration of the hardware of containment had to be postponed for lack of time.

The final two speakers of the morning were introduced as molecular biologists who represented bipolar attitudes on the severity of the Guidelines.

David Hogness

David Hogness's experience included working with *Pasteurella pestis*, a most virulent and pathogenic bacterium, under P2 conditions, and he had cloned and handled tens of thousands of recombinant DNA molecules from fruit flies in P2+EK1 conditions. Hogness commented mainly on the formation of the Guidelines, in which he had played a principal role.

> I have been struck by certain trends . . . the general tendency to increase the levels of containment . . . this trend . . . would clearly not be of concern to me were it based on data indicating that these experiments [recombination of *E. coli* DNA with that from cold-blooded vertebrates] were potentially more dangerous than we have previously imagined. However this was not the case. Indeed, the data that has accumulated since the *Science* letter was published [Berg's letter of 1974] argue in the opposite direction . . . the strain of *E. coli* used in these experiments is highly attenuated . . . does not colonize in normal bowels of humans or other tested mammals, and . . . does not even become pathogenic when containing certain plasmids that normally confer pathogenicity to more robust strains of *E. coli*. . . .
>
> Why, in the face . . . of this evidence were the containment levels increased? I suggest that one of the important factors has been a preoccupation and speculation as to the conceivable hazards of these experiments without a corresponding concern about their benefits . . . By the use of these shot-gun experiments one can isolate individual genes and examine in molecular detail their structure, function, and arrangement in the complex chromosomes of higher organisms. [p. 201–204]

Roy Curtiss

Roy Curtiss, professor of microbiology at the University of Alabama School of Medicine, was next introduced. He described himself as having "a rather consistent conservative attitude . . . believing that the potential biohazards were far more likely to be real than unreal." At the time of the Berg letter, however, Curtiss had renounced all of his laboratory dedication to making safer host vectors for recombinant work and had only recently returned to this aim, after being convinced that the last version of the Guidelines had corrected errors he saw in the earlier Asilomar version, including the Hogness version. Here he described his current efforts and goals.

> [Biological containment involves mutations] that preclude colonization and survival in the intestinal tract . . . preclude biosynthesis of cell wall in non-laboratory-controlled environments . . . [and] lead to degradation of DNA . . . . [With potential EK2 host strain x1776 and its even safer derivative, x1876 . . . the] data must still be subjected to the scrutiny and evaluation by scientists selected by NIH . . . [The last host-vector system would reduce transmissibility from 10^{-8} with robust *E. coli* K12 to $<10^{-23}$ . . . reducing danger to an astronomically small number.] [p. 204–211]

Judge Bazelon—Comment on Process

> Of course I wouldn't dare to express an opinion. I have none on the question of whether or not the guidelines are right, wrong, or in the middle . . . Whatever decision is made, Dr. Fredrickson, I think you can't go wrong . . . in making sure that the decision that you render has within it great specificity for the reasons why you came to the conclusion that you did. . . . The healthiest thing that can happen is to let it all hang out, warts and all, because if the public doesn't accept it, it just isn't worth a good damn. [p. 321–322]

Such respected advice reinforced a decision we had already begun to implement, the collection and preparation for publication of all the transcripts, letters, correspondence, and other data relevant to the decision process. Thus, the full transcript of the DAC meeting initiated the first of a series of yellow volumes under the general title of *Recombinant DNA Research*, the number of which had reached volume 20 by December 1995, when the archives moved to a website (http://www.nih.gov/od/oba/) with the greater volume and access of the Internet. Because the series is in many libraries around the world, the interested reader has access to detailed coverage of much of the correspondence, minutes, and documents of this continuing debate.

Some Dissenting Views

Thus far, the voices prevailing at the table of the DAC meeting had been those who had no doubt about proceeding with research. There were also dissenting views to be heard.

Robert Sinsheimer

We were eager to have the views of Sinsheimer, a world-class molecular biologist on the "other side" of the recombinant DNA controversy. The discoverer of chi1776 single-stranded virus and a former editor of the *Proceedings of the National Academy of Sciences*, Sinsheimer gave the DAC the opportunity to hear from the scientist whom Nicholas Wade had labeled—and correctly so—one of "America's most distinguished critics of continuing rDNA research."[10] It had taken courage at Asilomar to announce one's belief that science might be moving too swiftly toward some unseen chasm—promoting genetic intercourse between the prokaryotes and the eukaryotes at this period of evolution. It would still require another year of research before it was clear that such intercourse was drastically limited. It was not known until 1977 that prokaryotes could not "read" eukaryote genes.[11] Thus, at the time of the DAC meeting, one had only intuition to counter Sinsheimer's cautions of an "evolutionary divide" that could be accidentally breached with possibly fateful consequences.

Was there false comfort in the assumption of others that during all the preceding years all the possible recombinations had already been tried and that human ingenuity would not succeed in upsetting the established equilibria? If this mere opinion were wrong and some "Andromeda strain" resulted from rDNA experimentation and escaped to foray from an unhealthy niche in our environment, the morality play that would depict this dismal folly would find this California biologist illuminated as a melancholic Hamlet while the rest of us would be barely visible in the shadows of purgatory.

Samplings of Sinsheimer's considerable oeuvre relating to this debate during the 1970s[12] demand respect for the eloquence with which he could movingly place in perspective what he saw facing us. At the DAC meeting, Sinsheimer largely restricted his comments to stressing these views. (See below.)

There were others who felt that the desire of the majority of molecular biologists to end the moratorium was a dangerous attitude.

Susan Wright

Susan Wright, a history lecturer at the University of Michigan and sympathetic to the activist strains of the Science for the People movement, was a leader at the manifestation in Ann Arbor in 1975. A tireless attendant at the NIH meetings, she provides a critique of the process, including the actions of the NIH director.

> I would advocate a slower and more cautious approach . . . and the maintenance of the moratorium until more is known about the risks. . . . There should be two committees to consider the separate questions of the nature and the acceptability of the risks. . . . [a second] . . . should be drawn from

the . . . humanities, philosophy, ethics, policy areas such as law and administration, medicine and science . . . no member of that committee could be engaged in the experiments under consideration . . . many of the claimed benefits are dubious, the risks seem relatively clear. . . . [p. 266–270]

Charles Madansky

. . . To me there are . . . no pressing social benefits to be gained from this research. If they are to come at all, what matter if it takes 20 or 25 . . . or 105 years? [p. 254–256]

Others Leaning Favorably

Daniel Callahan

. . . I think there can be no moral obligation to do this research. . . . There is, however, a moral obligation to do no harm . . . now, having said all that, I think that the guidelines are going in the right direction . . . it seems to me it is possible to be an excessive worrier . . . I think the guidelines are suitably worried. [p. 329]

Peter Barton Hutt

. . . I am not as persuaded as Dr. Callahan that there is . . . no moral obligation to go forward with this research . . . this may be [a case] where inaction could be of greater detriment to the public than action. . . . the scientific community deserves enormous praise . . . for bringing this issue to the fore. . . . [p. 332–338]

Philip Handler

. . . The nay-sayers [here are] so very young as compared with the aye-sayers . . . in the past, conservatism has been the role of the elders of the tribe . . . the scientific benefits of this work are something I do understand . . . I think that you have no choice but to see to it that scientific work does indeed go forward. . . . [p. 324–326]

David Baltimore

. . . One could argue that the only appropriate response to this unknown potential hazard is to completely ban the experiments . . . that would provide the highest degree of protection of all, so there must be an argument why these experiments should be done at all. . . . There are two critical reasons why they should be allowed to go forward with appropriate protection. One reason is because of the scientific interest in the experiments . . . the ability to form recombinant DNA molecules opens up the opportunity to utilize the enormous resources developed in the study of bacteria to probe how higher cells function. The medical justification is two-fold . . . a potential for manufacture of biologicals, and there is a potential for understanding the complicated diseases that arise from the malfunction of cells. [p. 248–252]

Roy Hudson

. . . We are standing very much at a point where Columbus might have stood facing the Flat World Society. . . . I think we must take the concept. We must believe that the world is round. We must sail forth . . . but we must sail forth with precaution. [p. 347–348]

Questions

I called for an hour's respite. After lunch it was announced that Barkley's planned tour of facilities had to be put off for lack of time. Nevertheless, multiple questions hovered about the room. There were no unequivocal answers.

"Would someone want to insist that all the research be limited to P4 facilities?"

Robert Sinsheimer: . . . the research we are talking about at this meeting marks the advent of a whole new era, the real turning of the corner in biological research . . . from an analytical phase to a much more synthetic phase . . . [and] what we are doing is almost certainly irreversible. . . . knowing human frailty, these vectors will escape, they will get into the environment, and there is no way to recapture them. [p. 322–323]

Marion Koshland: I don't think there are data to indicate the nature of the risk. There is no way of assessing it right at the moment . . . but a guessing game . . . . [I'd like] NIH or some other organization to take on some of these risky experiments, perhaps at Fort Detrick . . . so that we have some data on which to make an assessment of risk. [p. 275]

"Could the guidelines cover all the rDNA work in this country?"

Someone (was it Peter Hutt?): Could these experiments be done by a high school biology teacher?

Alan Ladwig: What happens if somebody goes beyond the guidelines? Is he getting involved in a criminal type area? [p. 341–343]

"What about Escherichia coli?"

Joseph Melnick: As I listen to these sessions today, . . . [I realize] there is a big gap in knowledge as to whether the recombinant DNA is expressing itself in *E. coli.* [p. 240]

Richard Goldstein: We feel that [*E. coli*] is the wrong choice because everyone knows it is a human inhabitant. [p. 299]

Roy Curtiss: We know more about *E. coli* than any other organisms on the face of this earth . . . something about one-third of its entire genetics. We know how to disarm *E. coli* so that humans are no longer an ecological niche for it. [p. 304]

Maxine Singer: . . . We know now that *E. coli* K-12 does not colonize normal human bowels . . . we also know a fair amount about the kinds of organisms which will accept DNA from *E. coli* K-12. [p. 304]

Peter Hutt: The very fact that leads [Dr. Petersdorf] to conclude that it is relatively safe to use *E. coli*, leads the Boston group to conclude that it is quite the wrong thing to use, namely that if you have that level of *E. coli* that can reside in humans, then it is a very poor thing to use because it can easily be found in humans, and could carry those dangerous forms. [p. 241]

Robert Petersdorf: I take great issue with what you say, Mr. Hutt, because if your arguments carry through to their logical conclusion, there would be in fact no biomedical research, there is no completely safe research experiment. [p. 243]

Margo Haygood: . . . have already commented on my uneasiness about *E. coli*. I'm even more uneasy about how things are going to be when another organism is implemented. [p. 339]

Hutt's final advice on the matter of guidelines—including a lengthy and informative letter after the meeting[13]—was that the promulgation and revision of the NIH Guidelines would require publication in the *Federal Register* and full respect for the provisions of the Administrative Procedure Act. In closing, Hutt suggested that we consider using Section 361 of the Public Health Service Act. (See later chapters for more regarding Section 361.)

Closing

At the end of the day, I thanked the participants and asked them to send me a letter within 2 weeks, covering their impressions and further thoughts about the important matter they had been considering. As I looked about the room, I noticed how the committee members and all the other participants had rearranged themselves while seeking consensus on this important matter. Largely to myself, I commented, "We are not now an inner circle and an outer circle. We are all one in this. . . ." [p. 349]

Summing Up

The chairman of the RAC had been assigned the job of summing up. No one could have done it better.

> ***Hans Stetten:*** Dr. Fredrickson, you have been advised that the guidelines that were drafted were too permissive. On the other hand, you have been advised that they are too restrictive. On the one hand, you have been advised that we have moved too slowly. On the other hand, you have been advised that we have moved too rapidly.
>
> On the one hand, you have been advised that the language of the guidelines is too strong in that it embodies law, rather than guidelines. On the other hand you have been advised that it is too weak, that it uses the conditional mood, "should," where it should have been "shall."
>
> On the one hand, you have been advised quite recently that *E. coli* is the most desirable organism for this kind of study. On the other hand, you have been advised that it is certainly not the most desirable, and perhaps far from the most desirable organism.
>
> In trying to provide you with a starting point, I would like to mention that up the coast from here, if one sails, he soon encounters the Island of Newfoundland, which is separated from the mainland of Canada by a very narrow strait, The Strait of Belle Isle. This is a fog-ridden part of the water, yet the sailors up there sail it in dense fog with impunity.
>
> By listening to the surf on the left, and listening to the surf on the right, and when the surf on the left and on the right are precisely equal they know they are charting a safe course. This . . . is the best advice I can give you. [p. 317–318]
>
> ***Donald Fredrickson:*** That explains the roaring in my ears.

Commitment

Three days after the DAC meeting, Stetten, Joe Perpich, Leon Jacobs, and Bernard Talbot sat down to consider the next tactical moves. Stetten urged that we proceed as expeditiously as possible, for he was best tuned to the anxieties of the scientific community. Issuance of the Guidelines was the most urgent concern. Perpich therefore drew up a memorandum for me describing "the administrative law model" of how we should proceed. Examination of its approach is worthwhile because it shows how we were going to accommodate to FACA and APA while keeping the Guidelines in hand. It would also set the pattern of handling commentary that we would have to do over and over again for many years. The memorandum began:

> The DAC meeting was, in effect a public hearing where you allowed public comment that would be taken into account in issuing proposed guidelines. . . .
>
> 1. At the meeting you noted a 2 week period to allow for further comment by interested public witnesses . . .
> 2. During this period you could distill the comments . . . and analyze their contents. . .
> 3. With all of the comments assembled, you could then review the proposed guidelines . . . and get the comments of the RAC on the assembled comments. . . .
> 4. . . .You could then issue the guidelines with any modifications you find necessary . . . you could briefly explain which views were accepted and which were not and why . . .

Perpich concluded,

> The forum demonstrated wide differences of opinion among scientists present on both substantive and procedural issues, as well as differing views among representatives of the public. Because the scientific evidence is not conclusive in support of any particular position on the guidelines, I believe that how you reach your decision may be as important as what your decision is.[14]

And this is the pathway we committed ourselves to walk. We did not have an inkling how many trials of comment and response would eventually be required. Or how many afternoons and evenings some of the Kitchen RAC would join me, all bent over mounds of papers on the carpeted table in my office, vigorously discussing parsing of sentences, details of P3 containment, or other serious principles of public responsibility. When we were through, someone, most often Bernard Talbot or Joe Perpich, would spend many more hours in redrafting what had been torn apart. The group would begin tearing the product down again on the next day, and the cycle would continue. Although the directorship of NIH was itself a full-time job, I estimated later that I had to devote at least half of my time to recombinant DNA during 1976–78.

I have often revisited the archives of this time to preserve my perceptions of what we did. I especially marvel at the sincerity and quality of the advice and support we received. There were also not a few revelations of the seriousness with which most scientists took the new kinds of responsibility thrust upon the community of inquiry.

A few days after the DAC meeting, I received a long, thoughtful memorandum from Maxine Singer, who stated that she had thought it inappropriate for her to comment on the substance of the proposed guidelines during the meeting but added that "you will not be surprised to learn that

I do have some specific views concerning the decisions you must now make."[15]

Revision of the Draft Guidelines

A few weeks after the DAC meeting, we sent a letter to the RAC[16] with some suggestions for the Guidelines, amounting to changes in 30 specific items suggested at the DAC meeting, to be considered at the next meeting of the RAC.

The detailed changes that resulted shored up the structure of the implementation sections and strengthened the descriptions of the containment and safety procedures. We considered that the DAC review had significantly improved the Guidelines.

Not all were convinced of this. Participant Susan Wright later recorded skepticism:

> Despite the range of views the hearing solicited, it was immediately clear that the NIH leadership did not contemplate disrupting the policy paradigm its scientist clients largely favored. [In his summation . . .] Stetten advocated compromise, which did not deviate from the policy paradigm all along, a course with which Fredrickson was in evident agreement.[17]

Attitudes of other participants were more favorable.[18] All comments were included and responded to when the NIH Guidelines were officially released.[19] Balancing the technical input of the RAC with the response of the DAC, we had achieved a two-tiered review structure. We were on our way to setting a template for the "guidelines-type" regulation that we proposed to pursue.

EIS before Release?

A story of the relationship of recombinant DNA research to the National Environmental Policy Act of 1969–70 (NEPA) is told in rich detail in a later chapter. Actually, by the time of the DAC meeting, we had made a decision to postpone NEPA discussions until the public meeting was finished. Within a week after the DAC meeting, however, we were put on alert by bulletins from the HEW General Counsel's Office that the forces from within the department would be on our heels about whether the Guidelines could be issued without an environmental impact statement (EIS).

Outside opinions on this subject were also becoming more needling and accompanied with warnings about risks of litigation. Jeremy Stone of the Federation of American Scientists wrote a pessimistic opinion: "I do not believe that your legal staff has given sufficient consideration to the danger that the entire process NIH is following may be legally irrelevant if a suit arises under NEPA."[20] Several faculty members of the University of Michigan, commenting further on various aspects of the Guidelines, also inserted the injunction that NIH should write an EIS as required by NEPA,[21] and Professor Richard Andrews explained in gratuitous detail that "the National Environmental Policy Act of 1969 (42 U.S.C. §4321 et seq.) has established a framework of substantive policy goals that all agencies of the federal government are to pursue. . . ."[22] A month later, the Friends of the Earth wrote us further instruction.[23]

NEPA was a law well meant, whose purposes we could all endorse as intended to save our environment. However, NEPA was also a powerful tool to obstruct federal process. It was brandished against molecular biology with great vigor. One can pursue in chapter 6 why we bowed to pressure to compose an EIS and how we struggled to get its approval, even though the consequences of the "major action" were utterly unknown.

Striving for the Release

The lines between Bethesda and the laboratories of the molecular biologists were filling with anxious queries as we entered the final weeks before release of the Guidelines. Hans Stetten sent me a memorandum early in May.

> Having listened to Professor Green [Harold Green, Esq., a volunteer legal advisor to us on environmental affairs and regulation] and having read Mr. Riseberg's memorandum of April 19 on DNA research-NEPA requirements, having also received feedback from the community of impatient investigators on the outside, I am deeply concerned lest there be further delay in the issuance of the guidelines.
>
> . . . It is the clear and stated intent of the guidelines to recommend a series of constraints designed to reduce the hazards to the environment to a minimum, asymptotically to zero. This condition can only be approximated, however, if the guidelines are promptly published, assiduously read, and rigorously adhered to.
>
> . . . The community of scientists interested in this kind of research in the United States is clearly becoming restless. . . . It has recently come to my attention that the restlessness with regard to the delays that have already transpired may manifest itself in cynicism with regard to any NIH regulation in this area . . . also dampen the enthusiasm . . . of [Asilomar] . . . and the preceding months when American scientists behaved with remarkable responsibility and self-constraint.[24]

I have kept a copy of this memo. At the bottom is my scrawled answer: "As we have earlier discussed, this course will be followed: early release with statement on EIS to follow. First, have to alert Cooper/Mathews."

Key Stops on the Way to Release of the Guidelines

Perpich and I had to make two strategic moves and two tactical ones before my promise to Hans could be redeemed.

Creating the FIC

On March 30 we alerted the heads of some 16 relevant federal agencies by phone and sent a letter inviting them to "an informal meeting to exchange information and discuss forthcoming NIH guidelines on research involving recombinant DNA molecules." On April 8, representatives of 16 agencies filed into NIH's Building 31 to savor the challenge of rDNA. Assembled was most of the government's regulatory power and the National Science Foundation and Department of Agriculture, who sponsored genetic research.[25] We described the imminent promulgation of the Guidelines and hoped that all agencies agreed to come under the same Guidelines. Second, we urged them to an ecumenical examination of their authorities to see if any had jurisdiction over recombinant DNA research. We agreed to try to obtain a presidential charter for formation of a Federal Interagency Committee on Recombinant DNA Research (FIC). We planned on the next meeting later in the year.

Meeting with the Secretary of HEW

The secretary of HEW was now David Mathews, Caspar Weinberger having gone to head the Department of Defense. We told Mr. Mathews that we were about to promulgate guidelines for recombinant DNA research and that our goal was guidelines, not regulation. We also said we would appreciate it if he would send a letter—of which we had a draft—to ask the president to form a new interagency committee on this revolutionary new science. The secretary agreed to the second. For the first, we would have to consult his general counsel, William H. Taft IV.

Years removed, in 1989, I ran into a youngish-looking, pale man at the Cosmos Club, whom I recognized as William H. Taft IV. He did not recognize me as one who, for an instant in time, had been in and out of his long career prior to his accompanying Caspar Weinberger from HEW to Defense. I reminded him that as general counsel of HEW he had made two decisions that ultimately had an important effect on the speed with which the "new biology" was translated into practice. In the notes at the back of the book, I quote from his formal response following our visit to him on June 1, 1976.[26]

Richard Riseberg, the extension of the General Counsel's Office in residence at NIH, had insisted that rule making was the required route because:

> 1. That a court construes guidelines as an action not in accordance with the law or without the procedure required by law. 2. Publishing the Guidelines as a notice in the Federal Register would not fulfill the requirements of the Administrative Procedure Act that general policies and interpretations be published in the Federal Register nor apparently does it satisfy the Department on policy of such publication.[27]

The chief of the Environmental Safety Branch, an appointee of the HEW office overseeing NEPA, also had adamantly advised us that ". . . the current guidelines are not legal under NEPA."[28]

This advice notwithstanding, and to our great relief, Taft had left open the door in stating that the decision of taking the risk was, in his opinion, "a matter for policy and administrative judgment." The hazard of litigation, of course, encompassed the secretary as well as ourselves, and the secretary had already indicated his willingness. A memorandum from his assistant secretary for health in support of the NIH position provided further encouragement.[29] We felt we could continue as we had started.

Meeting with Industry

It was clear that issuing guidelines for recombinant DNA research that were applicable only in academic laboratories would soon be an untenable situation. It was more than time to inform industry what was coming and to test their temperature for voluntary guidelines.

Letters were sent to 25 American companies on a list prepared by the Commerce Department inviting them to meet with us at NIH on June 2. Most of the companies indicated an interest in reviewing the Guidelines. They were concerned about the limitation of cultures to 10 liters and possible analogs of the EK system for nonpathogens usually used in industry. An overriding consideration for industry was the need to protect patent rights and prevention of premature disclosure of research.

Telling Congress

June 14, 1976. A week earlier, we had sent 40 letters of invitation to congressmen and staff. Many came. Maxine was exemplary in her explanations. The "hearing" was uneventful.

Release of the NIH Guidelines, June 23, 1976

> Today, with the concurrence of the Secretary of Health, Education and Welfare, and the Assistant Secretary for Health, I am issuing guidelines that will govern the conduct of NIH-supported research on recombinant DNA molecules. The NIH has also undertaken an environmental impact assessment of these guidelines . . . in accordance with the National Environmental Policy Act of 1969 . . . I expect a draft of the environmental impact statement to be completed by September 1 for comment by the scientific community, Federal and state agencies, and the general public. . . .[30]

The press conference in Wilson Hall of Building One was attended by 20 or 30 reporters, many of them familiar faces like Harold Schmeck of the *New York Times*, Stuart Auerbach and Christine Russell of the *Washington Post*, Warren Leary of the Associated Press, and Nicholas Wade of *Science*. In the *Post*, I was quoted as answering a question about the risks of rDNA research with the statement:

> You can conceive an endless list of possibilities of hazards. . . . It is really the unknown that is the source of the fear, not certain or specific hazards that could happen."[31]

Colin Norman, writing from London for *Nature*, was bemused:

> How did Fredrickson arrive at his final decision when confronted with such a wealth of conflicting advice? His first, and most fundamental consideration was simply whether the research should be allowed to go on at all, in view of the potential hazards. . . . In a lengthy statement published along with the guidelines [he] recognized that the breach of genetic barriers might pose a hazard . . . [and] nevertheless argued that the research can be controlled so that it is carried out carefully. . . . Fredrickson pointed out that there is, in any case, no way to prohibit the research throughout.[32]

The preamble to the Guidelines was accompanied by 12 pages of introduction that described in close detail all the decisions we had made after the public comment obtained at the DAC meeting and afterward.[33] These included policy decisions, implementation considerations within and beyond the NIH, and actions taken by the RAC to constrain certain other experiments. In addition, there were several other new proscriptions achieved by the months-long deliberations of the RAC over my comments.

> I have decided that the guidelines should for the present, prohibit any deliberate release of organisms containing recombinant DNA into the environment. . . .[33]

It was apparent that no products that could be certified as releasable into the environment would be available before the Guidelines would be revised.

It was wiser to avoid any more issues than were already presented by the Guidelines as initially promulgated.

The mechanisms for modifying and adapting the Guidelines to the emerging new knowledge were set forth:

> The Recombinant Advisory Committee in conjunction with the Director's Advisory Committee shall continue to serve as an ongoing forum for examining progress in the technology and safety of recombinant DNA research. Their responsibility, and that of the NIH Director is to ensure that the guidelines, through modification when called for, reflect the soundest scientific and safety evidence as it accrues in this area. Their task, in a sense, is just beginning.[33]

Around the World

Word about the Guidelines went first to every American chief of station with a telegram from the Department of State.[34] Later, Francine Simring of the Friends of the Earth commented:

> On June 23, 1976, flouting the Federal law which requires that an Environmental Impact Statement precede such publication. . . . And in the face of mounting controversy over the Guidelines and the hazardous new technology, Fredrickson had shipped the document by diplomatic pouch for distribution by United States embassies and consulates over the world. An explanatory telegram from the office of Secretary of State Kissinger requested a "missions alert."[35]

Copies, accompanied by an explanatory letter, went without delay to the majority leader of the House of Representatives, to the members of the U.S. Senate Subcommittee on Health of the Committee on Labor and Public Welfare, and the House of Representatives' Subcommittee on Health and the Environment of the Committee on Interstate and Foreign Commerce, as well as to the chairmen and ranking minority members of the Subcommittees on Labor-HEW Appropriations.

Press

The national press and the science journals gave the event the kind of coverage that Asilomar and the Michigan manifestations had missed. This was an editorial event.

The *Washington Post* of July 2 carried an editorial, "A Scientific Breakthrough," in which the commentary included:

> . . . Not everyone is happy with the guidelines. Some scientists feel it is idiocy to attempt the curbing of scientific pursuits. Others believe that all experiments that might lead to genetic manipulation should be prohibited.

> The guidelines steer a narrow, but apparently safe course between these extremes. In the end they only extend the age-old medical admonition of *primum non nocere*—first of all, do no harm to a new frontier of potential discovery. The prognosis is good.[36]

While it had the initiative, the NIH was already under way to put in place the second stages of national use of the new rules. In the tale of the events, the mishaps, the gains and losses which follow, many run in parallel, for in the era of political management of biomedical science, things took place in several arenas at once.

5

Recombinant Ramblings

1976–77

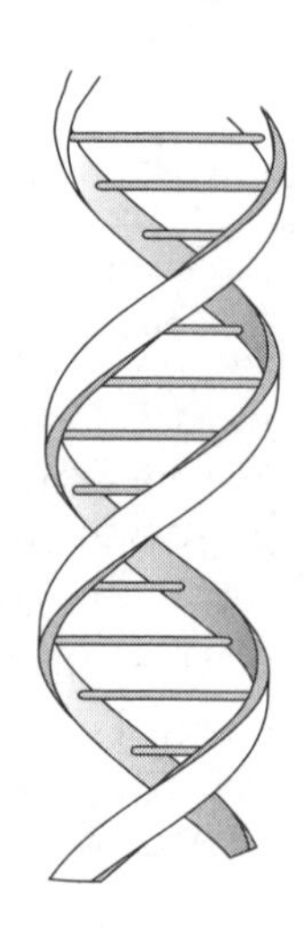

What is it about genetic knowledge that creates such ambivalence in us?[1]

Agonists

The number of persons disturbed by the awareness of recombinant DNA research will never be known. Nor were the sources of their anxiety ever studied and arrayed in a database. Some undoubtedly were alarmed about the possible spread of contagion of an unknown sort, an "Andromeda strain," or other chimeras raising havoc in the ecology. Some were fearful of scientists coming to possess the powers to threaten the genes of vulnerable persons. Many were angry at those who tried to usurp powers to bring more turmoil to an uneasy world. There were others who were jealous of the usurpers. Some of these latter may have been scientists, but other scientists were alarmed by their own perceptions of how little they knew of what their brethren were doing.

Identified as one of the two most distinguished opponents of pursuit of recombinant DNA research was Erwin Chargaff, professor and chairman emeritus of chemistry at the College of Physicians and Surgeons at Columbia University of New York.[2] He was fitted with a classical education in Austria and Germany, with further postgraduate studies in the United States. In 1974, he received the National Medal of Science, largely for his development of highly accurate micromethods for determination of purine and pyrimidine bases, from which he discovered that DNA contains a constant ratio of these bases and that in the DNA of all species there are equal amounts of adenine and thymine and of guanine and cytosine. These important findings were at the bridge when the Watson-Crick model roared across and the tide of molecular biology rudely washed over many of the earlier landmarks that marked the trail. Chargaff aired his dim views of the NIH in strong language: "Our time is cursed with the necessity for feeble men, masquerading as experts, to make enormously far-reaching decisions."[3] In June, I gave a little talk before the American Biochemical Society in San Francisco, in which I expressed shock at the "apostasy" of one of our scientific curia.[4] Chargaff's many articles cast in the tones of a wrathful Goethe supplied verbal tinder to light the flames that rose in the House at the beginning of the 95th Congress, when statutory regulation of recombinant DNA research was sought. Only cinders remained by the time I first met Chargaff, when I moderated a debate between him and Bernard Davis at Oberlin College in April 1980. I made it a point to have lunch with him, but our conversation could not find any common plane.

Agonists of rDNA were not always so remarkable as individuals. They were often more effective in groups, persistent at finding ways to block the course of revising the Guidelines—the longest pursuit of this entire story. Groups of environmentalists, adept at using process as weapons, usually launched their opposition with the most extreme demands. For example, members of Science for the People, particularly alarmed by the use of *E. coli*, began by wanting all the first experiments to be conducted in P4, the maximum physical containment.

But the vigilantes were not all small groups. There were municipal manifestations, uniting townspeople. One of these took place in one of America's greatest university towns in 1976.

Cantabrigian Counterpoint

> Stop DNA Research. There are already more forms of life in Harvard Square than this country can stand.[5]

P3 Lab at Harvard?

On the day of our briefing for the congressional staff, and just a week before the release of the Guidelines in 1976, Harvard's Dean Henry Rosovsky announced approval of plans for a P3 laboratory to be set up in two rooms on the fourth floor of the aged Biology Building on Divinity Avenue in Cambridge. That night the Cambridge City Council called a meeting and took a unanimous vote to hold special public hearings on Harvard's plans.

Biologist Mark Ptashne's desire to have a P3 laboratory was strongly opposed by some scientists, particularly some other occupants of the building, including biologists Ruth Hubbard and her husband, Nobel Laureate George Wald. Word of the disagreement and fears of the research soon reached city hall. Mayor Alfred E. Vellucci and the city council quickly became aware that 2 years earlier, the scientists had declared a moratorium on the very research proposed, and here it was now taking place at both Harvard and MIT. They also were aware that the NIH was promulgating guidelines—rules which some of the scientists complained were being developed by the very people whose interest was in continuing the work. The mayor took a politician's delight in this twist toward his position in the delicate balance of the endless struggles between "town and gown" at Cambridge. In fact, Hizzoner had never been handed a cause that could raise him so high above the walls of the disdainful and privileged properties within his domain. The Cambridge City Council proposed a series of public hearings to inform them. At these appeared scientists at Harvard and MIT, including Richard Goldstein, Jonathan King, and Jonathan Beckwith, who joined Wald in support of the mayor and in opposition to the laboratory. Their arguments were opposed by other Harvard and MIT faculty members, scientists such as Ptashne, Matt Meselson, Bernard Davis, and David Botstein, who just as vehemently supported the new laboratory. Leaders of the Asilomar conference also participated, including Maxine Singer and David Baltimore. Paul Berg was also among those quick to realize that pernicious barricades were rising in the road just opened by the arrival of the Guidelines and wrote the mayor and the city council of his views that the Guidelines were more stringent than any scientific evidence indicated was needed, and he urged an alternative of cooperation to suppression.[6]

Calls from Cambridge

On the morning of July 7, I received a call from Mayor Vellucci's assistant, Mr. McKinnon, urging me to make an appearance at the meeting of the city council. Shortly after that, Mark Ptashne called with the same suggestion. Walter Rosenblith, provost at MIT, who was gravely concerned about a shutdown of experimentation in two of the strongest biology departments in the world, at first urged me to come and then changed his mind later in the afternoon. Both Joe Perpich and the lawyer Harold Green were strongly against my making an appearance before the city council, reasoning that passage of a moratorium in Cambridge after the appearance of the NIH director could be very damaging to the federal role. Second, I could hardly condemn the council's inquiry, it being symbolic of the public participation which we were encouraging. I took their advice and did not attend. The council meeting produced a resolution calling for a 3-month moratorium on P3 and P4 experiments in Cambridge and passed an order that a Cambridge Experimental Review Board (CERB) be established to make recommendations. On August 10, the Cambridge city manager announced his appointment of nine nonscientists to the CERB.

CERB

During the fall, the CERB met every Tuesday and Thursday, hearing testimony from scientists on both sides of the controversy. One of those evenings, members of the Kitchen RAC and I sat before a speaker phone in my office in Bethesda answering questions posed by members of the CERB, whom we envisioned huddled over their speaker phone back home, admirably making an earnest search for wisdom. On September 27, the city council extended its moratorium 3 more months to allow CERB to complete its report. During this period, Science for the People and the Social Action Coordinating Committee held a teach-in at MIT; a few days after the city council's action, at a Cambridge forum, there was a formal public debate between George Wald and Matt Meselson.

CERB reported to the council on January 5, 1977, and the council converted its recommendations into municipal law, adding under the chapter on health and housing a new article II entitled "Ordinance for the use of recombinant DNA technology in the City of Cambridge."[7] This effectively lifted the moratorium at Cambridge, but some of its conditions were an egregious, direct assault on the autonomy of the cities' most prominent educational institutions and on the national efforts to cope with recombinant DNA with universal safety codes. Part of the ordinance had considerable merit in calling for all recombinant DNA research in Cambridge to be in strict accordance with the NIH Guidelines and insisting on mandatory training, including accident prevention, for all laboratory personnel, health monitoring of workers, and the addition of lab technicians and a community representative to the institutional biohazards committees.

But CERB then gave a demonstration of unlimited regulatory zeal. The EK2 level of biological containment was required for all experiments in P3, and impractical screening of all hosts for purity and resistance of experimentally created organisms to antibiotics was demanded. Finally, there was created the Cambridge Biohazards Committee (CBC). The CBC was a costly overlay of activities delegated to the institutional biohazards committees and the institution by the NIH and added site visits to laboratories, review of containment proposed for all experiments, policing violations, and many other duties representing such severe political interference with science that rDNA research would have to retreat to distant campuses.

The Cambridge affair did break down some traditions that made proud institutions more careful in their community relationships. The community, confused by faculty disputes and propelled by a zealous politician, acted as a united body. They worked hard, learned much more about the NIH Guidelines than some scientists, and demonstrated they could participate in an important public process—just what we had been asking them to do. Cambridge, moreover, played a positive role. It demonstrated the great danger and real limits of national policy when isolated jurisdictions attempt to cope with difficult global scientific problems. Clearly, it was another signal that the national effort must be capable of serving all communities in an effective way. Cambridge did stimulate environmental groups in other communities and local universities to respond to demands from students and citizens for more information. Very few communities went to the lengths of Cambridge. Some state legislators, particularly in New York, took their turns, threatening to balkanize rDNA research in the United States.

More Mundane Events

The appearance of the NIH Guidelines coincided with a number of other events, indicating the quickening of reactions at home and abroad. These can be briefly summarized before we return to find out what happened to our efforts to activate the Federal Interagency Committee.

ORDA

The Office of Recombinant DNA Activities (ORDA) was established (May 5, 1976) to facilitate the implementation of the new Guidelines. ORDA was the manager of the RAC. Its responsibilities were to administer and coordinate intramural and extramural rDNA activities at NIH. Its small staff had to review the institutional biohazards committees and certification statements and to monitor reports and information concerning accidents, containment, and safety innovation. A national registry of rDNA research, including that carried out by other agencies, was another of its responsibilities. William Gartland, who was a scientist administrator in the National Institute of General Medical Sciences, became the first director of ORDA. Along with Leon Jacobs, he had been the official representative of NIH at Asilomar. Gartland also had the able assistance of Charles McCarthy, head of the NIH Office for Protection from Research Risks to assist him in keeping track of the important memoranda of understanding between NIH and institutions housing research. If we add the one designated staff member who sometimes went out into the field to respond to cries of egregious violations of guidelines or fiducial rules, the total staff related to the Guidelines management was fewer than half a dozen persons.

Executive Recombinant DNA Committee

At the same time, an Executive Recombinant DNA Committee was formed in the office of the NIH director. It was chaired by the deputy director for science (Stetten) and included several associate directors of NIH as well as the director for research safety for the National Cancer Institute (Barkley) and the associate director for environmental health and safety for the NIH (Rudolph Wanner). The group was a necessary point of appeals on questions that could not be resolved at the level of ORDA.

Recombinant DNA Technical Bulletin

Since 1975, an NIH publication, *The Nucleic Acid Recombinant Scientific Memorandum*, had served to advise the scientific community about rDNA research. The traffic of such information rose in accordance with the need to hold together a network of interested committees and researchers, and a separate journal had to be sought. The Office of Management and Budget, however, has always been a strict accountant for the tons of newsprint used annually by the government. For months, permission was sought, and nothing yielded. Finally, we made a fortunate contact at the White House who realized that we had an unusual need. The first edition of the *Recombinant DNA Technical Bulletin*[8] arrived in August 1977 and enjoyed a very large circulation both in the United States and abroad.

These were basically housekeeping details. The next move, however, was of far greater significance in the steering of the NIH Guidelines through the rocky shoals that lay just ahead of them. The formation of the Federal Interagency Committee on Recombinant DNA Research (FIC) was, in retrospect, a signal achievement. The neutralization of tension among so many potential competing agencies of government made it possible for NIH to lead with a minimum of antagonism.

Formation of the FIC

From the Start

You will recall that HEW secretary Mathews had sent a memorandum to the president on June 18, 1976, requesting the formation of an FIC. On June 24, a return from the White House had questions for the secretary. What would be the nature of the charter? Who are the proposed members? The relationship to the Federal Advisory Committee Act? How does it relate to congressional action to place the responsibility in the hands of a presidential commission (S. 2515)?[9] On July 7, I had a call from Patricia Bowen, a staff person in the office of Senator Jacob Javits. Stan Jones from Senator Edward Kennedy's staff had just delivered a letter to the president for Mr. Javits's signature. Patricia wanted reassurance that it was something her boss would want to sign. Could she send it over for Joe Perpich to check? Knowing that Stan Jones was one of the "white hats" on the chairman's staff, and after I heard a bit of the message, I told Patricia that I would have Joe call, but that, yes, we "needed to prod the White House," indicating that I hoped the ranking minority member of the subcommittee would sign.[10]

Three weeks after its receipt, the White House memorandum was answered by acting HEW secretary Marjorie Lynch. Drafted by a member of the General Counsel's Office, the letter informed the White House that the proposed committee was to consist of federal officers and employees and was therefore specifically excluded from the Federal Advisory Committee Act. As for S. 2515, the acting secretary said that she was dispatching a letter to Congressman Harley Staggers, chairman of the House Committee on Interstate Commerce, in opposition to S. 2515 and added:

> . . . the study of Recombinant DNA research proposed by the Commission would duplicate the efforts of the NIH . . . the Committee that I am proposing would have a far broader mandate and a broader representation of interested parties.[11]

Lynch concluded that she "strongly urge you to recommend the President that I be allowed to proceed without . . . undue delay."

Senatorial Stimulus

On July 19, the three-page letter signed by Senators Kennedy and Javits was sent to the president.[12] The letter described the issues of rDNA research and the release of the Guidelines. It praised NIH for "the great care . . . taken . . . in obtaining the best scientific advice as well as advice from experts in law and ethics" in formulating the Guidelines and said that these steps taken by NIH "are a responsible and major step forward and reflect a sense of social responsibility on the part of the research community and the NIH. However, we are gravely concerned that these relatively stringent guidelines may not be implemented in all sectors. . . ." There was no mention of the stalled interagency committee, with the "kicker" in the penultimate paragraph:

> Given the high potential risks of this research, it seems imperative that every possible measure be explored for assuring that the NIH guidelines are adhered to in all sectors of the research community. We urge you to

> implement these guidelines immediately wherever possible by executive directive and/or rulemaking, and to explore every possible mechanism to assure compliance with the guidelines in all sectors of the research community, including the private sector and the international community. If legislation is required to these ends, we urge you to expedite proposals to Congress.

Senator Kennedy sent a copy of his letter to us with some gracious words:

> I want to offer my congratulations on the thorough and statesmanlike job NIH has done thus far with respect to formulation of these Guidelines. In particular I am pleased at your involvement of knowledgeable people from outside the medical research and scientific communities. . .[13]

White House Response

The White House responded a few weeks later in a memorandum to Secretary Mathews.

> . . . Since you would have the authority to establish the committee without the President's approval, according to the general counsel of HEW, we have no objection.
>
> If you decide to create such a committee and feel that a letter from the President to the heads of departments and agencies urging their cooperation would be useful, send me a draft letter."[14]

Letters promptly drafted at NIH, dispatched through the department to the White House, were sent from President Gerald Ford to Senators Javits and Kennedy 5 weeks later,[15] noting that Secretary Mathews was forming an interagency committee, the FIC. On the same day, the president sent a letter to members of the cabinet and independent agencies asking each to nominate members of the FIC to the secretary.[16] On October 16, 1976, Secretary Mathews signed the charter of the FIC.[17] One of the purposes enunciated in the charter was to facilitate a uniform set of guidelines for the conduct of this research in the public and private sectors.

The Climate as FIC Is About to Convene

High winds swirled through the public arenas as the FIC was about to be fitted with its charter. The city council in Cambridge continued its oversight hearings, and the assemblies of several states, including California and New York, were preparing to debate possible new state laws or regulations. In New York, Attorney General Lefkowitz held public hearings. A principal witness was Liebe Cavalieri, a molecular biologist at the Sloan Kettering Institute for Cancer Research. The *New York Times Magazine* of August 22, 1976, had carried his article about recombinant DNA research entitled "New Strains of Life—or Death."[18] Others were insisting the federal government get more involved, On November 11, a petition was addressed to the HEW secretary by the Environmental Defense Fund (EDF) and the Natural Resources Defense Council demanding that the secretary promulgate interim regulations to make the NIH Guidelines binding on all research with rDNA and that legislativelike public hearings be held.[19] Letters from Robert Sinsheimer and Alan McGowan, president of the Scientists' Institute for Public Information, were among those supporting the petition.[20] Preparations were being made for a forum on recombinant DNA to be held at the National Academy of Sciences.

Growing Congressional Appetite

The Senate Labor and Public Welfare Subcommittee held a hearing on September 22, 1976. Witnesses, in addition to me, included the assistant administrator of the EPA and representatives of the Pharmaceutical Manufacturers Association and the EDF.

Committee members Kennedy, Javits, and Schweiker (R-Pa) were present. Senator Kennedy reiterated his hope that the House would enact counterpart legislation to his resolution (S. 2515) supporting the creation of a Recombinant DNA Commission. Indeed, Mr. Waxman (D-Calif) had already introduced two bills (H.R. 15472 and, with Mr. Rogers, H.R. 15543), the latter proposing a permanent National Commission for the Protection of Human Subjects of Biomedical and Behavioral Research, including language to cover recombinant DNA research.

My testimony at this hearing was interrupted by repeated and pointed questions as to (i) why NIH had not received endorsements from all federal departments conducting recombinant DNA research and (ii) why the proposed Interagency Committee was so slow in getting started. The witness from the EPA was generally supportive of the NIH Guidelines, with the exception of a strong recommendation that recombinant DNA research not be performed with *E. coli.* A scientific panel consisting of David Baltimore, Robert Sinsheimer, Halstead Holman, and Norton Zinder commented on the Guidelines, with Baltimore suggesting they were too strict and Sinsheimer expressing the opinions that they were not strict enough and that our present state of knowledge did not permit us to say that the potential benefits outweighed the risks.

Burke Zimmerman of the EDF noted from the witness chair that the EDF would shortly petition the secretary of HEW to use the department's communicable disease authorities (Section 361) to promulgate the NIH Guidelines as regulations governing all recombinant DNA activities.[21] He further stated that EDF would seek a remedy by executive order or legislation to hold public hearings on the Guidelines and on whether *E. coli* was a safe subject for experimentation.[22]

The FIC Begins

The first meeting of the full and duly chartered Federal Interagency Committee on Recombinant DNA Research convened in Wilson Hall in Building One at NIH on November 4, 1976. Representatives of 22 separate departments or agencies were seated around the great table in this venerable conference room, the Department of Justice having been added since our preliminary gathering 6 months earlier.[23] The FIC, having a membership solely consisting of government employees, was not subject to the Federal Advisory Committee Act and thus held its meetings at the NIH, behind doors closed to the public. Spared advance notices and distribution of agendas in the *Federal Register*, the functioning of the FIC was flexible and allowed work to proceed in small groups when necessary. This might have caused much more criticism if the agencies had not actually served to increase the surface area through which its discussions diffused freely to their numerous constituencies.

At first, the press cared little that it was not invited to the FIC meetings, but as the proceedings picked up interest, it was agreed that they would be provided statements with a summary of the proceedings of each meeting.

November 4, 1976

As chair, I attempted to frame the significance of the FIC in my opening remarks, with a few new emphases beyond my inaugural comments in the preceding April. Our job was, I said, "to establish for the entire nation a workable system [for rDNA regulation] that protects adequately the legitimate interests of all. . . . If we are to serve usefully here, all of us must rise above those narrow interests of which we are often unjustly accused. . . . That is the test we are to undergo. Let us begin."[24]

Along with the mandate for the FIC, a checklist for the first assignment to each agency was handed out, laying forth the tasks that each had to consider with regard to their respective duties if they really wanted to have a regulatory role.[25]

Regarding the question of which agencies thought they had authorities that would assign to them one or more of the tasks ahead, I requested that at the next meeting (i) the "performer" agencies submit a statement analyzing the nature and extent of actual or planned recombinant DNA research and the role of the agency vis-à-vis the 11 functions or processes on the checklist laid out earlier, and (ii) the "regulatory" agencies submit a written statement analyzing the authority and role of the agency in the regulation of recombinant DNA.

The next meeting was set for 3 weeks away. A statement was also requested from the Arms Control and Disarmament Agency concerning the status of recombinant DNA as a possible source of biological weaponry. This request was answered in the negative a month later.[26]

November 23, 1976

At the next meeting, the full panel of participants was present, and the results were more than a little gratifying.

Reports of the Performers. Of the two major performer agencies[27] other than NIH, the National Science Foundation (NSF) reported it had accepted the NIH Guidelines, except for the role of NIH study sections. NSF had awarded 52 research projects involving rDNA, compared with about 125 such projects currently funded by NIH. The Department of Agriculture had earlier reached a decision to accept the Guidelines and reported on three or four projects in progress that met the definition of rDNA research. The Department of Defense said it had no such research, but if it did, it would follow the Guidelines.

Reports of the Regulators. EPA mentioned that Section 5(h) of the Toxic Substances Control Act provides an exemption from most of the requirements of the act for scientific research. (Much later when the question of releasing recombinant products into the environment became a national concern, there would again arise the question: is recombinant DNA a "chemical"?) The Center for Disease Control said that Section 361 of the Public Health Service Act could be interpreted as giving them the authority to regulate some recombinant DNA as an infectious substance. NIOSH, which sets standards for OSHA, suggested there be a central registry of laboratories, workers, and projects; medical examination of workers; informed consent of workers; and special health insurance. OSHA, on the other hand, the agent of enforcement of NIOSH standards, confessed the gap in its ability to cover state university laboratories. Only the 26 states that had a state OSHA plan were within their jurisdiction. OSHA also admitted that promulgation of a "standard" in the traditional regulatory process was a long and arduous process. This served to comfort those of us from NIH who had from the first a concern that process might triumph over substance in efforts to control the fast-moving, exceedingly complex substance of rDNA research. The Department of Transportation felt they had clear authorities to regulate transportation of harmful vectors—although definitions of which vectors were harmful would raise problems for all concerned.

The Department of Commerce reported a meeting with industry representatives on November 19. They related the willingness of commercial spokesmen to accept compulsory registration of recombinant DNA projects but recommended that compliance with guidelines must be voluntary. They were of the opinion that industry compliance with standards would be stricter than would obtain in most academic laboratories. Commerce's greatest concern was that of protection of proprietary data.

(On December 8, 2 weeks later, I had a second meeting with a group of industry representatives at NIH to explain the nature of the FIC. Their views to me were in accord with the Commerce report, but I had detected a note that some in industry would view the NIH Guidelines as affording them a measure of protection against the liability for untoward events. Thus, a willingness to follow the Guidelines in the private sector might eventually occur.)

Representatives from Agriculture, OSHA, EPA, CDC, Justice, Commerce, the Office of Science and Technology Policy, and NIH were asked to form an FIC Subcommittee on Authorities to resolve the question of which, if any, agency statutes represented a clear source of regulation of research. Attorneys from the Departments of Justice, Agriculture, HEW, Labor, Transportation, and the EPA dedicated their attention to four particular laws: (i) the Occupational Safety and Health Act of 1970 (Public Law 91-596); (ii) the Toxic Substances Control Act (P.L. 94-469); (iii) the Hazardous Materials Transportation Act (P.L. 93-633); and (iv) Section 361 of the Public Health Service Act (42 U.S.C. §264).

February 25, 1977

The Subcommittee on Authorities promptly reported that it concluded that none of the existing legal authorities of any federal agency would permit them to oversee all rDNA adequately by regulation. Further, the lawyers concluded that regulatory actions taken on any combination of existing authorities would probably be subject to legal challenge.

Summing Up

The FIC had amiably and thoroughly done the trick. NIH so far was holding the position in charge, and one set of guidelines would do for the country—provided more local versions did not appear. (They would, and the issue of federal preemption of the locals would be a hot coal to handle in the Congress.)

The Patent

> United States Patent 4,237,224, an application for a patent on a process to produce biologically functional molecular chimeras, by inventors Stanley N. Cohen and Herbert W. Boyer, was filed November 1974.[28]

The story was legendary.

> The experiment was conceived at a Waikiki Beach bar after a plasmid meeting. I [Cohen] had talked about plasmid transfer and Boyer about a restriction enzyme. We then decided on the next logical experiment and published the results in the *Proceedings of the National Academy of Sciences* in 1973 without a thought of patenting. It was David Baltimore who mentioned the experiment to a reporter from the *New York Times*. The latter did a story on the experiment. Niels Reimers of the Stanford Business Office read the article, and urged us to apply for a patent.[29]

In mid-June 1976, as we were readying the Guidelines for launching, I received a letter from Stanford that added more to the already excessive burden on my desk.

> Dear Dr. Fredrickson:
>
> From Paul Berg and others I know that you are aware of discussions taking place at Stanford over the wisdom of proceeding (on behalf of Stanford and the University of California) with an application for patent protection for discoveries in the area of Recombinant DNA. As you know, we began to move in this direction with the knowledge and consent of NIH and NSF; as you also know, the whole matter of patent protection is now the subject of a lively debate here. The purpose of this letter is to solicit your views.
>
> As further background to what you already have, you might be interested in the enclosed memorandum in which I have attempted to summarize some of the major questions . . . speaking for myself only, and not trying to articulate University policy. . . . I would guess that most of the University's senior officers would agree with my conclusions (though there is dissent), as do many, although not all, faculty. . . .
>
> Let me emphasize that we do not yet have conclusions. We are proceeding with the necessary steps in the patent application process and we have had discussions with a prospective licensee. We have taken no irrevocable steps, but we are rapidly approaching the stage at which binding decisions will need to be taken. . . . Your contributions to our deliberations would be extremely valuable. . . .

The issues we are dealing with are complex, interesting and important, and the way they are resolved is likely to have a lasting effect on science and education. I think this is not too strong a statement. I hope you agree and that we will have the benefit of your views.[30]

Sincerely,

Robert M. Rosenzweig
Vice President for Public Affairs

The accompanying memorandum contained more than a demand for a review:

> Before the decision is finally taken to press for patent protection, the leaders of the most relevant public agencies, e.g., NIH, [and] the President's Science Advisory Council, should be consulted. We should seek their agreement that the decision is a proper one *and their willingness to say so publicly* [italics mine].[31]

No matter where one was positioned in the early part of the recombinant DNA era, "the patent" was widely perceived as a modestly seismic event, a nervous shift at the conjunction of the academic/not-for-profit and commercial tectonic plates sustaining the crust of the biomedical research enterprise. To some scientists of my generation and fairly cloistered experience at NIH, it also heralded a certain loss of innocence. An unfamiliar entrepreneurial instinct now intruded upon the free competition for the discovery of nature's mysteries with professional recognition as the principal compensation. The rise of the biotechnology industry attending discovery of recombinant DNA technology had to be accepted as an inevitable culmination of years of basic research in molecular biology that could mean a quantum jump in understanding of both disease and health. It would also require capital investments leading eventually to extremely profitable commerce. Even by now, a quarter of a century later, all the long-term consequences of this change have by no means been realized. Astounding technical achievements, enormous growth in knowledge, and expansion of resources expended in both academic and industrial research are suspended on huge capital investments. At the time, one could not imagine that one of the more negative outcomes in the next quarter century would be a serious and growing conflict over patenting of even human genes.

HEW Patent Policies

The Stanford letter did not raise a question about the well-established arrangements of the HEW for allocating rights to grantees' inventions. The current HEW patent regulations were considered state of the art.[32] In 1977, the department had entered into an institutional patent agreement (IPA) with 167 universities, including Stanford and the University of California. Under the IPA, the sponsoring institution had the patent rights and could grant an exclusive license to practice an invention only if a nonexclusive license could be ineffective in bringing the invention into market. The shares of the royalties were also set by the agreement; normally, the inventor received 15%. The institution must grant the government a license to make or have the invention, and inventions could be used for research purposes without infringement and without payment of royalties.

Requesting that government take a stand and that the director of NIH make a public endorsement of the patent application seemed a somewhat callow challenge. In certain political corners at the time, such a stand could have snuffed out the usefulness of the NIH director as the neutral champion of the NIH Guidelines. After all, the acceptance of any patent application was the sole responsibility of the Patent Office, not likely to be influenced by what the HEW thought about it.

Patents in Biology in 1976

Given what might be coming in the way of new kinds of patents, one should catch a view through the narrow frame of history of public actions concerning biological patents in the years 1976 to 1978. In June 1978, the Supreme Court declined an opportunity to review an unprecedented ruling that a person can patent living things. The case involved a strain of bacteria found in certain Arizona soil. Upjohn scientists isolated the bacteria and used them to prepare an antibiotic named lincomycin. The U.S. Patent and Trademark Office rejected the application on the basis of a determination that Congress had not intended 35 U.S.C. §101 to permit patents on living things. The U.S. Court of Customs and Patent Appeals reversed this decision and held that such patents are permissible. The government then asked the Supreme Court to reverse this ruling, but the Court sent it back to the Patent Appeals Court without a further defining ruling.[33] The first decision awarding a patent for a life form created—in the court's words, "by what is sometime referred to as 'genetic engineering' "—was the ruling by the Court of Customs and Patent Appeals' decision on March 3, 1978. The new life form was a variant of *Pseudomonas* bacterium developed by Ananda M. Chakrabarty that was capable of eating oil spills.[34] The process used, however, was not recombinant DNA technology by the definition of the NIH Guidelines.

At the time of the Stanford letter, Congress was alert to questions raised by recombinant DNA. Senator Gaylord Nelson indicated that he was preparing to hold hearings on inventions derived from government funds for research that were "enriching" some universities.

NIH Deliberations

We were already having questions about any patenting of forms created by this "new genetic engineering." In conjunction with Joe Perpich and Norman Latker, the patent counsel for the HEW, we notified Rosenzweig that I was undertaking a careful review of our patent policies with respect to recombinant DNA research and would keep him informed of the outcomes. Latker was a staunch supporter of the patent policy of the department to facilitate the transfer of technology from the bench to the marketplace and was zealous to counter the prejudice of many members of the scientific community, where patents had a negative connotation. However, opined Latker, were I to sanction Stanford's proceeding to patent claims, it would open me to the charge by public critics that this policy is inconsistent with my statements in the release of the Guidelines and added, "While the guidelines signal a yellow light for proceeding with this research, allowing full use of patents signals a green light for research and development without concern for inherent risks."

Accepting the warning that any stance I took had to be carefully considered, the Kitchen RAC and I spent many hours considering a number of policy options that Latker had laid out. (See below.)

Latker worked hard to persuade us that the best approach was an option whereby Stanford would retain patent rights, but we would require clearance of all licenses. For the latter we could require certain conditions, such as having the licensee comply with the NIH Guidelines.

Concerns about Restriction of Research Information

A major concern of ours was the effect of patent laws and patent protection on the dissemination of research information. For domestic patent rights, once the research results are published, the inventor has a 1-year period of grace for filing a patent application. However, if one publishes first, one has waived all foreign patent rights. There was also a special desire for rapid dissemination of research results in the recombinant DNA area. The fear of losing foreign rights could cause investigators to delay publication of results pertinent to the RAC actions.

Latker also raised another element of the disclosure issue, the relationship between the patent laws and the Freedom of Information Act.[35] We would soon reach the point where consideration of industrial proposals in closed meetings of the RAC would become a tense period for NIH. Anticipation of this problem, however, could not allow us to oppose patenting of recombinant DNA inventions.

To our thinking that we could attach a condition to any license of a patent that all use of it should occur under the conditions prescribed by the NIH Guidelines, there were some who objected to NIH's forcing the private sector to use the Guidelines or demanding that the institution granting the license attempt to enforce the Guidelines.

The diversity of opinions led us to consider an expanded survey from the community, with request for judgments about certain options if recombinant DNA inventions were to be patented under IPAs. We began this process with the RAC in late August,[36] and in early September we solicited the opinions of the members of the DAC and all who had participated in its February 1976 meeting, the 22 representatives of private industry who had met with me earlier, and of all members of the FIC.

Letter from Director of NIH Soliciting Opinions on Patenting Recombinant DNA Inventions, September 8, 1976

The letter[37] began with an indication that we were soliciting the addressees on the "question of patent applications in the area of recombinant DNA research activity." The background explained that the filing of an application by Stanford and the University of California for a process for forming recombinant DNA, an invention generated in performance of an NIH grant, was in accordance with a prior agreement with the government (IPA) but that Stanford had solicited NIH's view on an appropriate plan for administration of this particular invention.

There followed a detailed summary of the allocation of patent rights by the department to research it had supported, and mention of some 167 patent applications having been filed in the period 1969 to 1974 under IPAs with millions of dollars having been committed to development of these patents; it appeared that the patent policy had both encouraged the development of new technology and demonstrated an economic benefit to the United States.

It was then emphasized that the research under the NIH Guidelines "requires a delicate balance between the need for rapid exchange of information unhampered by undue concern for patent rights and a potential for achieving uniformity in safety practices through conditions of licensure under patent agreements. " Noting that Stanford had indicated willingness to attach conditions to their IPA as it applied to recombinant DNA research, the letter proceeded to list five possible options for such conditions and requested opinions of the addressees.

Responses

Approximately 50 letters were received in reply.[38] The responses to Latker's five alternatives can be summarized briefly with interpolation of some individual responses:

Option 1. Institutions could be discouraged from filing patent applications on inventions arising from recombinant DNA research. There were no strict adherents for this proposal. Two members of the DAC, however, had specific comments about the proposal of any patents at all.[39]

Option 2. Institutions could be asked to file patent applications on inventions arising from recombinant DNA research and to dedicate all issued patents to the public. No respondents supported this option.

Option 3. Institutions could be asked to assign all inventions made in performance of recombinant DNA research to the department. The department as assignee of the invention could either pursue the licensing of whatever patent applications were filed or dedicate issued patents to the public. Some academic representatives, including several members of the RAC, supported this proposal but there were no industrial supporters. A member of the Asilomar organizers put his sentiments on this proposal on the record.[40]

Option 4. The department could continue to permit institutions to exercise their first option to ownership under the IPA but require that all licensing of patented inventions be approved by the department. The department could set certain conditions for approval, such as compliance with the *NIH Guidelines for Recombinant DNA Research*. This option was the overwhelming favorite of the majority of commentators. A few of the commercial firms voiced concern about the stipulation of any conditions by the government. One executive of a major pharmaceutical firm set forth detailed reasons why industry felt imperiled by any conditions.[41]

Option 5. The government could permit institutions to retain their first option, as in option 4, but approve only exclusive licenses. There was very little enthusiasm for this option. One correspondent not captured in the NIH poll offered another thought.[42]

Why Not Accelerated Patents on rDNA?

Without warning, there suddenly appeared a notice in the January 13, 1977, *Federal Register* signed by Marshall Dann, the Commissioner of Patents and Trademarks, bearing the title "Accelerated Processing of Patent Applications for Inventions." The announcement had been approved by Betsy Ancker-Johnson, deputy assistant secretary for science and technology of the Commerce Department, who was alternate representative to the FIC for Jordan Baruch, the assistant secretary for science and technology.[43]

Multiple Response

Four days later, Senator Dale Bumpers (D-Ark) sent a note to the new secretary of HEW, Joseph A. Califano, Jr., questioning the patenting of rDNA inventions. Califano summoned me to vent his wrath and threatened to "have a donnybrook at the level of the cabinet" if the order were not suspended. He shortly thereafter called upon Department of Commerce secretary Juanita Kreps to cancel the order. Part of the reason he gave was that he expected a report of the FIC on how to regulate firms entering this research area.

In a news account on February 10, Califano noted that the FIC was seeking some way to apply the Guidelines to all industrial concerns.[44] The same press report indicated that Secretary Kreps was also visited by two California state legislators who demanded that the order be canceled until an environmental impact study was issued. With the hair-trigger reflexes for which it had become known, the Senate Health subcommittee also announced that it would hold hearings on the subject in April as a follow-up to a National Academy of Sciences public forum on recombinant DNA research in March.

On February 14, the secretary of commerce reined in her troops.[45] On the 17th, Secretary Califano wrote Senator Bumpers that he and Secretary Kreps had agreed to suspend the order temporarily and that both agreed that the entire issue would be deliberated by the FIC, which was soon to recommend to him how the federal government should deal with this complex and important subject.[46] The FIC duly deliberated the proposal on accelerated patents for rDNA discoveries but never recommended the practice.

A Reply to Stanford (1977)

A year had gone by. There had to be an answer to Stanford's question. But this now was more problematic than ever, for the Senate Health Subcommittee was drafting a bill that would extend the Atomic Energy Commission's concept of government ownership of radiation patents to rDNA inventions.

Nevertheless, copies of the yellow-jacketed "Analysis by the Director"[47] were prepared for distribution to government officials and other interested parties by November 1977. After a description of the national query and the responses, the review and opinions of the FIC and NIH officials, there followed the reasoning behind our support of Stanford's position.

> This leaves the residual question whether the subject of the patentable processes (recombinant DNA techniques) is of such a peculiar nature that financial return to the inventors should be denied. . . . There are no compelling economic, social or moral reasons to distinguish these inventions from others involving biological substances or processes that have been patented, even when partially or wholly developed with public funds. Such inventions include vaccines for rubella and rabies, treatments for herpes infections of the eye, treatments for uremia, and prostaglandins—compounds that may have a number of possible medical uses. The argument that commercial development based on patent protection has or will assure maximum benefits of these inventions to the public applies as well to the putative benefits of recombinant DNA research interventions . . . It is recommended . . . that recombinant DNA research inventions developed under DHEW-NIH support should . . . be administered within current DHEW patent agreements with the universities. But each agreement should be amended to ensure that the licensees will comply with the physical and biological containment standards set forth in the Guidelines in any production or use of recombinant DNA molecules under the license. If legislation is passed these safety standards will be mandated by the law for all who conduct or support recombinant DNA research.[47]

At the end of December, the surgeon general and HEW general counsel Peter Libassi were convinced that the secretary's views about patents for DNA inventions now were such that a favorable recommendation would be accepted. A draft letter to Stanford had been waiting many weeks for approval. It was dispatched to Palo Alto on March 2, 1978. The key text was as follows.

> On the basis of the findings contained in the report and my discussions with Drs. Julius Richmond, the Assistant Secretary for Health, and Peter Libassi, General Counsel for the Department, it is my recommendation that at least for the present, recombinant DNA inventions developed under DHEW-NIH support should continue to be administered within current DHEW patent agreements with the universities. Each agreement, however, will be amended to permit the institution to grant a license under patents secured on any such invention only if the licensee provides assurance of compliance with the physical and biological containment standards set forth in the Guidelines in any production or use of recombinant DNA

> molecules under the license. In my view, the requirements set for NIH grantees and contractors will thus be honored by licensees as well. Accordingly Stanford may proceed to file recombinant DNA research patent applications.[48]

I didn't hear the last of this until a year later, when perched on a chair in the drafty old caucus room 304 of the Senate's Russell Building, I faced Chairman Gaylord Nelson, who had completed a statement about IPAs with a final comment: ". . . one unanswered question involves ownership of—and potentially enormous profits from—DNA research inventions."[49]

The worth of a year's struggle to answer Stanford's letter was redeemed. I gave the text of the report on their patent that we had so laboriously prepared. The chairman had no further questions and seemed to be mollified.

Announcement

The first press announcement of the Cohen-Boyer patent, now jointly assigned to Stanford and the University of California, appeared in 1980. This patent extended only to the gene-splicing and cloning technology. A second part left pending at the time was to cover products resulting from gene splicing. The research had been jointly sponsored by NIH, NSF, and the American Cancer Society. The announcement said that nonspecific licenses for "reasonable royalties" would be made and that both inventors had waived their rights to royalties.[50]

Not Far Behind

By March 3, 1978, others had made application for patents in the recombinant DNA area. These included Curtiss, University of Alabama ("Modified microorganisms and method of preparing and using same"); Wu et al., Cornell University ("Oligonucleotides useful as adapters and adapted molecules"); Rutter and Goodman, University of California ("Recombinant bacterial plasmids containing the coding sequences of the insulin gene"); and Goodman et al., University of California ("Purification of nucleotide sequences suitable for expression of bacteria").[51]

The Stakes

The Cohen-Boyer patent licenses were first marketed in December 1980; the sign-up fee was $10,000, with annual minimum royalty payments of $10,000 thereafter. Much later the sign-up fee was raised to $50,000 for large companies (over 125 employees). Initial royalty rates on end products were 1% on the first $5 million, 0.75% on the next $5 million, and 0.5% thereafter. Later, the standard end-product rates were raised, first to 1% at all levels and finally to 2%. The Cohen-Boyer patent expired on December 12, 1997. It is still earning, because all end products in inventory at the time of expiration still carry the royalty—thus in 1998 the two universities would realize over $30 million. Profits ended in April 1999. Total gross revenue will be around $240 million in all. After expenses of licensing, etc., each institution will have received around $90 million.[52]

Coda

At the time of this writing, 25 years later, I'm constrained to think what would have happened if the congressmen who abhorred the patenting of federally sponsored inventions had worked their will. I am certain we were right to cover technical inventions that would push the art as far as it has gone. I also know that had the question been my view of proprietary rights in human genes, I would never have accepted it.

First Recombinant Odyssey

The U.K. Guidelines having arrived in August 1976, I was itching for an opportunity to go see firsthand how they were being received abroad. Fortunately, the product of the Williams committee had not differed much from the NIH Guidelines. The British were not interested in biological containment and had slightly stiffer rules for physical containment. The greatest difference was their use of a committee (Genetic Manipulation Advisory Group, GMAG) to more or less develop rules as they went along. In September 1976, I was able to make a 9-day trip to Europe, beginning with attendance at my first meeting as ex officio member of the European Medical Research Council (EMRC) in Helsinki, Finland.

EMRC

The EMRC had been organized by a group of medical researchers from the research councils of all the European countries and had become a "standing committee" of the European Science Foundation (ESF). The council met annually, and the NIH director was accorded the courtesy of attending the meetings.

At Helsinki, the British representatives were very desirous that ESF adopt the U.K. Guidelines for Europe. I was eager to clear up some misconceptions about the NIH Guidelines but was determined to suppress national pride to avoid any War of the Roses over two similar guidelines that were, after all, both progeny of the "international agreement at Asilomar." The main thing was to get a system that worked in Europe, conforming to the important principles that had been agreed to and described in the Berg letter.

After the meeting, I flew to Munich to sign a research agreement between NIH and Germany for the HEW secretary. (Traveling secretaries were prone to return with a new agreement between their hosts and the NIH.) I next made a contact with the ESF by chance.

The ESF was founded in 1975. Its headquarters were in Strasbourg, France, where it had endorsed the formation of an Advisory Standing Committee on Recombinant DNA composed of lawyers and physicians as well as biologists. They had emphasized the need for uniform control of both public and private laboratories in Europe. The new executive secretary of ESF was Franz Schneider. He briefly received me at the headquarters of the Max Planck Society as he was preparing to leave to take his post in Strasbourg. He was courteous, but our conversation was brief.

"I am sorry to disappoint you, Dr. Fredrickson, but our committee has just decided to adopt the British system, and will shortly announce their findings."

I replied, "The British and U.S. guidelines are nearly equivalent, and I believe that the most important thing is that you recommend all your members to follow one or the other." Perhaps I was more piqued than I recognized, but I could not help adding, "I do have one question, however, Dr. Schneider. Are all your members willing to allow a British committee meeting behind closed doors to make decisions about their experiments?"

I left Schneider's office and strolled across the grass to Fitzi Lynen's Laboratory to pay a visit to the Professor.[53] He was not in, but in the corner of the laboratory, a scientist was translating the NIH Guidelines into German.

EMBO

The first world organization to compare the NIH and new U.K. Guidelines was the European Molecular Biology Organization (EMBO), formed by a group of molecular biologists in 1963 to promote educational programs and research opportunities to stem the tide of movement of young biologists to America. The activities were to be developed at the European and international levels and would include the establishment of a European laboratory. In 1969, 13 European governments formed a European Molecular Biology Conference to support these objectives. In 1978, the laboratory was established in Heidelberg, Germany. The excellence of the laboratory, educational programs, and organization of distinguished scientists made it a most prestigious source of opinion on recombinant DNA technology in the pan-European area. In February 1976, EMBO formed a Standing Advisory

Committee on Recombinant DNA.[54] A second meeting of this committee was held on September 18–19 in London for the purposes of comparing the NIH Guidelines with the report of the Williams committee. After a detailed comparison of both documents, the overall conclusion of the advisory committee was that the NIH Guidelines and the code of practice proposed by the Williams working party "provide adequate safeguards against the conjectural hazards of *in vitro* recombinant DNA research, although the two schemes differ in technical details, in the relative emphasis they give to biological and physical containment, and in the administrative procedures proposed for their implementation."[55] The sum of this meeting was a slight tilt toward the U.K. Guidelines, with the position that a nation might choose one or the other but not mix them. EMBO also had recommended that where the U.K. Guidelines were adopted, the country should have a committee like GMAG to interpret the Guidelines more locally.

Political Manifestations

For several years, there was no public reaction regarding recombinant DNA experimentation in Europe. In Germany, there were no public discussions or even a provocative *Der Spiegel* article. Later, when the Green party became much more powerful, political reactions and manifestations became intense in north Germany, and some of the great firms in the north would decide to move their biotechnology plants to either Bavaria or America.

Switzerland, for all its alpine serenity, may have been an exception to peaceful arrival of recombinant research. Complaints about genetic manipulations had been aired politically in Switzerland in the early 1970s. A question was raised in the Bundesrat by Oehen, head of the Republican Party in 1973. His party and the comparable National Action Party together constituted about 5% of the Swiss electorate but periodically contributed a disproportionate share of excitement from the radical right. When Oehen spoke against "genetic engineering" in the Bundesrat, scientists rallied quickly to help the government compose an answer, and the matter died without publicity. A year later, in the canton of Zurich, questions were raised about the work of Charles Weissman, a capable and entrepreneurial molecular biologist. His laboratory at the Technische Hochschule, in concert with the early biotechnical company Biogen, was cloning interferon genes from tissue culture cell messenger RNA.

The questions about the relationships between industry and the universities, or private and public funding, that were aired in the parliament would soon be common enough in America. The matter died after a flurry of discussion in the newspapers and in the Davos meeting on genetic recombination, attended by Paul Berg. The Swiss Society of Molecular and Cell Biology became involved, and despite lively debate and much philosophical discussion, no conclusions were reached.[56]

As I arrived back in Bethesda in mid-September, the Stanford patent question was still unsolved, and we had just started to fix the FIC.

6

The Environmental Impact Statement and the Polyoma Experiment

1976–77

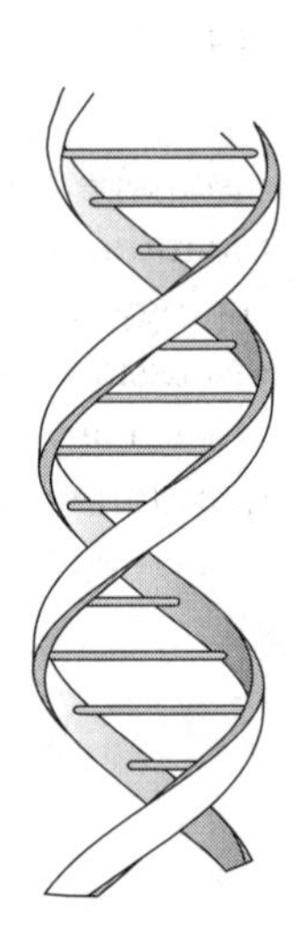

Assume that a new organism constitutes 90 percent of a population, but grows 10 percent less rapidly than its natural counterpart. The new organism will drop from a concentration of 90 percent to a concentration of 0.0001 percent (1 part in 1,000,000) in 207 generations. If the generation time of the natural organism is one hour, this decline in concentration will occur in about 8-1/2 days.[1]

Q: Does NIH plan to publish a NEPA document on the proposal to support DNA recombinant molecule research, both intramurally and extramurally?
A: Yes, NIH plans to utilize a systematic, interdisciplinary approach to assess the environmental impact . . . If necessary, impact statements will be published.

The Draft Environmental Impact Statement

At the time we released the Guidelines in June 1976, we had answered the critics within the public and our own General Counsel's Office with the promise that we would issue an environmental impact statement (EIS) in 1 or 2 months. At least, we planned to "attend" to it. We wanted to know if we could avoid what sounded like a useless burden, since no one had seen any of the mysterious side effects of recombinant DNA research.

The NEPA Episcopate and Its Rituals

Priests

Within HEW, responsibility for the National Environmental Policy Act (NEPA)[2] ran from the office of the assistant secretary for administration and management, who delegated his authority to the director of the Office of Environmental Affairs (OEA), a position occupied by one Charles Custard. His deputy was Paul Cromwell. The tree of responsibility extended from them through Boris Osheroff, principal environmental officer (HEW), downward to two people located on the NIH campus. They were Vinson Oviatt, chief of the NIH Environmental Safety Branch, and his deputy, Rudolph Wanner, agency environmental officer at NIH. The OEA ritual required all communications to go up and down the chain. Thus, to seek a consultation with Mr. Custard, one started with Dr. Wanner.

Rudolph Wanner was a good man, a thoroughly dedicated guardian of forests and streams. Eager to be loyal to NIH and accepted as one of its staff, he often was pained to be delivering the OEA ukases from on high. He received moral support from Richard Riseberg's insistence on the applicability of NEPA to recombinant DNA research. Wanner wrote the following to Joe Perpich in April 1976:

> On occasion, I have advised Dr. Fredrickson not to ignore NEPA on this question. It seems that he followed this policy on the DNA guideline development. Further, both PHS and HEW environmental officials expect NIH to produce an environmental document.[3]

As I review the record, I note the entry "environmental document" in Wanner's message and wonder what he really meant to say.

The EIA Gospels

Within NEPA are several variations that may permit escape from what is known to be a tedious preparation and submission of an EIS. There were contractors ready to take on the job for handsome fees. None of them had the faintest idea of what recombinant DNA entailed. We did know that a first station was performance of an environmental impact assessment (EIA). If the assessment judges that the impact is not likely to be significant, then the procedure goes no further than the filing of that opinion. The pressure had already risen too high, however, for us to escape the counting of "chimera on the head of a pin" if that was what we were trying to do.

The Generic Analysis

There was another option offered by NEPA. The generic analysis (usually an EIA) can be directed to the totality of an agency's activities. We knew that the FDA had achieved this sublime status. (Operating with its own independent general counsel, the FDA had been able to obtain a generic inapplicability statement without departmental approval. Any envy over such preferential treatment proved to be ill spent, because a federal court action in 1976 overturned the FDA immunity with respect to NEPA.)

In the case of NIH, with all its biomedical research and the operation of a gigantic hospital alongside, the history was not encouraging. In January 1974, NIH had applied for a generic analysis that potentially applied to all the agency's programs and activities.[4] The Asilomar conference was then a year away, and the Berg letter had yet to appear. Thus, recombinant DNA research was not a major subject of concern. Six months after the application reached HEW, in 1974, it was returned, described as insufficient, with a request that it be rewritten. A reworked document was submitted in February 1975 and returned with comments in May. Discouraged by the departmental responses, the generic analyses were in limbo. It was also uncertain if any EIS from any part of the department had ever cleared the OEA hurdles.

The NIH Decision

Meditation

Toward the end of April 1976, a full-dress meeting of the Kitchen RAC was held in Building One to decide (i) whether NEPA was applicable or inapplicable to recombinant DNA research; and (ii) if a statement of inapplicability was required, what would be necessary to fulfill this obligation; and (iii) if a statement of applicability was required, what were the next steps required to determine whether a full-blown EIS was inescapable. It had already been impressed upon us that to ignore NEPA meant a suit would be brought that would sweep us into court. We took another step.[5]

Test of Criteria

The guides to determining the requirements for "further environmental consideration" included a group of "initial criteria" in a manual provided by Wanner. Maxine Singer, an indispensable member of the Kitchen RAC, was asked to analyze the section of this administration manual[6] containing the "Initial Criteria for an EIS." She had been working with Joe Perpich on a collation of the Riseberg memorandum and material from her files relevant to how one should structure an EIA.

Singer concluded at the end of this extensive analysis that the several initial criteria relating to (a possible effect on) populations were relevant. One of these was "could bring about identifiable genetic change within a human population"; another, "the research is directed toward developing capacity to affect human populations. . . ." Although intent to alter human genes was not the objective of rDNA research, the feasibility of "developing a capability" for doing so elevated this to the level requiring consideration of an EIA. Recombinant technology in animal and plant populations was also sufficient to warrant an EIA.[7]

Perpich later described a meeting held on May 14 to review Maxine Singer's analysis. Boris Osheroff was present, and it was decided that an EIS would have to be done, based largely on Maxine's analysis. As a follow-up to that meeting, Singer, Emmett Barkley, Perpich, and Harold Green met with Vinson Oviatt and Mr. Dunsmore of the NIH Environmental Safety Branch. The first draft of the EIS was undertaken by Oviatt with the assistance of Singer, Barkley, and a scientific editor.

EIS for the Research or the Guidelines?

On May 7, Charles Custard sent word down the line to Wanner, discussing the application of NEPA to NIH recombinant DNA research. Custard inserted the following passage, which was read by us with great interest:

> If DNA research could bring about an identifiable genetic change within a human population, then one of the Department's Initial Criteria is met and would thus require an environmental assessment to determine if such a change could lead to a significant impact. *However* [italics mine] since NIH is adopting guidelines for conducting such research, the guidelines may incorporate safe-guards which would prevent such impacts from even occurring. If the environmental assessment and related documentation demonstrates this to be the case, then it may be possible for a statement of inapplicability to be prepared for all or parts of the DNA research program.[8]

A week later, Perpich met with Harold Green, our outside legal advisor. He told Green that at the May 14 meeting the director had come close to deciding to do the EIS on the research rather than the Guidelines.[9] Green felt the opposite, and, after further review, Perpich concurred that we had to forestall the impossible prospect of a NEPA assessment for each and every grant or contract involving use of this technology. This proved to be a correct and fateful decision, not the first in which I was saved from a catastrophic decision by able staff.

Guidelines Chosen

The first outline of a draft of the assessment/statement prepared in May 1976 was directed to the Guidelines.[10] The draft published on September 9, 1976, eventually would also be devoted to the environmental impact of the Guidelines.[11] But in May we were far from the printing presses. A note from Paul Cromwell indicated that there would be a 2-week delay in sending back comments.[12] Those comments never arrived.

It was no matter, however, for by June 15, Maxine Singer had begun to prepare draft no. 2, changing the format. On June 22, her outline was discussed with Warren Muir and Malcolm Bruce of the Council on Environmental Quality (CEQ), along with Harold Green. Custard and Cromwell were also present. Singer's outline was approved, and a release date of September 1, 1976, was agreed upon. Draft no. 2 of the draft EIS was circulated to Custard, Cromwell, Osheroff, and the NIH director's staff, on June 23. By June 30, we were back discussing Osheroff's lack of enthusiasm for draft no. 2. A week's further consultations, including Green, led to the decision to change the format back to that used for the first draft. By July 7, work had begun on draft no. 3. Draft no. 2 was sent back to us by Riseberg's office. The Guidelines had been promulgated—wisely, it was now clear—for this was going to be a long battle that scientists could never have tolerated had they known.

The Baffling OEA

The HEW Office of Environmental Affairs' strict protocol for communications was maddening. Cromwell's critiques went to Osheroff, with a copy to Wanner. In August, Cromwell informed Wanner elliptically that a structural change was now required in draft no. 3.

> The primary substantive problem with draft #3 is that it does not discuss substantive alternatives: it merely presents the one set of guidelines. It would be confusing, costly and useless to develop straw sets of complete guidelines merely for the purpose of comparison. However, the NIH guideline [*sic*] is based on the combination of six or eight variables. These should be carefully identified. Again the examination of various combinations would be confusing and probably arbitrary. However, the effects of altering any variable or set of variables appear to be the same: the more restrictive the variable(s): 1) the less risk involved in a relative sense and 2) the higher the "achievement" level (in the sense that less time will be required and . . . [fewer] . . . places will be able to conduct the research). The opposite is, of course, also true. Thus any substantive discussion concerning a comparison of effects between alternatives is very limited.[13]

A little knowledge of French poetry allows one to savor the dilemma in which we here found ourselves, in a reflection of Paul Valery's: "Ce qui est simple est faux mais ce qui est compliqué est inutilisable."[14]

Draft No. 4 in August

Early on August 12, Bernie Talbot, tireless scribe and reworker of texts, handed around copies of draft no. 4 to the members of the Kitchen RAC and numerous other contributors, who now numbered about 30, for another go at sharpening the "substantive alternatives." A copy also went to Cromwell, who discussed this new product with Singer on the telephone. I think now that we have to atone for calumny privately visited upon Custard and Cromwell. Theirs was an exasperating assignment in the smoky kitchen of NEPA, trying to instruct novitiates on the nebulous menus in the HEW administration manual. In a note commenting on draft no. 4 and pointing out that he could not meet the time frame for draft no. 5, Cromwell also discharged his impatience with the lot of us:

> In closing, I ask that the next time an NIH program decides to produce an EIS that you involve this Office in the earliest stages of the development of the draft. As you know, our input relates largely to the procedural aspects of the document, including the clarification of the objectives, the scope of the effects and the selection of the alternatives. It is difficult if not impossible to provide useful assistance on these points if we are not involved until the later stages of development.[15]

Halfway Home

On August 23, over my signature and Wanner's, draft no. 4 was officially sent through Osheroff to HEW with notice of our intent to publish it in the *Federal Register.* On the same day, Custard capitulated, in a letter to Osheroff:

> Thank you for your memo of August 23 forwarding Dr. Wanner's request of August 20. Given staffing limitations we will not be able to review the document within the established time period. However, given your approval of the document, we will transmit it to the Council on Environmental Quality. . . . Notice of its availability will be placed in the Federal Register by CEQ for publication . . .[16]

With a good deal of luck and a bit of bravado, we had reached the halfway point on our campaign and within the deadline promised on the issuance of the NIH Guidelines.

The comments began to arrive soon after publication of the draft EIS. Cromwell ordered 1,000 copies to be printed, 200 of which he would send to federal and state environmental officers. We also supplied a copy to all participants at the Asilomar conference; participants at the February DAC meeting; all representatives of private industry with whom we had met; all others who had addressed correspondence to us during the previous spring; and the chairmen and ranking minority members of the NIH authorizing and appropriations committees, as well as other congressional staff.

Public Comments on the Draft EIS

Within the 45 days allotted after the September 9 publication of the draft EIS, some 40 comments were added to the rising volume of the public record maintained by NIH.[17] A balanced summary of these, in terms of volume, is hindered by the appearance among the letters of five documents of 10 to 30 pages each from the Natural Resources Defense Council,[18] the Institute of Society, Ethics and the Life Sciences,[19] the attorney general of the state of New York,[20] Susan Wright and four other members of the University of Michigan,[21] and the Environmental Defense Fund.[22]

All five of these documents included some or all of the same criticisms. I have chosen to catalog a portion of these from the comments of the New York attorney general, who was holding public hearings on recombinant DNA during the fall of 1976. There was a tincture of arrogance in the paternalistic tone of legal impatience with the naiveté of the medical scientists.

> **Procedural Compliance with NEPA.** . . . Issuance of an impact statement after grants have been made to conduct recombinant research and after publication of guidelines was a *per se* violation of NEPA.
>
> **The Major Federal Action.** . . . The action taken by NIH which may significantly affect the environment is the *funding* of recombinant research.
>
> **The Adequacy of the Draft EIS.** . . . The DEIS does not satisfactorily discuss the ethical and biological implications of this research, the biological and environmental hazards inherent in the research nor the pertinent information on this research which is *readily available* in the literature.
>
> The full implications of using *E. coli* as the recipient host are not made clear. In the DEIS . . . no relevant information on the K-12 strain is given such as its rates of exchange with other wild type *E. coli* . . . There is no discussion of the fact that many pathogenic *E. coli* strains exist. Two out of every 1000 patients who enter Boston hospitals die from *E. coli* infections.
>
> **No Discussion of Ethical Implications of Research.** . . . The DEIS admits that future discussions of genetic correction of defects will require social forums because of the philosophical and moral problems involved. However, there exists an equal need for a moral and philosophical discussion of whether recombinant work itself should proceed.
>
> **Alternatives.** . . . As noted before, the "No Action" alternative is properly that the federal government will give no recombinant research grants. . . . Recombinant research supported by other federal agencies must be included in the DEIS as well.
>
> **Enforcement Procedures.** . . . The Guidelines place primary responsibility for their adherence with the "principal investigators." The concept of self-regulation when dealing with potentially pathogenic agents is not at all acceptable.[20]

Rebuttal

First, the *funding of research* was never considered by us as the major action. NIH had no formal program of recombinant DNA research; rather, it supported a number of uncoordinated studies of a technique adaptable to numberless explorations of molecular genetics. The variations in given basic research projects was such that requirement of a draft, then a final EIS, from each of potentially hundreds of investigators—and then passage through a filter moving with the speed of that provided by Messrs. Custard and Cromwell—would have diverted American research in molecular biology to other locations.

Second, the draft EIS was described as inadequate because it had neglected pertinent information on this research which was readily available in the literature. This argument neglected the fact that for 2 years a large fraction of the authors of that very literature had been laboring to create guidelines so that they might once more begin to contribute factually to the present debate. It was to be our experience throughout this period, and making a great effort to hear all dissident opinions, that we never heard any new fact not already considered publicly in the debate.

Third, the character of *Escherichia coli*, strain K-12, may indeed fool all of the world of bacteriology, but the chances of that had been debated at length and expressed in the construction of the Guidelines, which were the subject of the draft EIS. This does not mean that we were not taking a risk.

Fourth, more moral and philosophical discussions of whether or not recombinant DNA research should proceed were the kernel of this entire public exercise. Because we had no data at all on the hazards of this research—yet clearly could perceive great potential benefit of proceeding—what must be the coin of philosophical debate? Of the commentators, only Robert Sinsheimer presented a rational conjecture of violation of evolutionary law in the research. His comment on the draft EIS included the following:

> . . . It is probably *unlikely* that any single *mutation* will markedly improve the fitness or expand the ecological range [of organisms]. However, conversely I see no reason to believe that the evolutionary process has at any given time *exhausted* the possibilities latent in the existing mechanisms.[23]

This sober observation was the one philosophical element that we had to deal with, and we had chosen to proceed using a cumbersome web of guidelines to test this most important hypothesis. Not everyone else was content with the plan. Loretta Leive, a scientist in the intramural NIH program, not previously heard from in this debate, provided the following carefully considered comments:

> I believe that safety lies only in totally safe vectors, i.e., those with no chance of transferring or in any way permitting existence of their recom-

> binant DNA molecule outside the laboratory. While no vector *totally* safe may exist on this planet, I feel it is completely unreasonable and unacceptable to use a common saprophyte/parasite and viruses of possible mammalian oncogenic potential as vectors. . . .[24]

This commentator, fully aware of the "fervor and excitement with which scientists contemplate the experiments possible with recombinant DNA," felt that "after the great humanitarian effort in imposing a moratorium . . . it would be ironic and tragic if it were lifted too soon. . . ."

Fifth, the demand that recombinant research supported by other federal agencies be included in the draft EIS as well showed poor understanding of the range of a single government agency, especially when it suggested that barring all federal support of recombinant research was a feasible (or rational) thing for NIH to attempt.

Finally, there was the comment after the New York lawyers discovered that the Guidelines placed primary responsibility for their adherence upon the principal investigators: "The concept of self-regulation when dealing with potentially pathogenic agents is not at all acceptable." To whom, we might ask?

Other Comments (Acid)

Through the millstones grinding out public consensus, there always passes a varying grist.

One commentator, a physician and member of the National Cancer Advisory Board, found the draft statement "not useful . . . not valid . . . not in compliance with the law, and basically . . . a smoke-screen to lull the American people into an unjustified sense of safety."[25]

Another wrote that "It is my opinion that Doctors Erwin Chargaff and Robert Sinsheimer are basically correct . . . I hope you will see fit to establish a more or less permanent moratorium on such research and discourage its spread to other countries."[26]

Still another commentator concluded that ". . . I think your guidelines and EIS are pure Mickey Mouse in the face of hazards that may be even more likely to occur and more devastating than nuclear warfare."[27] A molecular biologist noted that "The September 9 statement makes no assessment of the damage that will be done to intellectual freedom and hence society."[28]

More Positive Statements

We appeared to have struck a balance. The executive committee of the Assembly of Life Sciences of the National Academy of Sciences Council "discussed the Environmental Impact Statement and strongly endorsed the contents of the document." They also believed that more contributions of experts in epidemiology and infectious disease would augment the discussion and recommended "no substantive changes" in the EIS.[29] From the corner of true NEPA expertise, the comments received from each of a half dozen departmental Environmental Office heads revealed that none found anything wrong with the structure of the draft or the conclusions of this NEPA exercise.

Back to the (NEPA) Kitchen

In the succeeding 2 months, the Kitchen RAC met regularly to perform two major tasks with which it was now becoming expert, if not enamored. First, all the commentary on the draft EIS and the Guidelines was sifted, organized, and responded to. Then, incorporating many changes suggested and recognized as we continued to work, we drafted and redrafted changes in both the EIS and, in some instances, in the Guidelines. All of these changes were detailed in the next draft of the final EIS, with continued keeping of the public record.[30]

A Major Transition

Into the cast of this melodrama, a new leading player moved to center stage.

In the election of November 1976, Jimmy Carter succeeded Gerald R. Ford as president. It was rumored, then soon confirmed, that Joseph A. Califano, Jr.. would become the 12th secretary of HEW.

NIH Director Is Placed on Ice

From my diary, I provide a verbatim account of my first acquaintance with this new leader and what transpired with respect to my position as the 11th director of NIH.

> *January 14.* Today I kept an appointment made for me on the 10th floor office of Williams, Connelly & Califano at 839 17th Street. Arriving at 11:00, I was told that Mr. Califano, the newly named Secretary of HEW, was about to go through Senate confirmation hearings and was busy. Could I wait? I could, even without the coffee brought as balm. About half an hour later, a tousled, short figure in shirt sleeves bounded into the foyer, picked up my file and papers, apologized for lateness, and sent me into his office ahead of him.
>
> "Tell me about yourself and NIH."
>
> One hour later, after much frank talk, I was asked to send a briefing paper on NIH, with chart of organization to add to a pile of charts on HEW in one corner of the room.

"What's your name? John?"
"Don."
"O.K. Don, [I'm] Joe—I'm sorry but I can't resolve your job just yet. I've got to ask you to remain in limbo for 2 . . . 3 . . . maybe 4 weeks. Will you do that?"
"Yes," I said.
"Prepare to brief me in about a week. How much time will you need?"
"About 90 minutes," I replied.
"O.K. Thanks for coming."[31]

Following James Shannon's departure in 1968, the position of NIH director had for the first time became politicized. Shannon's immediate successor, Robert Q. Marston, who had been head of an HEW bureau, once part of NIH, first encountered the new political coloration of NIH when he openly and courageously joined the large number of scientists who opposed the schism between NIH and the National Cancer Institute that was narrowly avoided in the National Cancer Act of 1971. This act, in making the director of the cancer institute a presidential appointee, had escalated the appointment of the NIH director to require Senate confirmation. Upon his reelection in November 1972, Richard Nixon requested the resignation of nearly 2,000 senior government officials. Marston's resignation had duly been handed in, and the White House indicated he had "expressed a desire to return to private life." As Richard Rettig states in his exhaustive history of this turbulent period, "Marston had indicated no such thing. He had been fired."[32] The term of Robert S. Stone, Marston's successor, whom I succeeded in 1975, had been similarly truncated. Thus, I had experienced firsthand the anxiety of the scientists and staff at NIH that accompanies any change in the administration. Before I met the new secretary, I was informed that while I had been out of the country on a skiing trip, Hans Stetten had urged a letter be written to Secretary Califano, requesting my reappointment. Many of the senior staff in Building One signed it. Some thought it dangerous to go on record, and most demurred at sending it. I was told that, when their opinions were sought, Jim Kelly, the politically wise comptroller of HEW, had said "Don't," and National Academy of Sciences president Philip Handler had said "Yes." The first opinion prevailed, and thank Heaven, the letter was not sent. Hans shyly gave me the signed but unsent letter on Monday morning before my appointment with Califano.[33]

House Cleaning

After President Carter's inauguration on Friday, there was a sense of tumbrels in the streets. Much of what I call "the cloud layer," the upper levels of HEW, seemed abruptly to have been cleaned out and everyone gone.[34]

At a National Academy of Sciences reception that week, Lee Shore, who was on the transition staff of Califano, said, "Don, it's fantastic, everyone wants you to be reappointed, there were no negatives." It was a nervous crowd though, wondering, "what will the new administration *really* be like?"

Limbo Renewed

Monday, January 24. I had my second appointment at 11:30 with Mr. Califano, this time in the office of the secretary. The vote on his confirmation that day had been 97 to 1. Senator Packwood was said to have opposed Califano because of the latter's opposition to federal payments for abortion. There was a cabinet meeting, and the secretary was late. At 11:45, I was called into the office of the new undersecretary, Hale Champion, about 50, short, cigar-chomping, ample in girth, refreshingly candid and blunt.

We talked, and I felt a similar empathy with him as I had had with Joe Califano from our first meeting.

Suddenly, Califano came in like a running bear. I unwrapped the briefing charts we'd prepared. I didn't use any fancy carrying case for I thought too much whiz-kid stuff wasn't wise at this stage. I explained later, in wrapping up, that "I didn't want this to look too professional." There was excellent give-and-take for 1 hour. We pulled no punches on what I considered the administration's abuse of appointments to NIH advisory groups in previous regimes. We covered all points, and I felt I had the most refreshing hour that I had experienced in the last year and a half on the job.

"Don, I've got to ask you to stay in limbo for a week or so yet—sorry."

The next week I was given a list of briefing memos to be sent, some on a 24-hour notice. The second trip to see Califano had provoked a number of inquiries, though there was no hint of what was to come.

On Tuesday, the 25th, I met with the directors of the NIH institutes to review what I had said to the new secretary. The charts I had used were shown, so all would know that their special causes had been pleaded.

On Wednesday, the 26th, Califano had a press conference and declared that the Carter administration would "depoliticize . . . un-Nixonize the NIH."[35] Dr. Gonzales, president of the Federation of American Societies for Experimental Biology, called and thanked me for the secretary's declaration and asked if they should "send a telegram urging my retention." I replied that it was not necessary (i.e., not a politic thing to do).

On more than one occasion during this hiatus, the new secretary had said, "O.K. I only want the *best* people at NIH and every agency." He once confided that Stan Ross, his friend who would be appointed to head the Social Security Administration, had advised him to get rid of all the agency heads and "appoint lawyers to replace them." I confided to my diary my impressions that

> Joe Califano is a product of the LBJ school [he had been President Johnson's special assistant for domestic affairs]—direct, candid, tough, expecting maximum loyalty, likely to be infuriated if "blind-sided." He says he wants direct contact with his operating chiefs and will keep channels open (the Assistant Secretaries have a hotline now—they must answer it themselves and he answers his). He wants one year extensions of the Cancer/Heart Authorizations this year. He knew I had conversations with Stan Jones (a staff member on the Kennedy Senate Committee) on these extensions and agreed that we could help prepare an omnibus research bill to go with reauthorization measures. He also suggested that I see Congressman Paul Rogers of the House Subcommittee on Health of the Interstate Commerce Committee to inform him of this cross-talk with the "Other Body." (It was one of the early lessons of many I was to absorb about the punctiliousness of Administration-Congressional relations.)[36]

Feather Reid, personal staff to Senator Magnuson (chairman of the Senate Appropriations Committee), also called to say that the chairman had requested my retention in Califano's interview with him.

> *January 29, 1977.* At 2:00 p.m. the phone at home rang.
> "Dr. Fredrickson."
> "One moment please for Mr. Califano."
> "Hello, Don. I want you to stay on, and it's O.K. with the President. I'd like you to keep it quiet, though, until I can announce it. It'll be on Thursday. I'm coming out to NIH . . . I want to do anything to keep the excellence of this place . . ."[37]

From the *NIH Record*:

> During a brief visit on February 3, HEW Secretary Joseph A. Califano Jr. spoke to a packed audience of NIH employees in the Masur Auditorium of the Clinical Center. A partial text of his remarks follows:
>
> . . . I recognize NIH as one of the greatest national treasures this country has; indeed it's one of the great treasures of the world. I will do what I can to help provide an environment in which you can pursue your work.
> . . .
>
> . . . I am happy to announce that we have completed our search for a Director of NIH. We have looked—as I have for every position it has been my responsibility to fill or to recommend to the President that he fill—have looked for only the best—only for excellence . . . The Presi-

> dent is announcing this morning and I am announcing to you that we found that person, and we found him right here at NIH in Dr. Fredrickson . . .
>
> . . . As far as politics is concerned, it's out of NIH . . . I will do my best to insulate these Institutes and this great institution from partisan politics . . .[38]

The EIS, Act II

This episode being resolved, I turned back to the "recombinant DNA controversy," along with all the other duties of the director. To the dramatis personae in act I of the EIS theatrical, a few highly significant adjustments had been made. Secretary Califano's expressed taste for the "most excellent" appointments was manifest in the galaxy of new young lawyers that surrounded him. All were superb draftsmen, tenacious advocates for the public, with tireless passion for detail. Thus, four new speaking parts now outranked Mr. Custard: Secretary Califano, General Counsel-To-Be Peter Libassi, Associate General Counsel James Hinchman, and Deputy Executive Secretary Rick Cotton. A brief cameo part was allotted to Tom McFee, the deputy assistant secretary for management, planning and technology, and in one tense scene, Custard and Cromwell were accompanied off stage by a consultant, Steven Ebbin.

We returned to the draft of the final EIS, which had been prepared after analysis of the comments received on the draft EIS.

Beware the Second "Final" Draft

Our initial casting of what was now to be the final EIS was delivered to the Office of Environmental Affairs on January 21, 1977. The comments had been thoroughly combed by the Kitchen RAC, and many revisions were made in the text of the EIS, which retained the organization of the draft statement.

At a speed quite foreign to its customary pace, the OEA returned the draft on February 2 with a detailed nine-page critique by Paul Cromwell. A shabby wreath of praise poorly hid the threat of devastation:

> The proposed final EIS is a distinct improvement over the draft EIS. However, we feel that it is not entirely responsive to the major issues raised by the commentators nor do we feel that the Decision Paper adequately reflects a serious consideration of those comments. We are also concerned that many critics of the Guidelines will feel that the EIS is merely an *ex post facto* justification for an action already taken. This in itself might en-

> courage litigation which could lead to the temporary suspension of support for recombinant DNA research.[39]

The OEA's demands for major revisions of the EIS included a number of ways to relieve confusion, particularly as to the difference between the Guidelines and the EIS. We could make these adjustments. Much more serious were the clear signs of a stalemate arising between the weary NIH drafters and an OEA doggedly determined that new bases must be established for "comparing impacts and probabilities associated with alternative Guidelines." Mind you, we were still a few months away from the Cold Spring Harbor experiments showing that prokaryotes could not splice DNA in the same manner as eukaryotes[40] and that shotgun loading of bacteria with the DNA of higher animals was almost certainly not going to create hazardous monsters. Thus, one must admit that the OEA at the time could not be condemned for the breadth of its concerns over the opening of Pandora's box. The obstacle to appeasing the OEA was the exhaustion of the ability of the molecular biologists to develop more imaginative scenarios of hypothetical disaster.

On February 7, I went to Cromwell's office to express these feelings to him as we went over the critique. "Enough is enough, Paul," I finally exclaimed and left the meeting to ascend to the top floor of the Humphrey Building to look in on Secretary Califano's gathering to search for consensus on whether to reactivate swine flu vaccine. It was a meeting I had helped to arrange.[41] At least here was a venue where consensus was flowing and not stagnant.

The Attack Intensifies

Our situation with the NEPA guardians was moving less well; indeed, we were in a crisis. On February 25, Custard sent the secretary a bulky memorandum[42] accompanied by lengthy commentaries from Cromwell and a critique prepared by OEA consultant Steven Ebbin, including the following excerpt:[43]

> **Violations of procedural requirements.** The proposed final EIS prepared by NIH . . . is severely deficient in a number of important, indeed, legally disabling respects. Clearly it is a promotional document which emphasizes "potential" benefit and deals with "putative hazards." There is no evidence to suggest that external commentaries and suggested alternatives were given serious consideration by objective judges . . . It is impossible to rule out, in absolute terms, the possibility that unpredictable evolutionary consequences would follow the inadvertent release of recombinant DNA organisms. Nor is it possible to predict whether such consequences,

> if any, would be trivial or substantial. Commentator Lappé notes ". . . 80% of all reported laboratory accidents have no known cause."
>
> . . . The EIS states that ". . . confidence in the Guidelines is derived mainly from the judgment of scientists," yet many scientists have evinced a lack of confidence in them (see, for example, Sinsheimer, Lappé, commentaries of the University of Michigan team, and the Attorney General of New York, et al., Appendix K).
>
> . . . The time to hold a public debate is now, before the technology and the public investment supporting it has gone too far. The horse will already have escaped the barn.

In his brief to Secretary Califano, Custard listed a series of six deficiencies, which

> . . . combined with an almost total absence of public participation in the Federal decision-making process with respect to recombinant DNA research will lead to NEPA litigation and a subsequent court order to restructure the EIS. (Potential plaintiffs include the Natural Resources Defense Council and the State of New York.) This might or might not be accompanied by an order enjoining all ongoing research supported by the Federal Government. However, in any event it will certainly result in a deterioration of public confidence in the Federal Government not to mention the scientific community. . . .
>
> . . . For these reasons we recommend that you require NIH to restructure the EIS in order to correct the major deficiencies cited in the attached memorandum.

Angry HEW Reprisal

A fortnight later, Thomas S. McFee, deputy assistant secretary for management, planning, and technology, which sheltered the OEA, projected a warning flare in the direction of the bridge:

> There is no question that this subject area [the final EIS] is one of the most complex and volatile issues that we have had to deal with since the National Environmental Policy Act has been in existence. My staff has reviewed the final impact statement proposed by NIH and believes that it contains the following [6] deficiencies [previously forwarded by Custard].

McFee stated that he too had no doubt that these EIA deficiencies uncorrected would lead to litigation.

> Due to the political and scientific significance of this issue, it will probably lead to litigation regardless of how successful we are in overcoming these deficiencies.[44]

NIH Retort

From the regimental headquarters in the embattled colonies, I shortly thereafter conveyed to Jim Hinchman in the General Counsel's Office my umbrage that the deputy assistant secretary and members of his staff had found that the final EIS prepared by NIH had "serious deficiencies."

> The portrayal of deficiencies seem to follow from the wholly erroneous assertion that "the EIS fails to properly define the action.". . . Because NIH has no program for supporting recombinant DNA research per se . . . nothing done by NIH with respect to support of recombinant DNA research could conceivably be regarded as a "major Federal action." On the other hand, issuance of the Guidelines constituted a major Federal action requiring compliance with NEPA. . . . Once it is recognized that the EIS deals with the Guidelines and not with the research itself, the alleged deficiencies can be viewed in proper perspective. . . . I recognize that the EIS is subject to legal attack, and that a court may find it wanting. But, as I have come to understand NEPA law, this is true of any EIS. . . . We are amenable to suggestions for improvement, but to "restructure" the EIS along the lines proposed by OEA would represent a substantial distortion of the situation dealt with in the real course of events. . . .
>
> Finally . . . I cannot permit to go unchallenged the assertion that there has been an "almost total absence of public participation in the Federal decision-making process with respect to recombinant DNA research". . . [There follows a recitation of the history of the issuance of the Guidelines and the invitation list for the public DAC February, 1976, meeting, including the *Federal Register* notices and the subsequent periods of comment.] [45]

After my memorandum, Hinchman reviewed our stalled EIS draft and reported that he now felt that it was legally sufficient, without predicting what a court might find in it if challenged. However, he continued, federal legislation was about to appear in the form of a recommendation developed by the Federal Interagency Committee for Recombinant DNA (see chapter 7), and he wanted me to know that "the department lawyers were now leaning" toward dropping the attempt to move the final EIS through the thicket, arguing that a new EIS would have to be done for regulations in any statute that might be passed. (I thought, "Another *major federal action*, of course.")[46]

Predictably, within a couple of weeks of this dismal news, the OEA, through its agent, Boris Osheroff, the NIH principal environmental officer, sent me a note indicating that he had taken the liberty of editing a recent draft statement of mine to try to arouse the secretary to move on the final EIS. Osheroff's proposed revision was contemptible. It would have me say that:

> In view of the pending legislation and, if enacted, its implementation: publication of the final Environmental Impact Statement on the guidelines is, in our judgement, inappropriate . . . accordingly the subject material are

> now issued in the form of a review of the NIH Recombinant Guidelines . . .[47]

Failing my signature, this counterfeit never made the trip downtown.

Judicial Entre Actes

Between presences now necessary at the Congress, I began also to make myself a nuisance at the Humphrey Building, bearing the message that we should stand naked before the bar in any suits against us involving violations of NEPA. The only defensive clothing would be a valid final EIS. Because this was evidently not forthcoming, a draft statement for the *Federal Register* was also being prepared to announce that the NIH EIS would be published as a revised draft document rather than a final statement.

Injunctions

On the weekend, when Congress took a brief recess, we received notice of the imminence of threatened legal actions. It happened suddenly when I found a note from Joe Perpich on my chair on returning to the office:

> Don/ The Friends of the Earth have sued the Department, naming Califano, Dickson and you to enjoin further funding for recombinant DNA research on the basis of: (1) a violation of the Environmental Policy Act from failure to comply with NEPA requirements in issuing the Guidelines and funding the research, and (2) failing to comply with the Federal Advisory Committee Act in establishing a recombinant advisory committee. The complaint will be filed in New York within a day or two/Joe.[48]

Friends of the Earth v. Califano et al.

On May 9, 1977, in the Second Judicial Circuit in Federal District Court, New York, a suit was filed by the Friends of the Earth, Inc., naming Joseph A. Califano, Jr., James Dickson as assistant secretary for health, and Donald S. Fredrickson, individually and as director of the NIH. The complaint follows:

> **VERIFIED COMPLAINT FOR DECLARATORY JUDGEMENT AND INJUNCTION**
>
> Plaintiffs, as and for the complaint, by their attorney named below, allege as follows;
>
> 1. This action is brought against the above-named defendants for the failure to comply with the requirements of National Environmental Policy Act of 1969 (NEPA), P. L. 91-190 . . . , the requirements of the Administrative Procedure Act (APA) . . . and numerous other laws and regulations (a) with respect to the funding of recombinant DNA re-

> search and its profound environmental impacts and consequences, (b) with respect to the Recombinant DNA Research Guidelines promulgated by the defendants to regulate the recombinant DNA research program, and (c) with respect to the proposals for legislation regulating recombinant DNA activities.[49]

After a long period of inaction, this suit was eventually withdrawn. But within a month after the Friends of the Earth had filed, another threat of litigation arrived, aimed to strike a death blow to the NIH's proposed earliest risk assessment experiment, the polyoma experiment, which was impatiently waiting its crucial stage of engagement. Quite unintentionally, however, this suit would spark the solution to the festering stagnation of our EIS.

The Polyoma Experiment

The polyoma experiment had been conceived, apparently, by Sydney Brenner and Wallace Rowe, as far back as the time of the Asilomar meeting. As the RAC was in its early stage of drawing up the Guidelines, comment was being collected on the concept, and many individuals were eager to sponsor experiments designed to measure the magnitude of risk involved in construction and utilization of recombinant DNA molecules. The RAC designated Rowe and Malcolm Martin at NIH as appropriate to carry out such an experiment. They selected polyomavirus DNA as a starting material. This virus, as well as its DNA, was infectious to mice and hamsters, in which it produced tumors and detectable antibodies. It was not infectious to humans. The circular polyomavirus DNA was to be split open with a restriction enzyme and inserted into the DNA of a phage or plasmid vector. The recombinant molecules were then to be introduced into *E. coli* K-12, and the resultant host-vector system was to be cloned and injected into mice by the oral or parenteral route to check for infectivity. The polyomavirus DNA would also be recovered from the host-vector system, purified, and tested for infectivity. The central question to be answered was whether the host-vector system bearing the foreign DNA might undergo lysis, scattering the DNA to loci where it might do damage. Specialized advice for these experiments was contributed by Roy Curtiss, Ron Davis, Stanley Cohen, Herbert Boyer, and others. The work would be carried out under the stringent conditions of P4. At its final (San Diego) meeting in 1975, the RAC voted to exempt these test experiments from the prohibitions in the Guidelines.

A Problem or Two Intervene

It was realized at once that the principal hurdle was finding a fully operational and tested P4 facility sufficient for this experiment in the United States. At that time, only one possible site existed. It was located at the Center for Disease Control and not available for this type of experiment. However, the NIH had earlier obtained a portion of Fort Detrick, Md., from the army and established there the Frederick Cancer Research Center. There were adjacent laboratories that had been used for production of vaccines. A contract was let for upgrading one of these units to the stringent requirements of P4 isolation, and it was finally nearing completion. In the intervening 2 years, it had been hoped to move the experiment closer to the NIH campus. Contractors were trying to convert a mobile laboratory to serve the purposes, but after months of work it had failed to pass the safety tests.

Comes now litigation to stall the opening of the laboratory and protract the experiment further at a time when it was so needed. Moreover, EMBO reported that their comparable experiment was under way.

Mack v. Califano et al.

On May 31, 1977, Ferdinand Mack, an attorney in Frederick, Md., filed a motion with the United States District Court for the District of Columbia, on behalf of his infant son, for a temporary restraining order and preliminary injunction to prevent an experiment testing the properties of polyomavirus DNA cloned in bacterial cells in the P4 laboratory under construction in Building 556 in the Frederick Cancer Research Center. The defendants were Joseph A. Califano, Jr., Donald S. Fredrickson, and John E. Nutter, chief officer of specialized research and facilities, National Institute of Allergy and Infectious Diseases, NIH.[50]

The plaintiff asserted that the nature of the organism to be created by the research was such that even a minuscule quantity, if released in the environment, would represent a threat to life and health. Furthermore, his brief alleged that the draft NIH EIS applied to NIH Guidelines, in which the very experiment was "prohibited."

As soon as the Justice Department attorneys assigned to defend us made themselves familiar with the cases, they looked to us to provide them with the core of the defense. Accordingly, the attorneys, under Frank Wohl, chief of the Civil Division in the U.S. Attorney's Office in New York City, examined the latest final EIS draft and strongly suggested its publication after a few modifications to strengthen its purposes for advocacy. They suggested that the format not be changed to extend the purposes of the EIS from release of the NIH Guidelines to the federal support of the research itself.[51] Justice filed its preliminary response to the New York suit in June, being unable to await the release of a final EIS.

The defendants described the importance of the research to understanding the new technology; how the Guidelines prohibited all experimentation involving highly pathogenic agents; furthermore, the experiments would be conducted using as a host-vector system a derivative of *E. coli* specifically designed to self-destruct in the colon and unable to colonize in the human intestinal tract, and which is not known to cause any human disease. Finally, they stressed that the P4 facility in which the experiments would be conducted was designed to avoid any outside contamination.

On July 18, both parties agreed to a court-ordered stipulation that all proceedings would be temporarily halted pending the finalization by the defendants of an EIS.

Department Reaction

At HEW, the general counsel, sensing a new pressure for solution to the EIS problem, directed Riseberg to draft an NIH memorandum to the secretary offering him two alternatives:

1. Publish the NIH's proposed final EIS, with minor revisions, and take the position in court that an EIS on the Guidelines in effect covers the research as well.
2. Publish the EIS but admit to the court that it covers only the action of promulgating the Guidelines and not the research.

Further, the General Counsel's Office would demand that text be added to the effect that "regardless of which alternative the Secretary pursues," NIH would begin work now on a comprehensive assessment of the environmental impact of the final standards to be issued after enactment of the administration's proposed legislation (the statute based on the recommendations of the Federal Interagency Committee).[52] We had to recognize that federal regulations arising out of a statute could be considered a NEPA matter, but I was strongly opposed to both the second alternative and the codicil implying that we might have to traverse again the Sisyphean path another year. One point of relief was the lawyer's operative word here of "assessment." As we had learned in the fray, a negative *assessment* could neutralize the need for preparing a necessarily more perilous *statement* and thus make the exercise relatively palatable. I was adamant, and this perfidious note was not sent to the secretary.

The Too Hard Hand of the Law

By late September, the Mack litigation being stalled while the final EIS was still awaited, Custard and Cromwell of OEA had lost control of the matter to the HEW lawyers. Rick Cotton had perhaps been ordered or simply challenged to end a dreary stalemate. His sharp pencil fell upon numerous perceived flaws.[53] When the critique was returned to the Kitchen RAC for more tuning, however, the first thing that met the eye was a demand that a preface be inserted that "announces the decision to prepare a subsequent [environmental impact] statement on the research that involves the use of recombinant DNA molecules."

For the HEW staff—and certainly for the government litigators—the sand was running out of the glass at the beginning of October. Patience and tolerance, too, had reached their lowest ebb among the Kitchen RAC at NIH.

The End of the Scientists' Tolerance

Maxine Singer

The decision of Maxine Singer not to participate further in the EIS process had been conveyed to me in a letter.[54] It was the hardest blow that I recall during this entire nightmare. It was also a serious loss for both the scientific and public communities, for Maxine's role throughout a controversy precipitated partly by her own sensitivity to the moral duties of science had been enormous and will someday merit its own telling.

Reply to Maxine

Maxine rejoined us after a few weeks, and I much later wrote her:

> . . . in your case it is time I formally replied to your letter . . . in which you resigned from participating in any further environmental impact statements. I remember thinking at that earlier time—It was during that dreary winter of what we hoped was our Little Big Horn (Custard's Last Stand)—that you were daring me, plumbing for the bottom of my tolerance for NEPA bureaucratics. I retain the illusion that I resisted this device. I recall vividly, however, the gratitude I felt when you quietly returned to your gun, restoring the critical balance of firepower without which we surely would have been lost . . .[55]

An Unfolding Revolt

Bernie Talbot, who had been with Hans Stetten almost from the first days of the RAC and was to be my "special assistant for recombinant DNA," had fully earned the nom de guerre "Stahkanovite of the Kitchen RAC." His hours and productivity as measured in countless drafts, lists, and assessments were phenomenal.[56] As this autumn deepened, Talbot continued to endure endless nights, piecing together the day's work on commentaries and drafts of the EIS, so they might be laid on the carpet on the director's table in the morning, usually to be torn apart and revised again and again. This October evening, Talbot was engaged with a response to criticism of the EIS from Rick Cotton, at this period effectively the executive secretary to the secretary. Suddenly, Talbot discovered he was fed up, and he composed a memorandum to the director, NIH, in the middle of the night:

> . . . When I first read Rick Cotton's September 28 memorandum on the EIS, my initial reaction was one of utter resignation. I thought, "I don't like the tone of his memo, but we can probably comply with all he requests. Where he wants passages deleted we can easily do that. Where he wants passages rewritten, we can rewrite, and if we bring the revised passages to him with the proper penitence and supplication, and we are lucky enough to find him in a good mood that day, perhaps . . . it will really finally get published. The more I thought about his memorandum over the weekend . . . my mood changed from resignation to anger. In my brooding I came to the decision to write you this memorandum, both as a way to express some of my rage and frustration, but also because I believe there is a major constructive action item which you should initiate.
>
> The action item to which I refer, and which I strongly urge you to pursue, is that you write to the Secretary to ask that he reconsider the "decision" he made that this EIS should be considered as covering only the Guidelines and that a separate EIS should be prepared on the research conducted under the Guidelines. . . .
>
> Rick Cotton in his memorandum says, "There is asymmetry in this discussion which is amusing but ought to be corrected." I also perceive an asymmetry involved in the preparation of the EIS; it is an asymmetry in the contributions of NIH on the one hand, and the contributions of the DHEW on the other.
>
> I know that you are painfully aware of the history of the preparation of our EIS. You will recall that throughout the summer of 1976, Maxine Singer and a whole team of NIH staff labored extraordinarily hard and long to produce the draft EIS, which was published in the Federal Register on September 9, 1976. Following receipt of comments from the public, which lasted until December 15, 1976, another major push again involved much overtime by many NIH staff, including Maxine, Michael Adler, and Bill Carrigan, led to the final proposed EIS which was sent to the DHEW on February 15, 1977.[57]

In short, Talbot was telling me that I had to stand up and stop this maddening circus, with my body if necessary.

Talbot's Chorus: Barkley

Next came a letter from Emmett Barkley, now the director of research safety at NIH and another paragon of public service in giving to the NIH

Guidelines structure and organization of containment and safety practices. Barkley had attended a meeting on another purpose, including Stetten, Rowe, William Gartland, and Talbot. The discussion changed to what was on everyone's mind. Barkley had come to some conclusions, which the others asked him to pass on to me.

> . . . Basically, our concern is that we are pursuing a treacherous course: a course for which we have no real mandate to follow. We took this course in a genuine effort to be responsive to any and all criticism. It was not a course we had to take and I am suggesting that we reverse our direction immediately. . . What we have done is involve ourselves in a bureaucratic struggle for which the boundary conditions are undefined, because no one (apparently at any level) actually understands either the purposes of NEPA or our responsibilities under NEPA. We are vulnerable because if our actions can be construed by DHEW as not being in compliance with NEPA, then certainly those who wish to take us to court will have a strong advocate in their corner . . .[58]

Barkley sent this letter along with a draft memorandum from me to send to the secretary announcing my decision to withdraw both from the draft EIS and from the NEPA process, "because of the distortions in the latter that did not serve the public."

Rowe

A note from Wallace Rowe seconded Emmett Barkley's plea. Wally Rowe was a person whose deep conservatism about the higher values of science and research I have mentioned before.

> This is to lend my strong support . . . that NIH reverse its position on finalizing an EIS on the guidelines. I have come to this position myself, feeling that the precedent will be horrendous. While failure to finalize the EIS would create serious complications with regard to the Mack suit against our polyoma experiment at Fort Detrick, I would be very willing to see NIH abandon that experiment rather than set the precedent that guidelines or experiments that in no way can affect the environment must be accompanied by an EIS.[59]

In refusing to take his advice, I wrote to Wally:

> . . . Among the small band of regulars who made it possible to assemble the proposed revision of the Guidelines, few scientists will have made more diverse contributions than yourself. You've not only been generous with your time; it's that special touch of conscience, and the willingness you have shown to sacrifice your own interests when you perceived the public interest to be different . . .This is not the time to be maudlin, or speak of special nobilities.[60]

Endgame

On October 6, Dick Riseberg floated another HEW draft for a decision statement from me to the secretary. It continued the blind course of promising another EIS on the research once the Guidelines had been taken care of. That evening I took up a pencil and drafted a letter to the general counsel, Peter Libassi. I referred to the Cotton critique that demanded a statement in the preface that announces the NIH decision to prepare a subsequent EIS on the research. I stated further that while we had been informed orally that the secretary was of a similar mind, we had never received written notification of so pivotal a change. In response to my requests for legal precedents for this, one of our staff had turned up a clip on a SIPI (Scientists' Institute for Public Information) case, which, though it involved breeder reactors, was more than instructive in our case. It provided a strong case that the extension of the NEPA processes to basic laboratory research with no known harmful effects would create an extraordinary precedent. I asked to meet with Libassi, Rick Cotton, and relevant general counsel staff to review the secretary's decision and the basis for proceeding to do an EIS on the research. I emphasized that we must have this meeting as soon as possible.[61]

I met with Libassi and Cotton on October 11. I then sent the secretary a memorandum noting that we had agreed to meet with him to "consider issues involved in the decision you make with respect to this matter." For purposes of that meeting, "I have added to the accompanying statement the reasons why I believe another EIS on recombinant DNA research is unnecessary."[62]

The quality of my appeal to the secretary's conscience was a testament to the effectiveness of the Kitchen RAC during what some at NIH considered an agonal moment. I regret there is an incomplete record of its preparation, save for a memorandum from Rick Curtin that provided the evidence of the lack of an EIS for basic research in the Department of Energy.[63] There is a legal gloss on my paper which was undoubtedly applied by Perpich.

Sweet Compromise

The meeting with the secretary occurred on October 18. When it was over, I took great pleasure in sending Peter Libassi a memorandum entitled "Approval of NIH Environmental Impact Statement."[64]

> . . . Enclosed is the Final Environment Impact Statement on NIH Guidelines for Research involving Recombinant DNA Molecules. This docu-

> ment, a revision of the Draft EIS published in the *Federal Register* in September, 1976, incorporates changes in responses to public comments. In accordance with our discussions of October 18, it also reflects suggestions by Mr. Rick Cotton and the Office of General Counsel. It is now ready for the Secretary's signature (see Proposed approval document, enclosed) and final disposition as required by the National Environmental Policy Act.
>
> Upon the Secretary's approval of the document, we will send the required copies to the Council on Environmental Quality and a notice will be published in the *Federal Register* stating that an EIS has been issued . . .

The participants at the crucial summit with Secretary Califano had been Peter Libassi, Rick Cotton, Julius Richmond, Jim Hinchman, and I. I recall little of what actually transpired, although I remember reminding the Secretary *again* that if the final EIS was not lifted from the dungeon of the departmental OEA, he and I would soon be standing bare before a U.S. court. Rick Cotton provided a delicately worded memorandum of the outcome of the meeting:

> Those present set out their views for the Secretary on the need for an environmental impact statement (EIS) to discuss and analyze the potential environmental impact of the recombinant research currently being funded by the National Institutes of Health.
>
> *The Secretary approved the issuance by NIH of the final EIS on the guidelines, without mention of NIH's future plans for compliance with the National Environmental Policy Act* [italics mine].[65]

The EIS and the comments upon it are presented in extenso in the NIH archive.[66] We had passed the biggest crisis so far in our custodial duties for the Recombinant DNA Guidelines.

Judicial Conclusion

EIS to the Rescue

Upon the successful completion of the breathtaking race to achieve a final EIS and to publish a notice in the *Federal Register* that it had been completed, the Justice Department was alerted. Copies of the EIS and a thick yellow document containing a glossary, description of the Guidelines, comments on the draft EIS, the Federal Interagency Committee report, summaries of recent legislation, and letters and article, were handed to the government attorneys of the defendants.

The trial of the Mack case resumed in February 1978. The plaintiff alleged that the EIS presented did not comply with the requirements of

NEPA. In its decision, the court noted that a recent ruling of the Supreme Court had "summarized the limited role of the courts in determining whether the agencies have complied with NEPA."[67]

> The only role for a court is to insure that the agency has taken a "hard look" at environmental consequences: it cannot "interject itself within the area of discretion of the executive as to the choice of the action to be taken."[68]

The court further observed that "The EIS does represent a 'hard look' by NIH at recombinant DNA research performed in accordance with its guidelines. . . ." The defendant's motion for a preliminary injunction was denied. The plaintiff immediately appealed the decision to the United States Court of Appeals of the District of Columbia. On March 8, 1978, this court denied the appeal.[69]

The government presentation had included an amicus curiae brief submitted by the American Society for Microbiology.[70] For the plaintiff, affidavits were received from Ruth Hubbard, Stephen Havas, and Christine Oliver, staff at the Massachusetts General Hospital. A supplement to the material considered by the court was an editorial by Richard Goldstein pointing out shortcomings of the NIH Guidelines.[71]

Back to Polyoma

P4 Ready

I note in my diary that on January 12, 1978, we opened our P4 Lab at Frederick Cancer Research Center. A few quiet demonstrators carrying banners reading "Laity and Clergy Concerned" were present for the morning press briefing. I was one of those participating during the afternoon meeting. Beverly Byron, wife of Congressman Goodloe E. Byron (D-Md), was the only congressional representative. After we all attended a movie about operations, I was invited to move to one of the metal tanks and put my hands in the clumsy rubber gloves that provided the only entry from the outside into the working chamber. I thought about the enormous, expensive effort it had been to assemble the stainless steel, vacuum locks, disinfectant dunk-baths—perhaps only to contain hypothetical chimeras.[72]

Scientific Conclusion

Once allowed to go forward, the series of polyoma experiments showed that a host-vector system containing polyomavirus DNA was totally noninfectious in the mouse and nontumorigenic in hamsters regardless of the route of administration, even though the original polyomavirus or cloned polyomavirus recovered from the vector was fully infectious.[73] A review of the results retrospectively concluded:

> In fact, it suggested . . . that the safest place to handle an infectious piece of DNA was in a plasmid contained in a suitable host. Under these circumstances, the safety was enhanced by no less than 10^8.[74]

While valid for only one type of DNA and host-vector system, the results were important—the first experimental evidence that dangers from this type of technology had been exaggerated.

The next chapter will take us into the chambers of the Congress, where the Guidelines and NIH will be dragged into a fateful "dance of legislation."

7

Acts of Congress

1977

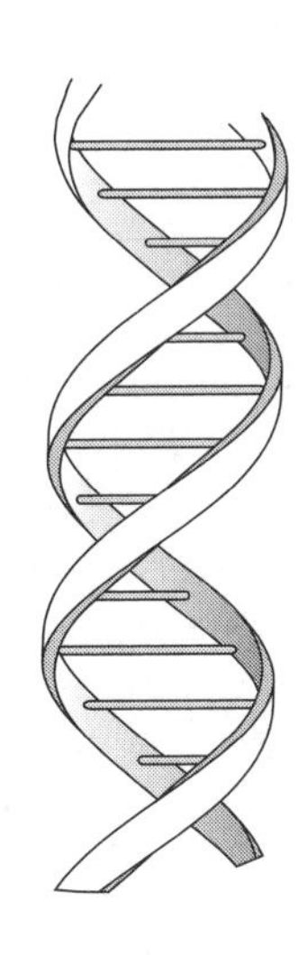

Congress and Medical Research

Host-vector systems in lower eukaryotes, such as the yeasts, *Saccharomyces cerevisiae* and *Neurospora crassa*—approved as HV1 systems—*S. cerevisiae* as an HV2 system—criteria for acceptance laid. Equivalence of lower eukaryote HV systems with those of *E. coli*, without viruses, with viruses, discussion leads to much amended resolution: "experiments involving complete genomes of eukaryotic viruses will require P3+HV1 or P2+HV2 containment levels." Because this is a "major change," from the proposed action as it previously appeared in the *Federal Register*, an additional opportunity for public comment will be given and it will be considered at the next RAC [Recombinant DNA Advisory Committee] meeting. Prokaryotes other than *E. coli* as host vector systems: *B. subtilis* accepted as an HV1 system, if meets detailed list of criteria. Requested addition of *Bacillus megaterium* to list proposed for exemption I-E-4: data judged poor, unanimously denied. Proposed exemption of *Streptomyces* species as chromosomal exchangers leads to multiple parliamentary objection, e.g., motion to table. Selected species broken into shorter list of non-pair-wise species, RAC votes 14 in favor, 2 opposed, 3 abstentions, recommendation not accepted by Director, NIH (due to uncertainty and disagreements in the votes) and deferred to next RAC meeting. Proposal for certification of new phage [lambda gt Aamal Lam439 Oam29.lambda B in DP50 SupF] as an EK2 system—unanimously recommended. Proposal for certification of a new plasmid pGL101 to be propagated in chi1776 as an EK2 system—referred back to Host Plasmid Subcommittee of the RAC. Proposal for certification of chi1776(MUA-3) as an EK2 host-vector system: RAC notes flaw in proposal, returns. Review of containment levels for shotgun cloning of primate and other mammalian DNA in *E. coli* K-12—multiple motions, eventually settled on lowering to P2+EK2 or P3+EK3. Next, three con-

> siderations of request for lowering of containment levels for three different clones of DNA, all rejected. Reports of risk assessment plans, including the results of the polyoma experiment.[1]

Of the branches of the federal government, it is the Congress which has long been the major proposer and disposer in bringing the federal government into medicine. This is overwhelmingly true when it comes to support of biomedical research, its epicenter for 50 years being the NIH. The creation of the NIH was due to numerous serial acts of members of both houses of Congress. The conversion of the Marine Hospital Laboratory into the National Institute of Health in 1930 was largely the devoted work of Louisiana legislator Senator Joseph Ransdall.[2] In 1944, deft rewriting of chaotic public health statutes by Congressman Alfred Bulwinkle (D-NC), with the eager support of Surgeon General Parran and his assistant Lewis Thompson,[3] inserted a Section 301 in Public Law 78-410 providing unlimited authorities to the surgeon general to support extramural research. When the Committee on Medical Research of the Office of Scientific Research and Development[4] prepared to close its operations in 1944, the unique Public Health Service (PHS) authorities paved the way for NIH director Rolla Eugene Dyer to remind the group of the new authorities allowing the PHS to continue the expiring medical research contracts.[5] He was able to take back most of them to Bethesda, where he lost no time in setting up the NIH extramural program in 1945. Representative Bulwinkle also inserted words that provided the PHS with the opportunity to create the Clinical Center. The next year, a visit by Dyer and several other PHS officers to the hometown of Congressman Frank B. Keefe (R-Wis), chairman of the House Appropriations Subcommittee for Labor and Health, resulted in a $60 million appropriation in 1947 to enlarge the campus and build the hospital. By 1948, numerous legislators had begun a process of pluralization of the NIH into today's large number of specific research institutes. The 10 years from 1955 to 1965 was a legendary period of growth in NIH support by Congress, which was mainly presided over by Senator Lister Hill (D-Ala) and Keefe's successor, Congressman Joseph Fogarty (D-RI).[6] The support of numerous congressional guardians to this day has propelled and maintained the NIH in its status as a world leader.

Congressional oversight is continuous and heavily involved in overseeing the distribution of support over the spectrum of research. It was inevitable that the news of the moratorium of the molecular biologists and the publication of *NIH Guidelines for Recombinant DNA Research* attracted the interests of the lawmakers. Several impromptu hearings had been held by

Senator Edward Kennedy, chair of the Subcommittee on Health and Scientific Research of the Senate Human Resources Committee, and at the close of 1976, Scientists' Institute for Public Information, a public interest group, promoted a 1-day seminar for members of Congress and their staffs in the Dirksen Building on the subject of "Recombinant DNA Research: Is Legislation Necessary?"[7]

In a period of relative calm, the explosive nature of the appearance of legislative interest in molecular biology at the opening of the first session of the 95th Congress in January 1977 was quite unexpected.

Representative Ottinger's Pique

On January 19, Mr. Richard Ottinger (D-NY) rose in the large hearing room of the House Subcommittee on Health and the Environment to propose resolution no. 131. His objective was to have the NIH Guidelines extend to any such research in the private sector, but the single "finding" in his introduction rang out with outraged conviction:

> Whereas unregulated research involving recombinant DNA is potentially devastating to the health and safety of the American people . . .
>
> The Secretary of Health, Education and Welfare under Section 361 of the Public Health Service Act . . . should [propose] . . . regulations governing all research involving recombinant DNA.[8]

At least he was for the Guidelines, but my fear of the consequences of invoking Section 361 had been clear from the first public hearing on the Guidelines at the DAC meeting last year. It would be a death blow to what we had been working slavishly for 2 years to accomplish. The core text of Mr. Ottinger's resolution had been lifted from the joint petition of the Environmental Defense Fund and the Natural Resources Defense Council circulated earlier.[9] The petitioners appeared to hope that they could throw both the Guidelines and NIH oversight into a fatal tailspin.

In Search of Animus

Congressman Ottinger was busy converting his resolution into a proposed statute 6 months later when I sought further information about his motives. At the secretary's insistence, I went to his office in the Cannon Building accompanied by Dick Warden, assistant secretary for legislation. On the way, Warden told me that Mr. Ottinger's district in upper New York state had been plagued with industrial wastes. Ms. Joanne Stoney, Mr. Ottinger's assistant, greeted us rather coldly. When Mr. Ottinger came in, he was not very effusive either. He told me that he had a thorough distrust of scientists, because "as the last issue of *Science* reveals, scientists don't give a damn about guidelines, restrictions, etc. . . ." The congressman was referring to the story in the *Smithsonian* by Janet Hopson, an intern who had spent several months in the San Francisco laboratories of Herb Boyer.[10]

I did not succeed in getting further in the graces of Congressman Ottinger. Having already submitted another bill, Mr. Ottinger was soon to return with a DNA Research Act of 1977 (H.R. 5020, identical to S. 621—see Bumpers below), with nine sponsors, which was referred to the Committee on Interstate and Foreign Commerce.[11] The findings of this bill were more moderate, but compromise is the name of the game if one seeks cosponsors of legislation. This time, "Congress finds that (1) Research related to recombinant DNA is of exceptional importance, with many potential benefits, but also major uncertainties regarding its possible effects on human beings and other organisms. . . ." Penalties for violation included a hefty $10,000 per day or up to 1 year's imprisonment.

Bumpers

Two weeks after Mr. Ottinger's first salvo (February 4), there was a roar from the "other House." Senator Dale Bumpers's (D-Ark) DNA Research Act of 1977 (S. 621) was the first bill drafted proposing regulation of DNA research by statute.[12] It would prohibit research until a license was obtained from the secretary of HEW; NIH Guidelines would be the rules for everyone. There would be civil liability for damages and penalties of $10,000 per day with up to a year's imprisonment for any infraction of the rules. Senator Bumpers's presentation remarks were earnest and no doubt well meant, but his phrasing was oratorical and hyperbolic, revealing that the senator and his staff had availed themselves of a very limited selection of scientific authority in arriving at the need for a new law:

> Mr. President, the thing that piqued my curiosity about this whole subject was an article in the New York Times magazine section of August 23, 1976[13] . . . written by one of the most knowledgeable science writers in the United States, in which he quotes Dr. Wald, a Nobel Prize winner at length . . . There is no question, incidentally, about some of the benefits that can be derived from this kind of research; but there, by the same token, there is no question about the devastating results that can occur if some of these microorganisms are allowed to escape from the laboratory.
>
> . . . We must ask, with Professor Irvin [*sic*] Chargaff, "Have we the right to counteract, irreversibly, the evolutionary wisdom of millions of years, in order to satisfy the ambition and the curiosity of a few scientists?" This world is given to us on loan. We come and we go, and after a time we leave earth and air and water to others who engage, under the leadership of the exact sciences, in a destructive colonial warfare against nature. The future will curse us for it.

> Mr. President, if I had my choice there would be a moratorium declared right now on all recombinant DNA research. I am convinced we are engaging pell-mell in one of the most dangerous kinds of research ever undertaken in the country, including the splitting of the atom. . . .

Despite a melodramatic introduction, S. 621 was constructed along more comfortable lines than other proposals that would follow. Under it, the HEW secretary would be the regulator. Also, in his introduction Senator Bumpers had declared that "nobody in this country can engage in this kind of research without complying with the NIH Guidelines." Facilities would be inspected and work suspended if the secretary found the licensed research activity constituted a public health hazard.

Three days after the Bumpers bill, Mr. Ottinger submitted a House version of his bill. More such bills would follow.[14] We, however, did not wait longer to rally the scientists with a most unusual warning.

NIH Informs Scientists

We swiftly invited a sample of leading scientists and academicians from across the nation to a meeting at NIH on February 19, to which members of the press were invited. Our scientific constituents needed to know what some in the Congress were cooking up for them, and we were gratified that, with only about a week's notification, over 40 prestigious opinion makers assembled in Conference Room Six in Building 31.[15] The several members of the national science press corps included Victor Cohn, the *Washington Post;* Harold Schmeck, the *New York Times;* Warren Leary, Associated Press; and a writer from *U.S. Medicine.*[16] Many of the scientists invited were not molecular biologists and were out of touch with the current status of the Recombinant DNA Guidelines. The majority were shocked when alerted to the aggressive mood of some in the Congress and listened in disbelief at the direction and tone of proposed legislation.[17]

Our purposes for the meeting were twofold. First, we would review federal activities concerning recombinant DNA research. This included the nature and composition of the Federal Interagency Committee (FIC), along with reassurance that, if we appeared to have "jumped into bed with the regulators," the NIH director was in the chair. We emphasized that the agencies had come to a realization that, when all the threads of their many regulatory powers were assembled, they could not be knotted into sufficient authority to regulate all such research, particularly its extension to the private sector. Therefore, the FIC was now reviewing what recommendations should be made in terms of proposing reasonable, minimal legislation, if it came down to that.

Second, we desired to expose some painful wording in the Bumpers bill, its definitions, requirement for licensure, civil liability of researchers and their institutions, as well as penalties for violations. The penalties were projected verbatim on a screen, with the suggestion that we were likely to be reading more of the same boilerplate in subsequent legislative drafts. The figures of up to $10,000 per day and a year's imprisonment for a single violation seemed medieval to the scientists, whose continuity of access to their laboratories and independence of action had never been so challenged before. The status of recombinant DNA rules in the United Kingdom and Europe were reviewed, and the absence of any legislation at this time in any country was noted.

This unprecedented and frightening move by Congress to regulate a field of biological research (other than for radioactive products) excited intense discussion. Fears were aired that this would lead to regulation of the search for new knowledge in other fields. Lewis Thomas felt the most important issue was the imputation of liability on the researcher in the Bumpers bill, which made the worker liable for any injury caused by his or her recombinant DNA research. Another deplored the term "research," a term without bounds. Sherman Mellinkoff felt "the way to guard against anti-intellectualism" was to "look for a process to show the scientific community is policing itself." The meeting went on after the 3-hour limit set for it. A portable camera from the local channel 5 television station (WTTG) took film, scenes from which were aired on the evening news accompanied by stills from the movie "Frankenstein." A sober but enlightened group of people filed out of the meeting. At least they would inform the community, and some might even add the *Federal Register* to their reading lists. We had prepared them for the possibility that, through the efforts of the FIC, the secretary might seek relief in a legislative proposal that would be concordant with the desires of the scientists and the public.

It was rumored that the "big guns" in the authorizing committees in the Congress, Congressman Paul Rogers (D-Fla) and Senator Edward Kennedy, were each preparing bills for presentation.

Glossary

While awaiting events in the center ring, the reader may wish to look over explanations of several major issues that will be featured in the history of this legislative period. These are (i) Section 361 of the U.S. Public Health Service Act; (ii) the "commission" concept as a replacement for the RAC, now appearing in one variation or another in the draft bills; and (iii) the FIC bill (administration bill). Each of these items played an important role in the hearings to follow.

Section 361

How ironic, in light of the role of NIH in the present controversy, that this troublesome bit of statute was a by-product of a thorough revamping and consolidation of all the old public health statutes in 1944 by Congressman Bulwinkle. His new Section 301 quickly propelled the NIH to the pinnacle of American biomedical research. There was also born a cousin, Section 361, written for fighting plagues and epidemics.[18]

An offhand recommendation that NIH use Section 361 had been made by Peter Hutt at the DAC meeting of 1976,[19] and it was the subject of a petition circulated by the Environmental Defense Fund and the Natural Resources Defense Council 9 months later.[9] The FIC had considered and rejected Section 361 in the search for existing authorities for regulating recombinant DNA research.[20] Section 361 became a "quick fix" for opponents of NIH's handling of the Guidelines, and a steady drumbeat of this demand reiterated for the next several years. To my chagrin, an assistant director of the Office of Science and Technology Policy in January 1978 appeared to succumb to the hypnotic simplicity of this remedy. He advised the HEW deputy counsel how easily this conversion could be done.[21]

> The plan included, *inter alia,* 1) Have the Surgeon General issue a notice of intent to issue regulations that would convert the NIH Guidelines to regulations . . . 4) designation of the NIH to review scientific questions and advise the Assistant Secretary for Health on the periodic updating of the regulations from time to time with full public participation . . . 5) The CDC to be designated as the agency to investigate reports from institutional IBCs, public interest groups, or others about possible breaches of safety regulations. . . . Activities in the private sector would be monitored with a mechanism under Federal Statute 18 U.S.C. 1905 to protect proprietary information.

On receipt of the plan, the HEW deputy counsel informed his superior, Peter Libassi, that in his opinion "the resolution of this issue involves the determination of whether recombinant DNA can be classified as a 'communicable disease.' "[22] When Peter Libassi inquired of my reaction, I hastily gave what had come to be known as "DSF's response to 361." It had become a department position, which I reiterated whenever I feared the secretary might waver. (To his credit, he never gave in.)

The "DSF Response"

- If members of Congress believe that recombinant DNA research is an infectious disease, they should have a roll-call vote on the record.
- Regulation of recombinant DNA research activities sets a precedent for Federal control over academic research laboratories. This Federal intervention is of such importance that a legislative solution—laborious as that may be—is preferable to a quick administrative solution without full public debate.
- Legislation with clearly specified authorities and duties to govern this narrow research sector is far preferable to invoking Section 361, which has been used to regulate brushes, plastic bottles and pet turtles.[23]

No Rancor

I should take care here to remove any idea that there was serious enmity between the Center for Disease Control (CDC) (which had responsibility for Section 361) and NIH beyond a family connection that often made them rivals. A famous "field man," Joseph Mountin, had been at NIH when the Clinical Center was rising. Mountin realized that most of the young men arriving in Bethesda had no appetite for "public health research" and that those who were adapted to field work in epidemiology and disease prevention would soon have no home. Mountin approached NIH director Dyer and asked his permission to take selected researchers and create a center adapted for this kind of work. Dyer is reported to have agreed, with two conditions: Put it somewhere outside of Bethesda, and "Don't call it an institute."[24] From the first, the CDC became an invaluable group of specialists who carried out successful hunts for new causes of infections and faithfully bore investigatory and regulatory duties to root out contamination. They were used to the powers that go along with regulation, such as right of entry, inspections, and seizure of suspicious materials. In the 1970s, the CDC had excellent microbiology but little molecular biology. Few felt that it would be capable of handling the first experiments in recombinant DNA.

The "Commission"

Essentially all of the bills introduced for regulation of this research during 1977–78 sported a "commission." Some of the proposed appointment mechanisms, composition, and duties—including replacement of the RAC—were alarming. Senator Kennedy, particularly, had early shown a predilection for the commission mode when he introduced legislation that created the National Commission for the Protection of Human Subjects of Biomedical and Behavioral Research in the early 1970s.[25] It was well chaired by Kenneth Ryan and considered to be effective in improving policies for the ethical conduct of clinical investigation, including that of children. It was unique in being freestanding and without a tie to the administration. A freestanding commission for recombinant DNA research falling under the oversight of a busy senator, however, posed a danger that the NIH Guidelines would become the responsibility of the Senate committee staff. If the commission were properly constituted and given a clear and limited charge, it might have performed a useful task in preparing for ethical issues that one day would come to complicate this powerful science. This was 1977, and by no means were there available sufficient facts to give much grist for the philosophers' mill. More important, the burden of the RAC (see the example on the first page of this chapter) was not for amateurs.

The FIC Bill

Although I had declared that I would do all I could to prevent a statute from interfering with management of the Guidelines, I found myself in the latter part of February carrying to the secretary the FIC's recommendations for model legislation. There was ample evidence that the Bumpers-Ottinger bill would be only the first salient of more bills to regulate recombinant DNA research. The need for the administration to move to set its standards for regulation under a responsible bill was imperative. I promised the secretary that the FIC would have a draft bill within the first week in March.

The FIC View of Necessary Elements of the Law

The FIC had failed to find any legal authority or combination of authorities to reach all such research. The key elements that the members agreed were necessary for minimal legislation included:

> (1) Primary responsibility for administration of the act to be vested in the Secretary, HEW; (2) A definition of "recombinant DNA molecules" consistent with the NIH Guidelines; (3) The NIH Guidelines to become the national standard, with such modifications as the Secretary may consider necessary [crucial to ensure future revisions]; (4) Bar any person within the U.S. and its territories from engaging in such research unless: a) permissible under standards set by the Secretary; and b) the licensing and registration requirements were satisfied; (5) Facilities would be licensed by the Secretary [the only efficient way to administer this requirement is to license the institutions, not projects]; (6) Allow the research to begin only after the project has been registered with the Secretary; (7) The HEW Secretary must have authority to enjoin production or research where he believes

> the activity constitutes an imminent hazard to health or the environment; (8) The Secretary is authorized to inspect facilities, make environmental measurements, and take other steps to ensure safety [thus a broad right-of-entry would be granted, sometimes necessary for enforcement of standards]; 9) Free disclosure of information, except for information exempted from disclosure by the Freedom of Information Act, or that needed to protect proprietary rights.[26]

Two more points: One was *preemption* of all state and local laws regulating the production or use of rDNA molecules. This issue was extremely important to prevent balkanization of the research. Second was a *sunset clause* fixing the law to expire at the end of 5 years.

Drafting Under Full Steam

The secretary's acceptance of the FIC proposal was quickly won, and the work now fell to those members of the General Counsel's Office, under Donald Hirsch, who were experienced in writing legislation. On March 16, the department released a press statement of the secretary's intent:

> New legislation is necessary to regulate the use and production of recombinant DNA molecules, according to a report transmitted today to the Secretary of Health, Education, and Welfare. In accepting the report from the Federal Interagency Committee on Recombinant DNA Research, Secretary Joseph A. Califano, Jr., said that the Department will immediately begin drafting legislation in the light of the recommendations made by the Committee. Califano noted that he had been closely monitoring the recombinant DNA issue since his confirmation and that he had been in continuous communication with Dr. Fredrickson, Director, National Institutes of Health and Chairman of the Federal Interagency Committee.

The secretary further said that he recognized the legislation would be an unusual regulation of basic science but that the potential hazards warranted the step, not only to safeguard the public but also to ensure the continuation of basic research in this vital scientific area. Secretary Califano added, "we are not saying that research should be halted," and he reaffirmed his "commitment to the principle of unfettered inquiry that applies in scientific research."[27]

OMB Rite of Passage

We arrived at the last hurdle in the executive branch of government on April 7. It may be useful here to record one of those ceremonies that Washington bureaucrats find routine but scientists find strange. All legislation proposed in the executive branch must pass the gamut of the fellow departments and agencies in the "federal family" in an aggressive, primal defense of turf. The venue was the offices of the Office of Management and Budget (OMB) in the Old Executive Office Building.[28]

The hearing was attended by members of the FIC. There was some dissension and arousal of issues not brought to the FIC. Agriculture sought to have the secretary of agriculture named as the authority to determine when plant recombinants could be released into the environment. Labor wanted inclusion of omnibus language to keep the Occupational Safety and Health Administration in the act. Defense insisted on national security exemptions. Justice talked long about the Freedom of Information Act, until HEW agreed to leave existing statutes to govern disclosure. HEW also conceded waiver of the National Environmental Policy Act (NEPA) on initial standards, a concession that I fervently hoped would not reappear to haunt us.

OMB staff members Joe Onek and Sue Wolsey and I engaged in a strong debate on that needling question of preemption. They nay, I yea, on the principle. Finally, HEW offered a compromise and added to the preemption section text regarding "Effect on State and Local Requirements."[29]

A Wary VA

The finale of the OMB hearing was an explosive objection from the representative of the Veterans Administration (VA), who shouted that the VA was not going to have "Joe Califano deciding what VA researchers were going to do with recombinant DNA." This was accompanied by a short statement.[30]

The administration bill as amended passed to the Hill.

I hope the reader will find these preliminaries have better prepared him or her for resumption of the history of these hearings. The date is still mid-March 1977.

The Hearings Resume

On March 9, Congressman Paul Rogers introduced The Recombinant DNA Research Act of 1977 (H.R. 4759) in the company of nine of his committee members.[31] A day later, he introduced an identical bill, H.R. 4849, to amend Title IV of the Public Health Service Act and add several new sections. The bill employed the definition of recombinant DNA from the NIH Guidelines, provided for licensing of all users by the HEW secretary, 10 centers for P4 containment, inspections, and civil penalties of $1,000. As a not uncommon practice, the House used some of the boilerplate from other bills—in this case the Toxic Substances Control Act—to provide a skeleton for the submission.

March 15-17: House Hearings Continue

As is necessary before a major hearing, our Division of Legislative Analysis (DLA) began preparing briefing books filled with answers to questions on recombinant DNA. This would be the first time that a thorough record was to be made in the House of Representatives concerning the history of why we now found ourselves before the Congress. Thus, my opening statement submitted in full would cover the scientific facts, a little about Asilomar, much about the RAC, the first congressional airing of the Guidelines, and the promise to complete an environmental impact statement (EIS) (not yet wrested from the reluctant hands of the Office of Environmental Affairs).[32]

A Philosophical Note. In the final half page of my opening statement, I tried to slip in a philosophical message:

> In conclusion, Mr. Chairman, I want to note that biomedical research is entering a new era in its relationship to society. It is passing from an extended period of relative privacy and autonomy to an engagement with new ethical, legal, and social imperatives under concerned public scrutiny. NIH has responded to this concern by requiring the formation of review boards to oversee human experimentation, animal care, and now genetic recombination experiments. Similar bodies may soon have to oversee other hazardous laboratory work. These responsibilities are inescapable adjustments to the rising demand for public governance of science, though this need not—and indeed, should not—go beyond what is clearly required for public safety lest we inadvertently impede successful research and hamper creativity. The progress of science will continue to depend on the initiative and insights—call it inspiration, if you like—of individual scientists.[33]

Other Witnesses

Thirty-three other witnesses shared the witness bench in those 3 days of House hearings. The witnesses covered a wide spectrum of attitudes about the research.[34]

Some scientists were perplexed at the legislative mores. Leon Heppel, formerly on the staff of the NIH and an expert on nucleic acids, was invited to a meeting in Washington a week after the hearings. He commented:

> I was rather embarrassed that I couldn't think of anything to say at the DNA meeting. . . . I felt the way I would feel if I had been selected by an ad hoc committee convened by the Spanish Government to try to evaluate the risks assumed by Christopher Columbus for what to do in case the

earth was flat, how far the crew might safely venture to the earth's edge, etc.[35]

The ranking minority member of the subcommittee and the only M.D. in the Congress was Dr. Tim Lee Carter (D-Ky), a country physician who had performed surgery on kitchen tables under candlelight. A person hard to get to know, but eventually revealing a great heart, Dr. Carter enjoyed the gamesmanship of a good hearing. He successfully humbled me, drawing me into discussions of cell fusion and the work of Beatrice Mintz, but he revealed nothing of his feelings about recombinant DNA research.

Congressman Ottinger attended the hearings and was permitted to question the witnesses. He was unfailingly courteous but skeptical. One of our brief exchanges is noted.[36]

Thus, the first skirmishes had occurred. There were neither casualties to be mourned nor victories to celebrate. Each side had taken the measure of the other. It would go on all year, and most of the next.

A Meeting with the Tiger

After we became well acquainted, Congressman Ray Thornton (D-Ark) and I would recall that he and I first met in a hospital room in the tower of the U.S. National Naval Medical Center across the street from NIH.[37] I don't recall the date; it was late February or early March. I had been asked to visit the bedside of a congressman who was a patient there. The patient sitting in bed in hospital whites was Representative Olin E. Teague (D-Tex), chairman of the House Committee on Science and Technology, a congressman of considerable seniority and prestige. The chairman, known to intimates as "Tiger Teague," directly asked me to explain just what was this genetic engineering that he had heard so much about. I spent about an hour there, answering questions about the technology and the putative dangers and explaining why I feared turning the Guidelines into statute. Mr. Teague was diabetic (and had suffered the amputation of one leg). I remember listing human insulin as one of the potential products of recombinant DNA. I have wondered since if this was fair politics—but eventually I was proved not to have exaggerated. I found out later that Congressman Thornton was one of the several other persons in the shadows in the room. As I recall, at the end, Mr. Teague spoke to Mr. Thornton and said something about "getting referral of any bill before the House."[38]

Hearings before the House Subcommittee on Science, Research and Technology

The day after the Rogers hearing, I received a letter from Congressman Thornton. The House Committee on Science and Technology had no jurisdiction over HEW and its agencies, but Mr. Thornton was taking his Subcommittee on Science, Research and Technology out on the first of a series of hearings on the science policy implications of the recombinant DNA research issue. It would give him an extensive tutelage, in no fewer than five hearings for a total of 12 days during 1977. I wondered if Mr. Thornton or Mr. Teague had been guided on this path by James McCullough of the Congressional Research Service, whose principal beat was this committee. He had prepared for Congress an excellent review of the basic science involved in recombinant DNA research. I admired McCullough as an example of the quiet, sometimes extraordinary talent that serves the public in such positions.

Thus, these were to be congressional sessions through which the public might be better informed, as well as the committee members and staff. On March 29 and 30, new faces, less familiar to the hearing room than many of their predecessors, were among the witnesses for the Thornton hearings. These included Roy Curtiss III, Daniel Nathans, David Baltimore, and Ronald Cape, president of Cetus Corporation, one of the first biotechnology companies that would be spawned by recombinant research. On the last day, in an examination of the actions of the federal government in this area, I was joined by Betsy Ancker-Johnson of the Commerce Department, who was a member of the FIC, and Dr. W. J. Whelan of the University of Florida, who was a member of the International Council of Scientific Unions.

Mr. Ottinger, a member of the parent committee, was present as a guest of Chairman Thornton.

I best remember a question by Mr. Hollenbeck (R-NJ) about the apparent conflict of interest that occurs when a scientist, using recombinant DNA techniques, participates in the development or revisions of the Guidelines. I replied that the Guidelines were so technical that the utilization of persons who were current experts in the field was an unavoidable necessity.[39]

Senator Kennedy Appears

The administration bill, as recommended by the FIC, was introduced by Senator Kennedy on April 1 and referred to the Human Resources Committee as S.1217 and by Mr. Rogers in the House on April 6 as H.R. 6158. It was the "minimum" bill as the FIC envisioned it. It had all the elements, from preemption to sunset provision.

As he was introducing this bill, however, Senator Kennedy abruptly gave warnings that he had some revisions in mind.[40] First, he surprisingly announced that he thought that one of the unresolved issues was the definition of recombinant DNA research.

Second, in order to give the public the right to participate in the decisions concerning this area of research, Senator Kennedy stated that he would like the witnesses at the coming hearing to address the question of "whether a commission within HEW should be established and be composed of a majority of individuals who are not involved in biomedical research."

Mr. Kennedy's Reservations

The Definition of Recombinant DNA? That the senator, in the act of introducing a new piece of legislation to control research, should immediately propose revision of the definition of that research at first appeared bizarre. Was there a reason? A message in code? A bid to appear au courant? Let's compare the definition patiently worked out by the RAC 2 years earlier (and appearing in the Guidelines in the *Federal Register* on July 7, 1976)—

> . . . molecules that consist of different segments of DNA which have been joined together in cell-free systems, and which have the capacity to infect and replicate in some host cell, either autonomously or as an integrated part of the host's genome . . .[41]

—with that prescribed by the senator on April 1, 1977:

> . . . molecules or segments of molecules of deoxyribonucleic acid—DNA—that are not known to be otherwise capable of being propagated in a particular species of living cell are rendered capable of propagation in that species by joining them by any method outside of living cells to another DNA molecule or segment of DNA molecules.[42]

We may find the key by selecting one clause in this substitute: "that are not known to be otherwise capable of being propagated."

The text of Senator Kennedy's definition was the child of Stanley Cohen of Stanford University, a principal agent in overturning dogmatic definitions where recombinant DNA was concerned. Cohen was working on some experiments dealing with recombinations that did not require the restriction enzymes in use in the classical recombinant DNA research up to that time. He had been in contact with Maxine Singer and other scientists. Singer quickly tested her opinions against those of others and conveyed them to me.[43] The Office of Recombinant DNA Activities (ORDA) and some members of the RAC were very unhappy about altering the already set definition, which had now become a principal model for many other countries. But Maxine Singer agreed with Cohen on the desirability of eliminating nonnovel or not exclusively manmade recombinants from the definition, because otherwise ORDA would soon have to process endless exemptions as more and more scientists began to use the technology.

In addition to performing research in molecular biology, Stanley Cohen, it turns out, was cultivating paths to various congresspersons. The route he had used thus far was mainly the conveyance of materials to the Kennedy committee via Lawrence Horowitz, the former Stanford surgical resident and now the principal scientific member of the committee's staff. There were other communications from Stanley Cohen. And it cannot be claimed that they were not useful.

Why The Commission? We have already discussed Senator Kennedy's predilection for handling legislative problems through long-standing committees or commissions whose members were sometimes handpicked by the senator or his staff; and regardless of their attachment or apparent delegation to some platform in the administration, they frequently had the appearance of reporting to, and being directed by, the senator. A suggestion in his opening statement of using a "subcommittee of the National Commission for Protection of Human Subjects" is a major case in point. What kind of commission did he have in mind?

Hearings on S. 1217

On April 6, Senator Kennedy began hearings on the bill he had so recently introduced, while inserting his "commission" question, with a hint that the National Commission for the Protection of Human Subjects of Biomedical and Behavioral Research would also have some role.[44] Secretary Califano and I, summoned as the administration witnesses, waited as Senators Metzenbaum and Bumpers, Governor Dukakis of Massachusetts, and Governor Byrne of New Jersey gave their views. Both governors were opponents of federal preemption as, not surprisingly, were two other witnesses, Daniel Hayes, chairman of the Cambridge Experimental Review Board, and David Clem, a member of the Cambridge City Council. The secretary brought no prepared statement with him but commented darkly on the proposal of another HEW commission, saying he "already had 322 of them." He suggested that, instead, the National Commission for the Protection of Human Subjects of Biomedical and Behavioral Research make an extensive review of all the ethical implications. In no part of my "long statement" submitted for the record, did I engage "the commission question."[45]

Better the Tiger Than the Fox

Three weeks later, on April 20, Allan Fox, the counsel of Senator Kennedy's committee, invited me and Peter Libassi, now formally installed as general counsel of the department, to come to his office at 8:30 in the morning to "discuss" the administration bill.[46] At the appointed time, Libassi and I made our way to the warren of decrepit office buildings behind the Senate's Russell Building, where overflow space for the ever-enlarging staff of legislators was provided. We found our way into what appeared to be a converted kitchen in which the shelves held trusses of papers, old coffee cups, and the impedimenta of an overworked office crew. We cooled our heels for some time before a staff person came out and handed us the corpse of a bedraggled draft, bearing the label "S.1217." From the moment we returned with the body to Libassi's office, we began a many months' apprenticeship in making our way through different sequential drafts of the administration bill as they appeared, now barely recognizable except for the label "S.1217."[47] These drafts came to be known by common names appended by legislative aides: "draft of 4/13/77," "merci" (4/23), "walnut" (4/27), "DH" (5/19), "subcommittee print #4" (6/14), and "subcommittee print #4 as amended" (6/16).[48] In these various guises, the drafts embodied the intentions of Senator Kennedy as revealed in his introduction of the administration bill on April Fool's Day. In print "4/13/77," a new Section 1801 called for establishment of a commission by HEW called the National Recombinant DNA Safety Regulation Commission, composed of 11 members, appointed by the secretary. Five members should be biomedical scientists; six, nonmedical scientists or laypersons.

The duties of the commission, summed from various sections, were to be challenging, indeed. It must promulgate regulations, hold hearings, hire personnel (to a maximum of five), obtain consultants, appoint advisory committees, disseminate information, register each project, make grants, enter into contracts, conduct or support training, inspect annually all facilities, starting with P4 facilities, allow any citizen to request an immediate inspection of any facility if he or she has reasonable grounds to believe that a hazardous product of recombinant DNA is present, and twice yearly publish a list of pairs of biological species known to exchange chromosomal DNA; an inspector for the commission may restrain, seize, or *destroy* (italics mine) hazardous recombinant DNA.[49] In later editions, destruction of the laboratory work required the permission of the commission chairman.

In its free time, the commission was to study the ethical and scientific principles relating to recombinant DNA activities. The National Commission for the Protection of Human Subjects of Biomedical and Behavioral Research was given the mandate to study other issues surrounding recombinant DNA.

ASM

It is the third of May. Harlyn Halvorson is on the phone. He relates that he's been talking to Allan Fox, who in turn has been talking to Stanley Cohen, relaying to Senator Kennedy his thoughts on new definitions of recombinant DNA. Halvorson and his organization represented an invaluable voluntary partner in the political processes that had become our daily chore.[50] The scientific society of which he was president, the American Society for Microbiology (ASM), was an organization of specialists in medical and basic science aspects of infectious diseases. Before Asilomar (1974), ASM had created a public affairs committee that later considered recombinant DNA issues. When the NIH Guidelines were issued, they were reviewed by an ASM committee headed by Harold Ginsberg, a professor of microbiology at Columbia University College of Physicians and Surgeons. In 1977, ASM took an active role in the debates and politics of possible legislation. I had received notice from ASM that the society would support legislation if absolutely necessary for licensing purposes, but in general they opposed it. Now ASM was in the thick of opposing faulty legislation and was a valuable and indefatigable source of information about the new bills and the revealing remarks of legislators and their staffs. There was also an Intersociety Council for Biology and Medicine, consisting of the executive officers of the major biological societies, who followed legislative developments closely. Their coalition finally expanded to include most of the major medical and academic associations.

House Markup

On May 3, the House Interstate and Foreign Commerce Subcommittee on Health and the Environment held a markup session on two DNA bills, the administration bill H.R. 6158 and Mr. Rogers's bill H.R. 4759.[51] At least 10 members of the subcommittee participated and discussed an outline of 28 major concepts or topics. As consensus was reached on each, the subcommittee moved on, leaving staff the task of cleaning up the bill. It was decided that not just recombinant DNA research but other rDNA activities would be covered.

Next came the question of who should administer the regulations, the HEW secretary or an independent commission. Representatives Ottinger, Maguire, and Markey argued for an independent commission. Mr. Ottinger felt that the secretary of HEW would be a "weak reed," and NIH was an interested party if it were delegated any regulatory duties. The chairman proposed that the secretary have ultimate authority but that there might be a DNA advisory committee *to help* the secretary. These proposals were narrowly approved. The role of local biohazard committees was discussed, and it was noted that they were given primary responsibility for inspections. Penalties for violation of the Guidelines were discussed. Mr. Ottinger suggested they be set at $100,000 per day. Finally, the conferees settled upon a structure of penalties:

Proposed Penalties for Violation of Guidelines:

1. Ineligibility for Federal funds for research for a specified period of time
2. Loss of license
3. Civil penalties: a) up to $50,000 per day for violations related to P3 or P4 facilities; b) up to $10,000 per day per violation related to P1 or P2 facilities
4. Criminal penalties: For a knowing or willful violation, upon conviction, a fine of up to $50,000 or imprisonment for up to one year or both.[52]

One may today pause and wonder at the magnitude of these proposed penalties. I suspect Chairman Rogers, a wily legislator, felt some "swagger" was necessary to assure others in the Congress, and especially some of the more anxious colleagues on his subcommittee, that the bill "meant business." Save for these excesses, the bill that emerged from these sessions and appeared in a print of May 24 was a "good bill." For example, one of the sensible features was Section 472, which relieved the secretary of complying with NEPA for revision of guidelines when the Administrative Procedure Act was followed.[53] This was important, as the RAC was in the process of proposing revision of the Guidelines. The NEPA waiver had an insistent tendency to creep in and out of texts of Senate and House bills.

Patents?

The House subcommittee had a desultory discussion of patents, with only one member being against patents in this area. Obviously, no one was aware that the Senate subcommittee was working in the opposite direction on this question. Indeed, Mr. Kennedy's draft bill was using as a model for S. 945 the script of the Atomic Energy Commission Act of 1946, with the words "recombinant DNA" replacing "atomic energy." An expert examining the draft concluded that

> The bill (a) puts ownership of all inventions "useful in recombinant DNA research" that stem from Government contracts or arrangements, in the

> United States, subject to specified waiver provisions, and (b) prohibits the patenting of all inventions "useful solely in recombinant DNA research."[54]

The longest days of the summer brought a brief legislative vacation. Meanwhile, the "routine" continued at NIH. The FIC was still considering the patent issue. The RAC was considering recommendations for revisions of the Guidelines. The Kitchen RAC and I fussed with the EIS in earnest and would be suddenly hit in the flank by the summons to litigation. The first quarter of the 95th Congress was over.

Kennedy and Rogers

The battle positions of the two principal adversarial authors determined to regulate recombinant DNA by law were changing. Each, at the outset, had carried the pennant of the administration bill. However, the Senate bill (S. 1217) had changed repeatedly and no longer pretended to represent any of the hopeful suggestions of the FIC. The principal House bill was also accruing new growths with each markup. It still carried the elements of a bill favoring the HEW as the focus of rules. Its sponsors, too, were determined to interpose more commissions of citizens, less informed about the complexities of recombinant DNA research and more concerned with projections of the ethical, moral, and philosophical problems associated with technology.

The Administration Position

HEW had invested much effort to protect the science that everyone felt was going to spark a revolution in knowledge. The secretary was on our side. Confronted, however, by two of the members of Congress who determined the fate of much crucial health legislation, Mr. Califano had a heavy investment in the outcome. He sent an all-points call for an update: "I am concerned about our Recombinant DNA Bill. . . . Let's get straight on responsibility and who's doing what."[55]

Joe Perpich assumed responsibility for directing exploration of how we might tie up loose ends and then prepare position papers for approval downtown. He had a useful dialog with CDC and assured the assistant secretary for health that CDC and NIH were not in conflict. If the secretary was ordered by a law to implement the Guidelines within HEW, we could meet CDC's need to satisfy its regulatory mandates while retaining NIH responsibilities. I sought support from Peter Libassi to act as a joint sponsor in approaching the secretary. Mr. Califano, of course, would immediately want to know what his attorneys thought about any arguments the scientists were using. We had answers for his anxiety, but the situation was fluid enough so that we frequently needed a sounding to be sure of our depth and position. Besides, we had to give him the opportunity to reconsider any of our previous administration positions, for he was now faced with differing legislative moods in the Congress. Libassi drafted an "action memorandum,"[56] to be backed by my analysis of the problems we and the scientific community saw in the commission now occupying a comfortable and more secure niche within each succeeding Senate draft.[57] We also sounded out others for their opinions. The economist Henry Aaron, assistant secretary for planning and evaluation, was becoming an apt pupil of the technology, and his views helped buttress ours.[58] Libassi's action memorandum restated the administration's major positions, with the little boxes for "agree" or "disagree." The key elements of the administration position were the following.

> *The Secretary should administer the recombinant DNA legislation, and in doing so, should avail himself of a strong outside advisory committee.* The current Kennedy draft still sticks with the use of a semi-independent regulatory commission because of Senator Kennedy's concern over what he sees as the conflict of interest inherent in entrusting the safety regulation of hazardous research to researchers. Both the Rogers Subcommittee and the scientific community support the Administration position.

In this important declaration, we also asked the secretary to reaffirm that the other positions of the administration were still the same, i.e., that questions of freedom of information (including proprietary material) should be left to general law, that other federal regulatory laws should not be allowed to cover recombinant DNA activities (until they emerged from the laboratory), and that a rigid inspection rule should not tie up a disproportionate portion of the resources allotted to DNA regulation. And finally that there should be a "sunset" provision to any law.

In my brief to the secretary, I concentrated on opposing the Senate committee's use of a quasi-independent regulatory commission in lieu of the secretary to regulate recombinant DNA activities:

> **Use of a commission obscures accountability.** Diffusing the bill's regulatory authority within a commission that would act by vote on the matters that came before it would obscure the locus of responsibility for its actions. Whom would the Congress call to account? Certainly not the chairman, for the bill assigns the chair no special regulatory powers. Certainly not the Secretary, whose only function under the bill would be to appoint the commission's members.
>
> **A part-time commission cannot do the job.** The subcommittee's staff bill would invest the commission with a combination of policy-making and

administrative functions that is beyond the capacity of a part-time body of private citizens. Merely the certification of hosts and vector is, alone, an enormous task. . . A commission structure promises delays in decisions inimical to both the public safety and to the usefulness of the research.

Even if restructured, the commission would duplicate existing regulatory capacity. The Department has the capacity to administer the proposed legislation through existing agencies. To establish yet another government agency . . . is wastefully to replicate current institutional capacity, even if the new agency were restructured to meet the previous objections to a part-time commission.

A strong advisory committee can meet the objective that a commission is intended to serve. We appreciate the apprehension of subcommittee staff at the supposed prospect of a DNA regulatory scheme wholly in the hands of individuals whose desire to see the research performed outweighs considerations of public safety. This fear can be met under the Administration bill by the Secretary's resort to a distinguished advisory committee appointed by him from among the lay and scientific communities, which would advise the Secretary on the policy to be pursued in regulating recombinant DNA activities. We do not question the need, as an aid to this regulation, of obtaining the views, on some regular basis, of persons free of any suggestion of bias in favor of conduct of those activities. . . . it is worth remembering that the impetus for the present NIH Guidelines, as well as the proposed regulatory legislation are derived from a concern for public safety initially expressed by biomedical research scientists.

Of course, this opinion was not shared by everyone. Ten years later, Diana Dutton would write of the revision of S. 1217 in this vein.[59]

The bill proposed by Kennedy had the most far-reaching role for the public: a freestanding commission, including a majority of non-scientists, which was to make regulatory decisions and also to consider the broader, long-term policy implications of the commission. Through his commission, the public would participate in the crucial initial steps of delineating the issues to be decided as well as the decisions themselves.

Joseph A. Califano, Jr., however, was far ahead of me in understanding the chaos that could arise should regulation fall back into the hands of a congressional committee. I perceived that he intended not to let this happen.

Senate Bill (1217) Prescribes Its Commission

In the committee print of S. 1217, on May 19, our already tragically abused administration bill was adorned with the following:

Finding (5) It is necessary to establish a Commission to be known as the National Recombinant DNA Commission to assure that recombinant

> DNA activities be conducted in a manner to protect the public health and welfare of the American people . . . composed of eleven members: Six who are not and have never been professionally engaged in biological research . . . who are qualified to serve . . . by virtue of their training and experience in medicine, law, ethics, education, physical, behavioral and social sciences, philosophy, humanities, health administration . . . or public affairs . . . Five who have been professionally engaged in biological research . . . [60]

The First Amendment and Research

As a relief from the rising tension, Mr. Thornton's House Subcommittee on Science, Research and Technology held hearings on May 25 and 26, another of his engagements with the issues while poised for a sequential referral of Mr. Rogers's bill, if Mr. Teague could swing it. The first panel of the Thornton hearing included four professors of law whose most consistent theme was that the First Amendment includes scientific inquiry as a form of speech.[61] However, all parties agreed that even the First Amendment has limits and that freedom of inquiry may be limited where there is a need to protect society from harm. When protection is needed, all four professors agreed that federal regulations should preempt state or local requirements.

Meanwhile, on the Senate side, Kennedy's staff was working on a draft which they chose not to reveal. To Peter Libassi, it looked as if Kennedy would ultimately seek a DNA regulatory commission within HEW but not under the secretary's control, stressing a very heavy participation of the lay public in development of the regulations governing DNA research.

By March 27, Califano had decided to call Senator Kennedy. Accordingly, Donald Hirsch prepared for him a "talking paper," a script that laid out the secretary's objections to the present features of the many-headed hydra that was S.1217. An especial objection was to the regulatory commission, particularly its inspection provisions, in which inspectors could enter licensed facilities and destroy DNA materials without judicial approval. The paper suggested that the secretary remind the senator that the scientific community would view administrative action, particularly ex parte, as a poor substitute for judicial proceedings to protect their property.[62]

Another House Commission

On June 6, Secretary Califano sent the standard courtesy message reassuring Congressman Staggers that H.R. 7418, the House version of the administration bill, still had the blessing of the administration.[63] It was de rigueur not to complain to the chairman that the administration creation had undergone some odd transformations after adoption by its House sponsors. It had, for example, acquired a burdensome new rDNA committee of 15 persons.

The day after the secretary's letter, Chairman Rogers and at least 10 of the 14 members of his subcommittee sat down to mark up H.R. 7418 another time.[64] Several times, Mr. Rogers brought down the gavel with his customary touch for compromise in handling a flurry of amendments. The maximum fine of $50,000 as the civil penalty for a Guidelines violation had fallen in just a month to $5,000. The chairman added two other amendments, proposing yet another definition of recombinant DNA, and stipulated that the secretary must approve each local biohazards committee. (Apparently no one thought that there would be over 100 of such bodies in another year.) The number of members on the 15-member advisory committee to the secretary not engaged in, and not financially related to, recombinant DNA activity, was increased to nine.

The Ottinger Amendment

From our standpoint, the least desirable change in this House bill was precipitated by Representative Ottinger. Though not present, he had an amendment introduced on his behalf to create a Commission for the Study of Recombinant DNA Activities.[65] This commission would consist of 13 individuals appointed by the secretary. The commission would study ethical and scientific aspects of the research and terminate 26 months after the act became law. Representative Tim Lee Carter objected strongly to this new commission as redundant and an unwise use of the taxpayers' money. It was pointed out that the secretary had reportedly already contacted the National Commission for the Protection of Human Subjects of Biomedical and Behavioral Research and asked them to undertake the same tasks. Nevertheless, Mr. Ottinger's amendment was accepted. It appeared in Title II of the "clean bill" to be forwarded to the full Interstate and Foreign Commerce Committee.

An examination of the clean bill left no doubt that the Ottinger commission would be as effective as the Senate's version in idling the laboratories while the regulatory apparatus crafted by NIH, the FIC, and the HEW was replaced. Here are a few of the words in the new bill as dissected by a staff member:[66]

> The Secretary shall select members from individuals distinguished in the fields of medicine, law, ethics, theology, the biological, physical and environmental sciences. [The tilt was maintained to be sure that] only a minority, (6) of the thirteen members, shall have any experience with the science involved.

There were three pages of duties expected of the Commission. It would "conduct a study on the appropriateness of continuing recombinant DNA activities, conduct a comprehensive investigation and study to identify the basic ethical and scientific principles which should underlie the conduct, applications, and use of recombinant DNA activities. . . . a comprehensive review and critique of the regulations . . . and recommendations to the Congress and . . . HEW . . . on the safest and most appropriate uses. . . ."

Ryan's View

In deference to Senator Kennedy's remarks at the Senate hearing, Secretary Califano wrote Dr. Kenneth Ryan, the chair of the National Commission for the Protection of Human Subjects of Biomedical and Behavioral Research.[67] The commission had recently delivered recommendations dealing with federally sponsored research on the fetus. Despite the fact that ethics could be said to be the commission's metier, it was not ready to conduct a review of "the social, ethical and legal implications of recombinant DNA technology to alter the genetic character of man." Dr. Ryan sensibly declined the secretary's suggestion that his commission lift up such a burden. Congressman Ottinger asked for my views on his latest amendment. Observing that the national commission had a previous year's budget of $3 million and required the assignment of 15 personnel, I deplored the additional cost of mounting yet another independent survey of the relevant ethics. Mr. Ottinger answered that he'd be willing to compromise if the present national commission could be reconstituted to ensure appropriate public and scientific expertise in the area of genetic research. The General Counsel's Office said that any change in the composition of the national commission would require going back to Congress.[68]

The newly marked-up version of the House bill, the clean bill (old H.R. 7418 becoming H.R. 7897), arrived on June 20, headed for the full committee on June 21.[69] Arguably cleaner, the bill was still not without barnacles. A provision had been added that would have the secretary's advisory committee review and comment upon all proposals to work at P3. The Office of the HEW General Counsel commented that "the workload . . . could potentially be enormous and would not . . . serve the scientific community or the public interest."[70]

Senate Marks in Private

Meanwhile, over on the other side of the Capitol, the Kennedy staffers seemed to be breathing nitrous oxide. In the draft of June 14, the National Recombinant DNA Safety Regulation Commission had been escalated to *Presidential appointment of the members, and Senate confirmation of the Chair.*

The bill also commanded that its commission (while carrying out the continuous review, actions for evolution of the Guidelines, and overseeing and enforcing their use) also undertake an ongoing study of the basic ethical and scientific principles that should underlie the conduct of rDNA research. In this endeavor, the commission was to consult with the National Commission for Protection of Human Subjects of Biomedical and Behavioral Research.[71] The commission and its staff could easily devote all their efforts to conducting the ethics study alone. The amended draft Senate bill of June 14 also contained three alternative proposals for preemption. The commission was commanded to rule on any exemptions and publish any setting aside of local or state rules.

On observing the Laocoön state of legislation, Perpich sent a note to me concluding,

> . . . the various versions of the Rogers bill reflect in great part our working with the Committee staff. In the case of the Kennedy bill this has been a Horowitz production from the very beginning with very little opportunity for exchange.[72]

He urged me to carry our concerns to Libassi.

I wrote to Libassi on the 28th, after we had reviewed the latest markup of the Senate bill.

> Though there has been some improvement from the original draft—notably the elimination of the patent provisions—there remain a number of troublesome features . . . especially the establishment of the DNA Commission. I believe it is most important that the Secretary be informed and speak to Senator Kennedy.[73]

Libassi accepted a draft from the two of us to the secretary urging the latter to make such a call to Senator Kennedy to achieve these objectives.[74]

ASM in Action

Hal Halvorson, writing for the ASM, had sent two long letters to Senator Kennedy in May, indicating that the ASM supported national legislation to develop acceptable standards for recombinant DNA research, indicating a number of acceptable features and some others that the microbiologists considered undesirable.[75] These communications carried with them endorsements of the same sentiments at the recent annual meeting of the ASM. Ever resourceful, Hernandez from DLA sent me two documents.[76] The first was a copy of an ASM letter to all members of the House Interstate and Foreign Commerce Committee.

> . . . Although the Society is pleased with most aspects of House bill as reported by the Subcommittee, we are especially concerned about . . . the

> imposition of a penalty of $5,000 per day for each infraction, the National Advisory Committee review of P3 proposals, and the bioethics review by Mr. Ottinger's Commission. It would be imprudent to substitute ethical debate for excellence in science. (Regarding Federal regulations preemptions . . . microbes do not respect state or community boundaries . . .)[77]

The other document, Hernandez noted, was "a copy of suggested language for a minority bill report that may accompany S. 1217 . . . prepared by ASM and transported to Senator Gaylord Nelson's staff on June 27."[78]

Positions on H.R. 7418 were conveyed contemporaneously by the Intersociety Council for Biology and Medicine to the House Committees on Interstate and Foreign Commerce and Science and Technology and the Senate Committee on Human Resources.[79]

While NIH could not imitate the overt actions of the ASM, we could always give candid answers to members of congressional staff. Witness a note from Perpich about an inquiry from a member of Senator Chafee's staff, who said she had been led to believe that NIH had no objection to the commission in the Kennedy proposal. "I informed her we did."[80]

The "Tiger" Reappears

As evidence of an outside plot to hobble the Kennedy bill accumulated, we heard that Congressman Olin Teague had gained from Speaker O'Neill a sequential referral of H.R. 7418 as soon as it left the Interstate and Foreign Commerce Committee.[81] Teague's (Thornton's) Subcommittee on Science, Research and Technology would have 30 days to consider the bill, and staffer Gayle Pesyna said it would take all 30 days. In the meantime, the bill may not be considered by the whole House. On July 11, speaking in his deep Texas drawl, Mr. Teague urged his colleagues in the House to realize that

> . . . the DNA recombinant issue cannot be dealt with in isolation from the rest of the world. Just as it is necessary for the Congress to evaluate the relative merits of potentially different sets of State and local regulations to emerge as contrasted with Federal preemption of State and local option, the Western European nations are beginning to see the international implications of different national rules and regulations. . . .[82]

Teague then recommended that the members become familiar with the hearings his committee had held (10 since March) on this matter.

Desiderata: In a letter in mid-July to Mr. Rogers, the Central Intelligence Agency, over the signature of the director, Stansfield Turner, eschewed any desire to do rDNA research.[83]

Senator Kennedy's Views

On July 18, Senator Kennedy met for an hour in his office with half a dozen leading scientists.[84] The senator said he might be predisposed to modify the composition of his presidential commission, to review the civil penalties prescribed, and as well to reconsider the requirement that license applications must be published in the *Federal Register* along with a detailed description of the experiments to be undertaken. However, on the central questions of who shall administer the law, perform the regulatory functions, and who shall enjoy the right of preemption, there was no suggestion of a disposition to bend. The senator made it clear that he was also still opposed to federal preemption of more stringent local laws.[85]

Letter to Horowitz

I used part of the next day to compose a letter to Mr. Kennedy's principal staff member on recombinant DNA issues, Lawrence Horowitz, to place on the record my personal views on pending DNA legislation.

> I am writing at your request [on July 15] to explain more fully my views on the pending legislation . . . Some reasonable form of legislation is desirable to maintain a single set of standards for use of recombinant DNA techniques. . . . I am, however, also of the opinion that unless the new legislation enacted is sufficiently flexible to permit realistic regulation, the cost to the American scientific community and to the general public will be so excessive as to make it preferable that there be no legislation at all.
>
> The latest drafts of the Senate and House bills . . . differ significantly from points [in the Administration bill] in ways critical to the scientific community. . . . the Senate bill [wants to] create a commission with responsibility for developing and implementing standards and enforcing them through regulation. . . . such a commission will cost more than the research it intends to regulate. It will not be capable of handling expeditiously the great burden of highly technical matters, and will quickly find it necessary to create a bureaucracy collateral to that already in place . . . with inevitable delays, duplication, and diffusion of responsibility which could reach disastrous proportions. . . .[86]

The Gorbach Letter

That same day, I sent the secretary, through Libassi, a copy of my letter to Horowitz, along with a letter from Sherwood Gorbach, who had chaired a recent Falmouth workshop in which he reiterated that, from Asilomar onward, the hazards of using *E. coli* had been seriously exaggerated.[87] I summed up the evidence that potential risks of the techniques had been grossly exaggerated and that an increasing number of scientists believed that no legislation was necessary.[88]

Secretary Califano quickly requested letters for both Senator Kennedy and Mr. Rogers. The first informed the senator of new evidence that risks of rDNA may have been exaggerated. Saying he now thoroughly favored the House approach, the secretary requested an opportunity to discuss the matter with the senator at his earliest convenience. The letter to Mr. Rogers, a gesture of protocol, went with a copy of the letter to Kennedy, noting that he favored the House bill.[89]

Full Committee Markup

On July 22, when the full Senate Human Resources Committee reported out the latest version of S. 1217, the monotony of a vote was suddenly broken. All but one member (excepting the absent Senator Hatch), were recorded as "aye" to ordering the bill favorably reported to the Senate. The one "nay," Senator Gaylord Nelson, was a most significant sound in the history of DNA legislation.[90]

Two supplemental views appeared in the published report. A joint view of Senators Eagleton and Chafee considered that such a commission

> . . . structure was unworkable . . . and we question whether or not the Commission members would even be able to carefully review staff decisions. . . . as presently drafted S. 1217 simply produces an added layer of bureaucracy, which, in our judgement, will not regulate the safety of DNA research as effectively as the Department [of HEW].[91]

The supplemental view of Senator Gaylord Nelson supported his rejection of the bill. "S. 1217 is unnecessarily burdensome and detrimental to the future of this important biomedical research."[92] (We judged it probable that the senator's staff person, the competent Ms. Judith Robinson, had seen the ASM memorandum carried to us by Joe Hernandez.)

Senator Nelson's statement continued with a description of quite recent views from the scientific community as represented by the 1977 Gordon Conference on Nucleic Acids, a descendent of the 1973 conference that had sparked the creation of the NIH Guidelines. Now, 4 years later, 137 attendees dispatched to the legislators a concern that legislation should only strive for uniform standards. Nelson next noted that the prominent members of the National Academy of Sciences, including Robert Sinsheimer, had requested the president of the NAS on April 26, 1977, to relay to the government that the proposal of a national regulatory commission was "a wholly new and unfortunate departure." There followed a detailed analysis of the defects in S. 1217, noting that

> . . . a simple extension of HEW's authority to allow for enforcement of DNA guidelines in the private sector would accomplish the objective of establishing uniform national guidelines for recombinant DNA activities. . . . The potential for obstructing research and impending progress in conquering diseases appears to be much greater than the benefits accruing to the public through the provisions of this legislation.[93]

The full report of the Human Resources Committee on the Kennedy bill contains quotable but hardly supportive information.[94]

> [page 8] . . . representatives of the Peoples Business Commission and the Friends of the Earth both stated that an immediate moratorium on all recombinant DNA research should be imposed until the public had an opportunity to study and assess the situation. Dr. Halstead R. Holman of the Stanford Medical School expressed the view that less hazardous alternative techniques exist to accomplish the same objectives in many instances. He suggested these be explored.
>
> [page 12] . . . the Committee believes that there is an important role to be played by the National Commission for the Protection of Human Subjects of Biomedical and Behavioral Research.
>
> [page 14, under definition of recombinant DNA] . . . The burden of proof as to whether something should be exempted from the definition of recombinant DNA for the purposes of this act rests with the petitioner. The Commission then must make a determination that a given class or part of a class of experiments does not present an unreasonable risk to the health of the persons exposed . . .[95]

Visit with Senator Kennedy

On the afternoon of July 26, Secretary Califano, his legislative liaison, Dick Warden, and I met with Senator Kennedy—and Larry Horowitz and Daryl Banks of his staff—in Kennedy's suite in Room 431 of the Russell Building.

Asked to speak first, I presented our view that the regulatory commission would be so burdened that it could not function without crippling delays or delegation to a second bureaucracy. Moreover, not being subject to oversight, it eliminated all possibility of administrative appeal. The secretary stressed the last as the basis for his concern that it would not work. When we were about to leave, I gave Senator Kennedy copies of the "chessboard scene"[96] and received his signature on one previously signed by Congressman Rogers.

In a note made later for the record, I stated that "Senator Kennedy appears amenable to working out some solution as we head to conference . . . this was a friendly meeting and of value in establishing clearly our position."[97]

A letter from the highly respected Herzstein Professor of Biology at Stanford was on my desk when I arrived back at the office.

> I personally believe that if either of the current bills on recombinant DNA research is passed in its present form, research in the biomedical sciences will be plunged into a period comparable to the Dark Ages when inquiry was prohibited. This would come about despite the absence of any evidence that recombinant DNA research is harmful. We would appreciate your continued efforts.[98]

It was a poignant reminder of how the great majority of American biological scientists felt at this juncture. There was a gleam on the horizon, however. It was a sword in the hands of the Senator from Wisconsin.

The August Amendment

Gaylord Nelson broke away from Edward Kennedy on August 2 and offered a sweeping amendment to Kennedy's bill for regulation of recombinant DNA.

> Mr. President, I am today introducing an amendment in the nature of a substitute to S. 1217, a bill to regulate activities involving recombinant deoxyribonucleic acid (DNA) research. The substitute proposal reflects new information and views, and differs from both the Senate bill (S. 1217) and from H.R. 7897, a bill pending before the House Interstate and Foreign Commerce Committee, with respect to a number of major issues: the nature and the extent of the regulation necessary to combat potential risks from DNA research activities; the definition of recombinant DNA activities to be regulated; penalties for compliance; the authority of the Federal Government to preempt State and local regulations. . . . I ask unanimous consent that the amendment be printed in the Record following these remarks along with an article in *The New York Times*, July 31, 1977 by Walter Sullivan, science editor . . .[99]

A comparison of the "Nelson substitute to S. 1217" with its rivals, S. 1217 and H.R. 7897, prepared by our DLA[100] summarized the dismembering of S. 1217, with these strokes:

- A change in the definition of recombinant DNA, restricting recombinants covered to "novel" ones. [The ASM soon pointed out to Judith Robinson of the Senator's staff a defect in the new definition.[101]]
- The commission has been eliminated and the Secretary is to form a Recombinant DNA Advisory Committee of 17 members.
- The inspections of project sites will be performed by designees of the Secretary, including IBCs in institutions, removing the Kennedy bill's "citizen-vigilante" clause permitting anyone to demand an immediate inspection of a laboratory.
- Administrative seizure or destruction of recombinant DNA materials is deleted.
- Violations of the regulation, if wilful, will be penalized at a maximum of $2000.
- Preemption by States and localities allowed only if "necessary" and more stringent.
- Sunset of legislation after 5 years: only Nelson has this provision, as had been recommended by the FIC in the now forgotten Administration Bill.
- The requirement of the Kennedy bill that the IBCs shall consist of 1/3 nonscientists; 1/3 employees of facility; and 1/3 outsiders is softened

and the provision that the IBCs "shall review and approve each project and shall publicly notify the local community of each project approved" has been removed.

- The only major difference between the Rogers bill and the Nelson amendment is the retention in the House bill of a 13 member [Ottinger] commission to study the scientific and ethical aspects of rDNA activities and which shall review the regulations promulgated under this act.

Agonal Moves

On September 7, I had telephone calls from Larry Horowitz that indicated how much Gaylord Nelson's move had stalled the Senate subcommittee's march to capture Moscow before winter. Horowitz suggested "chucking up" the whole S. 1217 package for a "one-page" bill extending the NIH Guidelines to the private sector and a new commission for 9 months to review all activities and report to Congress and the executive branch about any further legislation. As for the commission, they were thinking about having members split 50-50 between the public and the scientific community and that its members could be chosen from recommendations from the Congress, the Institute of Medicine, and the Public Health Service, either the surgeon general or the NIH. The one-page bill would not mention the commission; however, Senator Kennedy would call a press conference to state his idea for a study group and propose how it might be set up. He said he had not yet discussed the proposal with the staff of either house.[102]

Horowitz called back on the 12th to report that the staff of other committee members were not all in favor of the idea. Some, he said, felt that the bill proposed by Senator Nelson should be defeated (by bringing the Kennedy bill to a Senate vote).[103]

I informed the secretary about these calls and said that I had told Horowitz that, while I could not speak for the department, I was not negative to his proposal, provided we could see the exact language before its submission.[104]

Further Thornton Subcommittee Hearings

Chairman Ray Thornton's House Subcommittee on Science, Research and Technology had 2 days of hearings on September 7 and 8 on "the science policy implications of DNA recombinant molecule research." I viewed that the importance of the Thornton hearings was the desire and ability of the chairman to invite strong and respected members of both the scientific and lay communities to testify. As a result, a broader representation of scientists was coming before the Congress, including many out of the apolitical "silent middle" who had been so far reluctant to put themselves on the records of this epochal controversy. Notably, Kenneth Ryan and Patricia King of the National Commission for the Protection of Human Subjects of Biomedical and Behavioral Research said they were only lukewarm in support of a DNA study commission.[105] Other witnesses included the Honorable Chief Justice Howard Markey (chief judge of the United States Court of Customs and Patent Appeals), Dr. John T. Edsall, Dr. Tracy Sonneborn, Dr. James Sorenson, Dr. Bentley Glass, and Dr. Lewis Thomas.

Meanwhile, Perpich was working with his staff toward preparing suggestions for the secretary as to how we might propose that the government implement DNA regulation in the absence of legislation this year.[106]

House Bill Readied for Full Committee Report

The draft of the full committee report on H.R. 7897 was made available to us by Burke Zimmerman, who, with David Meade, an expert House draftsman of legislation, had come up with a carefully crafted text.[107] Our having the report for comment ahead of time allowed us to make requests for revision in numerous areas, especially in wording that would have called for an EIS for every single project. Mr. Ottinger's commission remained intact. All in all, however, Bernie Talbot and Joe Perpich concluded that the new draft had taken into account most of the concerns of the scientists.

At another markup session, evidence of the perilous course of attempts to write legislation over science emerged as more members eagerly scrambled to tack a few more amendments onto the bill. Congresswoman Mikulski wanted to change "necessary" to "reasonable" as the secretary's option to accepting state or local rules for preemption. Mr. Markey also wanted a study commission with presidential appointment of members.[108]

Horowitz Returns

Meanwhile, word was received from Larry Horowitz that he had now gotten "a good reaction to the 'one-page' bill idea" from the rest of the committee but that Paul Rogers wasn't at all enthusiastic about such flimsy legislation.[109]

As the support for Senator Kennedy's last draft of S. 1217 decomposed (further catalyzed by Senator Javits's queries about a compromise on preemption to add to Nelson's defection), we learned from a Halvorson editorial that

> In a surprise move, Senator Edward M. Kennedy (D-Mass) announced on September 27 that he would withdraw his controversial bill S. 1217 on the regulation of recombinant DNA research. In a speech before a Medical

> Writers Association, Kennedy referred to recent scientific evidence suggesting that the risks associated . . . have been overstated . . . He proposed that the NIH Guidelines, as revised and updated, be the basis for new legislation.[110]

On the opposite side of the Capitol, Congressman Rogers was also encountering resistance from some members of his committee in regard to the amendments most recently proposed.

The coming of autumn lifted our hopes that we might keep the management of the Guidelines free of entangling statutes. The Thornton hearing could be counted on to project an image of more rational dialog between scientists and the legislators.

October 4, 1977: Visit to the Humphrey Building

As the fortunes of S. 1217 and H.R. 7897 vacillated, we seized the opportunity to brief and to test the reactions of the secretary to the current ferment in the Congress. Present were Warden, Libassi, Carole Emmott, Julius Richmond, Perpich, and I. The secretary began with a strong statement that he favored the House bill against the Kennedy one-pager. During our discussions of about an hour, the tone of his remarks contained suspicion that we were helping to orchestrate the intense science lobbying that had been credited by the press (Victor Cohn, Judith Randal, and others) to the changing of Senator Kennedy's mind. The secretary was also suspicious of the similarity of Nelson's bill to our FIC report and finally asked me if we had written Senator Nelson's recent remarks before the Senate.

All agreed that the current scene provided opportunity to return to the original bill, or one close to it, and that this ought to be the department's position. The secretary indicated that he had not told Senator Kennedy that he "was with him" but that Larry Horowitz had told others as much. He agreed to call Kennedy that day and Rogers 2 days later. We prepared briefing documents for him on the differences between the two bills.

Aftermath

At 6:20 p.m. the secretary called me at home. He said that Senator Kennedy, with whom he had just talked, was quite unhappy. Moreover, the senator had told the secretary that Richmond and I were solidly behind him (Kennedy). What Horowitz had apparently conveyed to the senator was that all of our statements (David Hamburg's, Richmond's, and mine) had placed us in the position that if Congress could pass no legislation and a study were deemed to be the best holding action, we would support his commission for the purposes of study but not regulation.

The secretary signed off glumly, and I put down the phone uneasily. I began to realize that we both seemed to be victims of conflicts between the ego of a senator and the survival tactics of a powerful but slipping person on his staff.

The Friendlier Body

The House of Representatives was usually a more congenial place. But suddenly it was less friendly.

The House Interstate and Foreign Commerce Committee began marking up H.R. 7897 on October 12, 1977. But a quorum deficit allowed only a brief exchange of words, demonstrating true congressional hardball by highly unusual language from two experienced parliamentarians. The powerful chairman of the full committee, Mr. Harley Staggers, revealed that he, too, had his contacts among the scientists—including the ubiquitous Stanley Cohen—who were suggesting that the potential hazards were much less than earlier considered. Mr. Rogers, sensitive legislator, author of a carefully crafted bill for which he had politely and persistently collected a record and stirred through compromises to keep a majority of his committee behind him, was unhappy at this. He insisted that a bill was needed to balance the need to protect the public health and the desire to allow research to continue. His ranking minority member on the subcommittee, Tim Lee Carter, however, was saying that while he generally supported H.R. 7897, he agreed with Mr. Staggers.[111]

A few days later, the House committee met to further mark up its bill. Except for some amendments to protect proprietary data, the preoccupation of the members was the wording of the preemption section. Representative Mikulski again proposed to substitute the word "reasonable" for the word "necessary" in Section 484(b)(2) of the bill. This prompted an elegant example of the art of debate. Representative Eckhardt (D-Tex) offered an amendment to Ms. Mikulski's amendment, adding placement of the words "If the Secretary finds that" before the word "reasonable." The effect of this pirouette was to shift the decision of what was reasonable back to the secretary. The committee approved the amendment unanimously, a small but hopeful gesture toward maintaining a national policy for conduct of the most revolutionary science of the century. But it was not to be the end of the struggle.

8

Returning to the Witness Table: the Remainder of the 95th Congress

1977–78

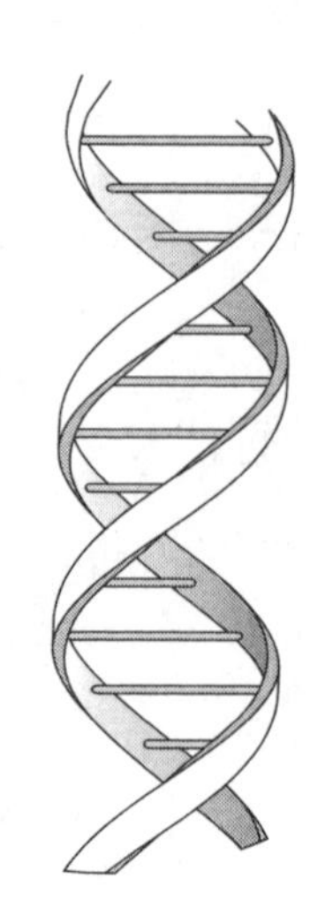

Summary

As described in chapter 7, for 10 months of the first session of the 95th Congress (January through October 1977), the efforts to regulate recombinant DNA research by statute caused the introduction of 11 bills and resolutions in the House of Representatives and two resolutions and one bill in the Senate. One of the bills was an administration bill constructed according to recommendations of the Federal Interagency Committee (FIC). This bill was introduced in the House in April as H.R. 6158 by Congressman Paul Rogers (D-Fla), chairman of the Subcommittee on Health and the Environment of the Committee on Interstate and Foreign Commerce. In this month, the same bill was introduced in the Senate by Senator Edward Kennedy (D-Mass), chairman of the Subcommittee on Health and Scientific Research of the Committee on Human Resources, as S. 1217, the identifier carried by the remaining bill(s) sponsored by Mr. Kennedy. In August, a chilling amendment, in the nature of substitution for S. 1217, was suddenly introduced by Senator Gaylord Nelson (D-Wis).

In November, action in the Senate had turned to the Subcommittee on Science, Technology and Space of the Committee on Commerce, Science and Transportation (see below). The subcommittee chairman, Senator Adlai Stevenson, joined by Senator Harrison Schmitt, took testimony on November 2, 8, and 10, as oversight of the NIH management of the regulation of recombinant DNA research. They laid particular emphasis on a reported violation of the Guidelines at the University of California at San Francisco (UCSF). The principal witness was the NIH director, with appearances by numerous other witnesses, including Philip Handler, William Rutter, and

Herbert Boyer. Senators Stevenson and Schmitt would leave for the next year the completion of their report and conclusions.

A summary of the second session of the 95th Congress begins with the Yuletide recess, where only two vessels of the flotilla of would-be DNA laws that had been assembled during the preceding months remained in the shallows of the congressional lagoon: Senator Kennedy's battered S. 1217 and Mr. Rogers's H.R. 7897. The latter had been passed by the Health and Environment Subcommittee, but the chairman of the parent committee (Interstate and Foreign Commerce), Mr. Harley Staggers, had refused to release it in the last days of the session.

As the second session is opening, Mr. Rogers is trying to rewrite H.R. 7897 when he learns that a new House bill (to be cosponsored by him and Staggers) (H.R. 11192) is appearing and will supersede H.R. 7897. In this same time, Senator Kennedy quietly amends S. 1217. It now resembles the Staggers-Rogers bill in naming the HEW secretary in charge of regulation, but it pointedly does not support federal preemption of local regulations and would activate several commissions. On March 14, H.R. 11192 is marked and sent on sequential referral to the Committee on Science and Technology (Mr. Teague and Mr. Thornton). The bill is returned at the end of the allotted month (April 21) with a few amendments, which are then passed by the full Interstate and Foreign Commerce Committee. Mr. Thornton and his committee are also completing a report of their extensive oversight hearings in 1977. This is released by the parent Science and Technology Committee in April, accompanied by a definitive analysis of "Science Policy Implications of DNA Recombinant Molecule Research." In July, Mr. Staggers asks the Speaker to schedule his bill before the Rules Committee to get it out on the floor; however, summer doldrums have taken hold of legislation. The Senate is idled by Mr. Kennedy's apparent loss of interest in legislative rule over the research, stranding his committee, Human Resources. Throughout the year, Senators Stevenson and Schmitt have pursued answers to questions remaining from their oversight hearing of the preceding November. They, along with members of the Human Resources Committee, now eye Section 361 as the only viable kernel of possible regulation. Senators Kennedy, Javits, Williams, and Schweicker, along with Stevenson and Schmitt, write to Secretary Califano, asking him how he might apply Section 361 to regulation. The director of NIH and the general counsel are asked to prepare answers for the secretary. They oppose use of Section 361. The replies are ready in July, but the secretary elects to send them in September. It is then too late for further attempts at legislation, and Senators Stevenson and Schmitt issue a divided report.

The climax of the drive for legislative control occurs at the start of the 96th Congress in January 1979. Mr. Staggers announces to the House that it will not engage in any further legislation regarding recombinant DNA. All the bills from the preceding Congress are dead. So is the hope of legislation to regulate DNA research.

Adlai Stevenson Oversight

Early in September 1977, Senator Adlai E. Stevenson, chairman of the Senate Subcommittee on Science, Technology and Space, reminded OSTP director Frank Press that his committee had jurisdiction for comprehensive science policy oversight and was watching DNA legislation to be sure it balanced carefully the duty to protect the freedom of scientific inquiry with the government's interest in the public's safety.[1] Shortly thereafter, Stevenson inserted into the *Congressional Record* a lengthy statement supporting the passage of flexible legislation for regulation of safety in recombinant DNA experimentation.[2] He then scheduled several days of hearings over 2 weeks. I was to appear during the second week.

November 2, 1977

The hearings were held in Room 5110 in the Dirkson Senate Office Building, the quarters of the "Committee on the Hard Sciences."[3] The room was filled with pictures of rockets and outer space and a glass-walled cabinet containing model airships, boats, and other toys of commerce. Chairman Stevenson was joined on the presiding bench by Senator Harrison Schmitt (R-NMex), the engineer-astronaut, whose remarks revealed a prejudice for science that attempted to cushion the sometimes litigious tones of the chairman. Stevenson, however, had sent us an excellent letter of invitation, implying sensible questions with uncomplicated answers and no eagerness for dramatic climaxes. The NIH Division of Legislative Analysis had prepared the usual briefing books for me days in advance. The team had temporarily lost its edge, however, for there were no marginal notes to warn the director that when he came up, he could be left to hang out to dry.

On the first day, November 2, Philip Handler, president of the National Academy of Sciences, was the leadoff witness. Handler was always candid and sometimes could be churlish at the witness table. I later experienced this when we formed a team for joint appearance before Congressman Albert Gore, Jr., on ethics in science.[4] This time, however, I thought Handler took just the right tone:

> Mr. Chairman, you who consider such legislation will appreciate that you have embarked on a thorny trail. The risks of violating that which we cherish under the First Amendment surely loom as large as the hypothetical risks against which these safeguards are being erected. . . .
>
> In his book on Congressional Government (1885), Woodrow Wilson said, "If there be one principle clearer than any other it is this: That in any business, whether of government or mere merchandising, somebody must be trusted."

The colloquy turned briefly—and somewhat abstractedly—to the question of whether science can be trusted to police itself.

> ***Senator Schmitt:*** But if there is a violation, do you believe that within the scientific community there can be sufficient capability of the examination and punishment of the violator?
>
> ***Dr. Handler:*** Within reasonable limits, yes sir. But at the extremes, no, sir. As a nation, we are full of people who have used their own discretion as to what is reasonable . . .
>
> ***Senator Schmitt:*** . . . What I am trying to get at is whether in the general realm of research the Government should be the regulator or whether the research community could regulate itself . . . do you think it is possible for the scientific community to police itself?
>
> ***Dr. Handler:*** I believe the community can police itself much better than the Government can, as long as the Government gives general guides. . . . But the effectiveness of that procedure will lie in whether or not those engaged believe that the guidelines are rational and that they conform to reality as they understand it, and that the risk they are averting is a real risk.

Handler was succeeded on the rest of that day by a group of witnesses with differing views.[5]

The first witness a week later was Frank Press, chief science advisor to the president and director of OSTP, who appeared together with his assistant director, Gil Omenn. Both Senators Stevenson and Schmitt soon swung the axis of the testimony to what condition regulation of recombinant DNA would be in if legislation was not passed this session. From there we proceeded to penalties if any violation occurred. I was soon brought in to answer some of the questions; my answers included pointing out that we had not yet had a question involving sanctions but that NIH might lift funding from the project in which an infraction occurred, but not arbitrarily extend this to an entire institution.

By now I had become the only witness. After multiple pages of testimony, the following occurred:

> ***Senator Stevenson:*** Are you comfortable with NIH's role as a regulator? You adopted your guidelines and you also enforce them. Is that a proper function for NIH? Is it a realistic one?
>
> ***Dr. Fredrickson:*** No. I am not comfortable with it, Mr. Chairman, in the sense of enforcement. I think our role is to sponsor research, to propose standards, and to interpret these highly technical matters. I think these are very appropriate things for us to do. On the other hand, I do not think we should be the policeman, and that is a difficult role for us to play . . .

The discourse moved onward, through infections, pathogens, where lay the different expertise for regulation in HEW, the role of the institutional biohazards committees (IBCs), the Stanford letter, institutional patent agreements, and overall patent policy. Suddenly, a prosecutorial edge appeared in the senator's voice.

> Dr. Fredrickson, how will the NIH policy with respect to proprietary interests be applied to the recent insulin gene experiments at UCSF, which were funded by NIH?

A curtain mercifully fell on this scene before I had to answer immediately. I was asked to make way for Margaret Mead, the distinguished anthropologist, who had to catch an airplane. Dr. Mead, leaning on her stick at the podium, her fine head crouched over a hastily prepared and confusing text, seemed to be flailing at too much research, then, too little research . . . and a world in danger. As she discussed her views that there needed to be more prioritization of what research the government was supporting, I was turning over in my mind the incident of the "insulin gene experiments." At the end of September, the reporter Nicholas Wade had obtained the Office of Recombinant DNA Activities (ORDA) records through the Freedom of Information Act and reported in *Science* a breach of the NIH Guidelines that had occurred earlier in the year in the Department of Biochemistry and Biophysics at the University of California at San Francisco.[6]

The previous evening, I had received a telephone call at home from the chairman of that department, William Rutter, who informed me that he and Herb Boyer had "straightened out the paper work on their somatostatin experiment with the university's IRB."[7] This served notice that I should likely see Rutter and Boyer at the hearing on the next day. Previously, Rutter had dealt with Stetten on issues related to their use of plasmids, and I knew very little about the incident that was looming ahead as the testimony proceeded.

Maxine Singer and Paul Berg, DAC meeting 1976 (DAC '76)

DeWitt ("Hans") Stetten, Asilomar

David Baltimore

James Watson, DAC '77

Wallace P. Rowe, Asilomar

Sydney Brenner and Roy Curtiss III, Asilomar

Academician A. A. Bayev (director of the USSR Academy of Sciences Institute of Molecular Biology), Asilomar

Herbert Boyer, Asilomar

Donald S. Fredrickson and Joseph Perpich (Associate Director for Planning and Program Evaluation, NIH)

Director's Advisory Committee meeting, February 1976 (DAC '76)

DAC '76. (*Front, left to right along table*) Leon Jacobs, DeWitt Stetten (*leaning forward*), Donald Fredrickson, Ronald Lamont-Havers (Deputy Director, NIH), Judge David Bazalon (U.S. Third Circuit Court of Appeals; *behind, hand on chin*), Joseph Perpich, Robert Sinsheimer (Chairman, Division of Biology, California Institute of Technology, Pasadena; *obscured*).

Esther Peterson (Consumer Advisor, Johnson administration)

From left to right: (*rear*) John Tooze (European Molecular Biology Organization [EMBO]); (*front*) Walter Rosenblith (Provost, MIT); (*rear*) Anna M. Skalka (Committee on Genetic Research); (*front*) Sir John Kendrew (Director, EMBO)

Robert Petersdorf and Peter Hutt, DAC '76

Margo Haygood, DAC '76

Daniel Callahan (director of the Hastings Institute of Society, Ethics and the Life Sciences), DAC '76

Harold Ginsberg (*left*) and James Neel, DAC '77

Donald S. Fredrickson and DHEW Secretary Joseph A. Califano, Jr.

DHHS Secretary Patricia R. Harris, visiting NIH in 1979

Fredrickson and Peter Libassi (the Libassi Hearing)

Protesters at the National Academy of Science forum on March 7–9, 1977

The NIH Recombinant DNA Advisory Committee (RAC), 1979: (*front*) Robert Mitchell, Susan Gottesman, William Gartland (ORDA), Ray Thornton (chair), Elena Nightingale, and Elizabeth Milewski (ORDA); (*standing*) David Baltimore, John Scandalios, Abdul Jarim Ahmed, Patricia King, L. Albert Daloe, Mark Saginor, Werner Maas, Robert McKinney, James Mason, Gerard McGarrity, Bernard Talbot (NIH), Richard Goldstein, Stanley Barban (ORDA), Ramon Piñon, David Martin, David Friedman, Myron Levin, Nina Federoff, Kenneth Berns, King K. Holmes, and Arthur Landy

The events leading to the revelation that recombinant DNA technology was capable of opening the way for production of human insulin—as I had suggested to Congressman Teague might someday happen—entered into this particular hearing because they had involved a technical violation of the NIH Guidelines. These events have since been the subject of an entire book by Stephen Hall[8] and also described in a detailed report of this particular hearing, released in 1978. The story has since reappeared, 20 years later, in reports of litigation concerning the University of California's patent on insulin, which was obtained in the 1970s.[9] The central issue involved the use of a plasmid before the NIH had formerly certified its use. In January 1976, the RAC had been asked to review a particularly promising new vector, pBR322, developed in Boyer's laboratory. RAC approved it, pending receipt of further information on its safety before submitting it to the NIH director for certification. The first edition of the 1976 Guidelines had specified that the certification of EK2 and EK3 host-vector systems was to be made by the RAC, with no referral to the director for action.[10] Subsequently, the director was included in the loop, and ORDA dispatched this information to all biohazards committees and principal investigators and clearly stipulated that the use of new vectors was not permitted until the NIH director had certified each one.[11] However, prior to its certification, pBR322 was being used experimentally with rat genes to make insulin in Rutter's laboratory, despite warnings from ORDA's chief that this particular plasmid had not yet been certified and against the advice of Boyer to the laboratory staff. Confusion between "approval" and "certification" was blamed for the use of pBR322. When the mistake was reported to NIH, Hans Stetten advised the experimentalists they would have to destroy the materials. They reported they did so and successfully completed the experiments using a third, newly certified plasmid.

In illustrating the great difficulty of imposing complex regulation on scientific experimentation, the events showed the damaging collision of two forces—one, the attempt to maintain the promise of uniformity in the protection of society from unverified risks; the other, the pressing drive of every scientist to be first. Moreover, investigation showed the IBC at the university to be functioning below expectations and made it evident that if this quality of institutional monitoring were to serve for regulation by the Guidelines, the system would prove unsatisfactory. The extremely exciting evidence that the power of technology was greater than expected was mixed with the seriously depressing message that NIH might therefore lose its bid to be the successful regulator of the technology.

Senator Stevenson returned to questioning me. After pages of further interrogation, we came to his questions about the UCSF IBC:

> ***Senator Stevenson:*** The . . . report from Dr. Cleaver [UCSF IBC] says that Dr. Boyer informed members of the team on February 4 that pBR322 had not yet been certified, yet work continued on the clones until March 3. The clones weren't destroyed until attempts were made to determine if the vector wasn't going to be approved soon. Are you [Dr. Fredrickson] undisturbed by these circumstances? . . . Isn't it true that the biosafety committee was informed that the vector had been used in June, 3 months after the experiments had been halted?
>
> ***Dr. Fredrickson:*** Yes, it is disturbing to me . . .
>
> ***Senator Stevenson:*** Now, if these events are disturbing, do you plan to take any action?
>
> ***Dr. Fredrickson:*** I plan to take some general action. I am sure that NIH must in its revision of the guidelines be more explicit and more demanding of the institutional biosafety committees and also of our communications with them, and to develop administrative practices that are more consonant with appropriate adherence to the guidelines. I am distressed.
>
> ***Senator Stevenson:*** But Dr. Fredrickson, in the period I am referring to now, there wasn't any confusion. They knew that the plasmid had not been certified. Dr. Boyer informed members of the team on February 4. So your communications, it would appear, could be improved, but in the situation I am referring to now, there wasn't any communications problem. There was a failure to respond. What do you do about that?
>
> Let me ask you another question to give you more to ponder. It is not just the lack of authority at NIH that is disturbing. It is the lack of authority, of an acceptable chain of command, within the institution that is disturbing. Shouldn't responsibility for the conduct of this research be clearly identified in the institution?
>
> ***Dr. Fredrickson:*** Yes.
>
> ***Senator Stevenson:*** And that those people be held to account when it is abused and neglected?
>
> ***Dr. Fredrickson:*** Yes, I agree with you, Mr. Chairman. I think that it is a joint responsibility. Of course, the scientists, the institution and NIH should make sure that these responsibilities are clear and that they are appropriately tended.
>
> ***Senator Schmitt:*** . . . This concerns me. I don't know how you enforce guidelines or statutes when reputable, highly respected scientists can convince themselves that it is a non-problem.
>
> ***Senator Stevenson:*** You would prefer not to have the enforcement responsibilities at NIH. Is that correct?

> ***Dr. Fredrickson:*** Yes. NIH has neither the desire nor the capability to be a regulatory agency. Our role is to set and interpret the guidelines.

Rutter, followed by Boyer, later proceeded to the witness table and underwent demanding questioning about the events surrounding the unauthorized use of plasmid pBR322. I left the hearings before they completed their testimony. According to Hall,[8] the witnesses' "comments steadfastly reflected the university's position, which in effect blamed the National Institutes of Health."

Both scientists emerged intact, if chastened. Joe Hernandez reported that a parting note was sounded after them by Senator Stevenson, who noted that both witnesses had indicated they did not favor legislation. "If there is legislation, it will be because of scientists such as you," said the senator dryly.[12]

Postmortem

Joseph Perpich's combination of psychiatric and legal training and—more important—his experience in the halls and committee rooms of Congress was at no time more valuable than in the postmortem sessions that usually followed the more bruising encounters, particularly in the Congress. Here he opined:

> The Stevenson hearing was extraordinarily difficult for you as the NIH witness as it raised important questions about our regulatory role in regard to recombinant DNA activities. I believe the hearing was a fair one and we should respond constructively by examining our policies. . . . I do think the Committee staff prepared the worst-case scenario involving the researchers in California. I do not believe it was a question of malfeasance on their part nor on ours, but simply a groping with informal processes to make them more formal but yet far short of overt regulation. The hearing does emphasize the importance of process and providing written notice to people promptly for an opportunity to respond in a timely fashion.
>
> The grilling you received from Senator Stevenson is an important reminder of how correct your course is to move carefully and cautiously in the interpretation of the Guidelines and the certification of host-vector systems. I have no doubt that your prudence in this regard has prevented many other problems that could have surfaced at this hearing or future hearings. As the hearing showed, you bear the ultimate responsibility for actions and will be held accountable no matter where they occur.[13]

Angry Postscript

In his opening testimony, Philip Handler had made a characteristic nod to the First Amendment privileges of science. However, when he returned to the NAS, he came into a meeting of the executive committee of the

Assembly of Life Sciences and angrily urged them to dispatch a resolution to NIH demanding a full inquiry of events at UCSF. The assembly came very close to doing so, and Hans Stetten was asked to obtain my reaction. I advised him to counsel against it, for it might have resulted in a public inquisition and a resulting auto-da-fé of Boyer, NIH, and volunteerism in the conduct of scientific research.[14]

95th Congress—Halftime Score

Thus, the battle over recombinant DNA in both houses in the first session of the 95th Congress had ended without passage of any bill. Between January 4 and December 22, 1977, 14 bills or resolutions had been introduced, most debated, and all but two discarded. Three different committees had held a total of 25 hearings or markup sessions, listening to nearly 100 witnesses.

Near the end of November, Hal Halvorson paid me a visit. We sat at the "Rembrandt" table in my office and reminisced over the legislative year. I complimented him on his performances as a leading scientist-lobbyist in the Congress this year.[15] He declared he was holding steadfast for reasonable legislation when Congress returned, and I agreed, suggesting that we should contact Burke Zimmerman (an environmentalist scientist who was now Representative Rogers's only staff member with knowledge of molecular biology). "We must get him to rake through the ashes," I said, "and find a coal to warm up Paul Rogers—maybe get Mr. Staggers with him—to do a more realistic bill."

Second Session, 95th Congress (1978)

Last Days of H.R. 7987

Burke Zimmerman had been busy during the holiday. He had found Congressman Rogers amenable to paring his bill to an interim statute that would fill the need to extend the Guidelines to government and private sector alike. It would have a strong federal preemption provision to capture the support of Harley Staggers. During early January, several of us were invited to a drafting room in the basement of the Cannon Building in the company of Zimmerman and Steve Lawton of the House committee staff, where we could kibitz the bill as it came off the pen of the drafting clerk, David Meade. The slightly musty cellar seemed to hold multiple ghosts of statutes unborn and of many others admitted to the arena only to perish there. One of the latter would be the new version of H.R. 7987, for word suddenly arrived that a new House bill would be introduced. The principal sponsor was Congressman Harley Staggers. Joe Hernandez said the bill would be introduced early to serve as a basis of negotiation of compromise between Staggers and Rogers and that the model for it was promulgated last year by Nan Nixon and Dan Moulton, two "DNA lobbyists from a consortium of northeastern universities headed by Harvard."[16]

A quick study of a first draft of this Recombinant DNA Safety Assurance Act reveals:

- All recombinant DNA activities shall be conducted under the requirements of the NIH Guidelines.
- Revision of the Guidelines need not comply with the National Environmental Policy Act (NEPA).
- A commission, appointed by the secretary of HEW, is "to study, review and recommend on the Guidelines . . ."
- The National Commission for Protection of Human Subjects of Biomedical and Behavioral Research will review ethical, social, and legal implications.
- Preemption of regulations of states or other jurisdictions, with exceptions for secretarial permission . . .
- A two-year sunset of the legislation.

H.R. 11192

Mr. Staggers ("for himself and Mr. Rogers") formally introduced the new bill in early February. The press release said the bill was developed in consultation with Senator Edward M. Kennedy (The inclusion of the national commission was evidence of the collaboration.) I was quoted as saying: "Although I am speaking only for myself and not the Administration, which now has this bill under review, in my opinion, this measure offers the most promising solution yet proposed for establishing national standards."[17] On February 16, Burke Zimmerman at the American Association for the Advancement of Science meeting commented that "the draft bill contains a strong Federal preemption position. It also exempts the research from regulation under the National Environmental Policy Act." From the floor of the House, the draft was criticized by Representative Richard Ottinger, who said the bill was "miserable." Ms. Pamela Lippe, spokesperson for Friends of the Earth, is reported to have charged that the provisions omitted most of the safeguards that environmentalists had tried all year to have incorporated into legislation.[18]

Copies of a series of Hal Halvorson's letters to Staggers, Rogers, Senators Nelson, Harrison Williams, and Kennedy arrived. Halvorson had developed, with the aid of Harvard lawyers, an argument for preemption based on (i) the worldwide ubiquity of species of microorganisms, (ii) the need for international agreement on what dangers might conceivably exist in technology, (iii) the uncertain hand of local authority, often uninformed, and (iv) the need for national evaluation of whatever dangers there may be from experiments. There was also detailed concern of scientists about bothersome quirks in new definitions of rDNA, illustrated by Halvorson, Roy Curtiss III, and Frank Young,[19] who urged a change in the definition of rDNA.

In a conspicuously silent Senate on the first of March, the Human Resources Subcommittee quietly "laid on the table" an amendment 1713 to S. 1217. Title I of the amendment was identical to the Staggers-Rogers bill, with two missing features: the NEPA waiver and preemption. Title II described a commission of 11 members differing in details from that described in the House. A copy of a letter for Mr. Staggers from the secretary was passed to us, accompanied by a note saying that Rick Cotton would be asked about the NEPA waivers and the strong preemption clause in the new bill.[20] That same day, Cotton called me at home. He and Dick Beattie (an assistant general counsel) questioned the constitutionality of the preemption mechanism. I replied that "we are calloused by a year of futile inventions on preemption; I urge the Secretary to support preemption for a single set of standards as far as he will tolerate." I then remembered I had also asked myself, "Has Rick overlooked the NEPA position?"[21] The answer came 2 days later; Dick Riseberg, our cautious counsel, had informed Libassi that he believed we could publish the initial regulations emanating from this bill without satisfying NEPA.[22]

March 10. The secretary called me in to his office. "Doooooon," he intoned, "are you opposed to the NEPA waiver in the Rogers bill?"

"No," I replied.

"I thought so. They [Senate staff] have misrepresented your views."

I retorted, "I'm in favor of the waiver, because you as secretary are in danger of being stoned to death under NEPA while you are trying to regulate this research sensibly."

Califano, red-eyed, flu-laden, listened as I told him that the Council on Environmental Quality and other environmentalists likely would kill that waiver anyway, because it wasn't in Kennedy's "new amendment" to old S. 1217. "Therefore," I said, "I'll withdraw my demand for your support of the waiver, in order to retain your support for preemption! But thanks for thinking of my interests, Joe."

"You are always in my thoughts, Don."[23]

March 14. During the markup of H.R. 11192, there was a flurry of amendments from the more conservative members. After Mr. Rogers gave everyone his or her chance—exercising one of his great legislative talents—he then offered three amendments so fast that they were accepted without objection. The first amendment did away with the NEPA waiver; the second changed the definition of rDNA as scientists had requested; the third changed the language of preemption so that the authorities of other federal agencies were not impinged. The vote to report the bill out of the House subcommittee was 17 to 6.[24] As the bill was passed for sequential referral to the House Committee on Science and Technology,[25] it bore a minority report summarizing strong dissenting views.[26] These were collectively expressed by Representatives Waxman, Mikulski, Maguire, Markey, Marks, and Ottinger, mourning losses from the older H.R. 7897. They had fought for:

- a Recombinant DNA Advisory Committee—a "commission" (9 of 17 members to be engaged in the research);
- local biohazards committees in each community where research was present;
- designated inspectors;
- a commission enabling mandatory involvement of the citizenry in rule making and in local enforcement and study of research issues at the national level;
- a preemption provision favoring local authorities;
- promulgation of worker safety rules; and
- opposition to authority of the secretary to exempt any of the established rules for anyone.

Thornton Committee

During 1977, the Subcommittee on Science, Research and Technology of the Committee on Science and Technology had prepared itself in the event that "Tiger" Teague, the chairman of the parent committee, was successful in obtaining sequential referral of any bill reported out of the House Committee on Interstate and Foreign Commerce. An Arkansas lawyer, Thornton had acquired, in his hearings, considerable understanding of recombinant DNA and acquaintance with a generous cross section of the friends and foes of this science and the NIH. His committee, with the aid of able staff, notably Dr. James McCullough, senior life scientist for the Science Policy Research Division, Congressional Research Service, and Dr. Gail Pesyna, a science consultant to the subcommittee, had prepared a report encompassing the testimony of 50 witnesses heard in 1977. A portion of this report, released March 1, "Science Policy Implications of DNA Recombinant Molecule Research,"[27] was one of the most cogent congressional documents written during this remarkable period and merits the attention of scientists today.

Sequential Referral Report

The Committee on Science and Technology returned their review of H.R. 11192 to the authors on April 21. The bill was amended as suggested in the sequential review. It is noteworthy that one amendment was relevant to provision for a national committee to examine long-range questions of genetic manipulation. The Teague committee changed the provision for the secretary to form this group to presidential appointment. The subsequent history of genetic research provides retrospective grounds for believing that this might have become a useful arrangement for working on the ground rules for a hippodrome of human gene manipulation in the coming millennium.

Committee Views

In their legislative critique of H.R. 11192, the reviewers from the Committee on Science and Technology were both candid and probing. One example is the "general section" reflecting strong opinions of the members of great interest to scientists:

> Legislation in this area would represent an unusual regulation of activities affecting basic science. . . . The regulation of recombinant DNA research may be an opening wedge into future regulation of other areas of basic research. . . . This fact is viewed by many people—especially by scientists—as a major attack on the freedom of scientific activity. . . . The real threat to the freedom of scientific inquiry may be less likely to come from those who would restrict or regulate research to protect the public health, safety and other values we cherish than from those who believe there are some things man just should not seek to know. . . . Thomas Jefferson once stated that "There is no truth on earth that I fear to be known." . . . Equally crucial is the very thorny question: should government become the arbiter

> of knowledge? . . . As in any area of regulation, the government must guard against excessive zeal to protect society from all risks, however minor. Such activity would particularly cripple science.[28]

Today, when I reread the critique of the Committee on Science and Technology, I feel the sense of exhilaration it gave me when I first read it over 20 years ago. This report also contained several other observations of comfort to our point of view. One was applicable to the study commissions: "If the Commission is to function effectively in its evaluative tasks, persons should be appointed to it who have a working knowledge of the field of genetic manipulation . . ."

There is also a reference to my earlier testimony on the bill, "We strongly favor preemption," to which the members recorded an ambivalent reaction. They predicted that a broad range of comment on this delicate subject would be heard when the bill met the floor. Finally, the report was commendatory in its evaluation of the NIH procedures but had a concern that "turning NIH into [a] regulatory agency would do great damage to its principal function."[29]

Senate Reaction

With the House DNA bill now moving smoothly in the month of May, there was little activity in the Senate. Some of the Senate staff said that the environmentalists wanted no bill so as to avoid preemption getting in the way of local lawsuits. Warden wrote the secretary that "Kennedy has now taken the public position, however, that legislation is unnecessary. Horowitz reports that the Committee would permit legislation to move only in response to a major, visible push from the Department."[30]

A letter from Senator Stevenson had arrived at NIH before the New Year.

> . . . Your testimony, responses to questions, and information to be supplied for the record will be very helpful in formulating recommendation for legislation and administration of existing statutory authorities.
>
> In that connection, I would appreciate your commenting further on four issues related to the unauthorized use of a plasmid vector by researchers at the University of California at San Francisco . . .[31]

The dozen questions were answered in detail[32] before the end of 1977. Through much of 1978, however, numerous further interactions between NIH, the secretary of HEW, and Senator Stevenson would follow. With the pertinacity of a skilled prosecutor, Stevenson remained focused on our discussion of the NIH Guidelines and an alleged violation. The ranking minority member, the former astronaut Senator Harrison Schmitt, was less critical of these allegations, but he would give his views in the fulsome subsequent report of this subcommittee.[33] Like the previously cited reports, this document was a rich lode of opinions and attitudes current in this controversy and further was witness to the seriousness that both senators felt about the responsibility of institutions for the research being carried on in their confines. I was sensitive to these views but not deterred in my insistence that the burden of oversight must be shifted peripherally from the NIH to the institutions and their safety committees.

Section 361

The popularity of Section 361 of the Public Health Service Act (42 U.S.C. §263) and our opposition to its application for recombinant DNA research have been covered earlier. The piece of statute was deficient, as witnessed by the advice of experts presented in full detail in the minority report of Senator Schmitt.[34] Nevertheless, this fragile piece of law was becoming the lifeline for numerous legislators who still felt the need for law.

In early May, I received a long letter from Senator Kennedy in which he asked me multiple complex questions expressing frustration and a curious sympathy for the state of recombinant DNA research. It was symptomatic of a groping for issues that might appease the various proponents of tougher laws: . . . Are the NIH Guidelines preventing development of vaccine to prevent viral diseases? . . . or research on development of antibiotics? . . . new sources of energy from the biomass? . . . Is the current administrative process at NIH bringing some research to a halt? . . . How many people are working in ORDA?. . . Have the newly implemented NIH administrative procedures taken away responsibility from the local committees?[35]

Concurrently, Senator Schmitt also wrote the secretary of HEW, noting the debate over aspects of the pending House and Senate bills. He had many requests of the secretary's opinion on the bills.

> If Senate amendment 1713 were enacted, what was the FIC opinion on the lack of an appropriate statute? What problems does the Secretary have with the use of Section 361 to regulate the research? There are also numerous questions about how the Secretary would use Section 361.[36]

HEW undersecretary Hale Champion wrote Senators Kennedy and Harrison Williams, chairman of the Human Resources Committee, identical brief replies, enclosing the secretary's letter to Senator Stevenson and my testimony before Stevenson's committee, expressing the department's strong support for H.R. 11192 and belief that Section 361 did not represent appropriate authority to achieve uniform regulation of recombinant DNA research. "We would urge the Human Resources Committee to consider passage of comparable legislation," concluded Champion.[37] Senator Kennedy apparently had planned to get the committee to send us a request to extend the Guidelines to the private sector under Section 361 authority. Hale's letter derailed this strategy.

Allied Forces Are Coalescing

Gail Pesyna sent a message to me on May 11 that she had been "specifically authorized" to inform me that the Science and Technology Committee would back me in opposing Section 361.[38]

OSTP adhered to the administration discipline in a letter from the director, Frank Press, to Senator Schmitt.[39]

> Q. Could Section 361 be used to cover the private sector?
>
> A. Yes. . . .
>
> Q. What problems might arise. . . ?
>
> A. Invoking Section 361 would seem to acknowledge risk of transmission of disease . . . Strictly conjectural. Was not meant to regulate research. Does not deal with risks to the environment. . . . The Secretary under 361 would not have the authority to preclude State or local actions.

Flanking Maneuver

A team of senators then determined to achieve what they were denied by incompatible bills by resorting to force majeure. They were now writing joint or conjoint letters to the secretary:

> . . . there is a need to correct deficiencies in the present system of regulation. Privately supported research activities are not subject to monitoring by NIH and there are no sanctions for failure to comply with the guidelines. Application of NIH standards by other Federal agencies is voluntary. As Director Fredrickson has stated on several occasions, it is doubtful this enforcement by the principal Federal sponsor of the research is appropriate. And the agency should not be responsible for the enforcement of the regulations nationwide. The Federal Government should anticipate commercial applications and the concerns they are likely to raise.
>
> In view of these developments and in view of the heavy legislative schedule of the Senate and Human Resources Committee, we are writing to inquire whether the deficiencies in the present regulatory system can be remedied through executive action *in the event final agreement on legislation is not possible* [italics mine].
>
> Specifically, it would seem advisable to shift monitoring and enforcement responsibilities from NIH to a more appropriate agency within HEW. It would also seem possible to remedy the problems of accountability and of coverage in the process of revising the . . . Guidelines. Finally it has been suggested that section 361 . . . provides sufficient authority to promulgate regulations . . . for research . . . conducted within the private sector. . . . We request you solicit a legal opinion from the Department of

> Justice to use the authority of section 361 in this manner. . . . We recommend that an appropriate group of experts and lay persons, such as the advisory committee to the NIH director continue to monitor the scientific evidence of the hypothetical risks of recombinant DNA research. The authors would appreciate prompt response.[40]

On June 6, Dick Warden wrote the secretary with advice to "make a strong legal case for use of existing authorities in order to urge the Senate not to act, or to draft perfunctory responses to the letters from Kennedy et al. and Schmitt that would continue to support the Rogers bill, but would fall short of forcing Senate action on legislation. . . ."[41] He argued that there is "a growing feeling that the research is not as dangerous as initially thought, an aversion to open old arguments regarding preemption, and a busy Hill calendar have combined to change the climate significantly."

I was asked to prepare a draft reply for the secretary to the senators. The substance of what I wrote (with some amendments by Dick Warden and Carole Emmott) is as follows:

> . . . In view of these scientific developments, you raise the question whether legislation is necessary. Specifically you ask whether existing statutory authority would be sufficient for purposes of legislation and you cite the regulatory authority of the FDA and the Public Health Service Act (Section 361). The virtue of legislation is that it may include a number of specific provisions that permit useful flexibility in implementing regulations. Such provisions might include:
>
> - preemption of State and local laws,
> - clear authority for the Secretary in relationship to other Federal laws,
> - APA waivers for initial promulgation of NIH Guidelines as regulations, and
> - authority for the Secretary to waive regulatory requirements that pose no significant risk to health or the environment.
>
> In recommending legislation, the Federal Interagency Committee on Recombinant DNA Research reviewed all existing statutory authority and found that none could provide for comprehensive regulation of these activities. In reviewing Section 361, the Interagency Committee concluded that "there would have to be a reasonable basis for concluding that the products of all recombinant DNA research cause or may cause human disease. Such a conclusion would be tenuous at best, and it is unlikely that resulting requirements could be effectively imposed and enforced." You suggest in your letter that we seek a legal opinion from the Department of Justice. However, Justice is represented in the Interagency Committee and participated in the review and recommendations concerning existing statutory authorities, including Section 361.
>
> Further, the authorities of the FDA were also reviewed by the Committee, but in as much as recombinant DNA has not yet reached the stage

> where it has yielded products to be regulated by FDA, it was agreed that FDA probably does not have authority to regulate such research at present.
>
> In light of the events that have led you to conclude that legislation may be premature, it may also be premature to invoke regulatory authority at this time. As you know the National Institutes of Health has been considering revisions to the Guidelines . . ."[42]

At this point, the secretary interjected that these Guideline changes (revisions) would include an environmental impact assessment, that there would be public review and comment, and that one of the proposed changes would be a proposal to permit, on a voluntary basis, registration of activities in the private sector, with protection of proprietary information. Indeed, a national registry of rDNA products could be maintained.

> In light of these revisions under consideration, it might be best at present not to invoke regulatory authority to reach the private sector.

The senators were further advised that the FIC would continue, as would the other advisory committees, including the RAC and the NIH Director's Advisory Committee, as well as the continuing close contact between CDC and NIH.

> In summary, I believe the Department, in the absence of legislation and without invoking regulatory authority, can continue to provide the standard setting and oversight on a voluntary basis for the present. The proposed revisions of the NIH Guidelines attend to the needs of the private sector, and if problems are presented . . . appropriate actions can be taken through existing regulatory authority or legislation.

At the end of July, Michael Goldberg, who had replaced Joe Hernandez as our man walking the beat of the legislature, reported what little news he could get. The Speaker of the House had delayed calling up legislation to clear a legislative backlog. Further, the DNA bill had a low priority. If action was not taken before the end of the month, the bill would likely die. Representative Rogers would only press for the bill if convinced that the Senate would also consider it. "Senator Kennedy continues to delay," opined Goldberg. "He realizes that in the absence of legislation the possibility for local control of the research exists. He also understands that Section 361, even if invoked, has no preemption authority."[43] Mr. Kennedy would apparently not move until he received a reply to the letter sent by him and other senators to Secretary Califano.

Actually, the secretary's reply was not sent until September 12. Senator Stevenson a month later printed the tardy letter as a rebuke with a major address on recombinant DNA in the October *Congressional Record.*[44] I was grateful that the text I drafted had survived more or less in whole cloth. On September 12, the secretary addressed a firm answer focusing on the department's opposition to the use of Section 361.[45]

Last of the Oversight Reports

The Subcommittee on Science, Technology and Space finally reported.[33] The delay was due to the considerable time taken by Senator Harrison Schmitt to collect the responses to his questions about regulation of science by statute. His was a minority view and partially explained the absence of his signature on the July letter signed by the six other senators to the secretary of HEW.

This is among the reports that deserve to be consulted in the future when new policy issues pertaining to public governance of biomedical science appear. The majority report of Senator Stevenson left no doubt that the members perceived the importance of technology and that it was moving quickly toward important application from the manufacture of drugs and industrial enzymes, waste treatment, production of food, and treatment of genetically related diseases. The full reports, majority and minority, are widely available.[33] The lengthy statement of Stevenson in the *Congressional Record* on "The Status of Recombinant DNA Research"[44] was a reasoned analysis, but the changes in the science would soon make it obsolete.

Scaling the Summit

On the last day of January in the first year of the 96th Congress, Mr. Harley Staggers, chairman of the Interstate and Foreign Commerce Committee, addressed the House of Representatives.

> . . . In addition to whatever agenda each Member may now have, I trust that all of my colleagues will also want to consider the critical importance of the studies and advances in the area of recombinant DNA which are being made by American scientists . . .
>
> . . . As the technique is refined and developed, and as our knowledge of the fundamental mechanisms of gene expression increases, further important discoveries and applications may be confidently expected in the immediate future. . . .
>
> . . . The technique of combining in the laboratory genetic elements from different species was first developed in the United States, and, of course, because of its vast potential, was soon extensively practiced worldwide. The principal American scientists who developed these . . . called for caution in plunging into an area of research which might pose hazards to mankind and the environment . . . [which] led to the development of safety guidelines by the [NIH].

> . . . Some scientists felt that the NIH guidelines were too restrictive or inflexible. . . . Other scientists contended that the guidelines were so restrictive as to demoralize young investigators and discourage them from pursuing careers in science.
>
> . . . Uncertainties over the protection of confidentiality on legislation to extend the scope of the NIH guidelines to private industry or fears that the Federal Government would prohibit any large-scale recombinant activities led several companies to pursue their industrial applications in their subsidiaries in Europe and Japan. . . .
>
> . . . It is time to reverse this unfortunate situation so that such extreme measures are no longer necessary. . . .
>
> . . . The Committee on Interstate and Foreign Commerce will continue to monitor the implementations of these [new] guidelines with care. . . . The committee will be particularly alert to the establishment of any unnecessarily restrictive State or local requirements. . . [46]

Another session of Congress had come and gone without the passage of legislation to regulate research.

9

The Revising of the Guidelines

1976–78

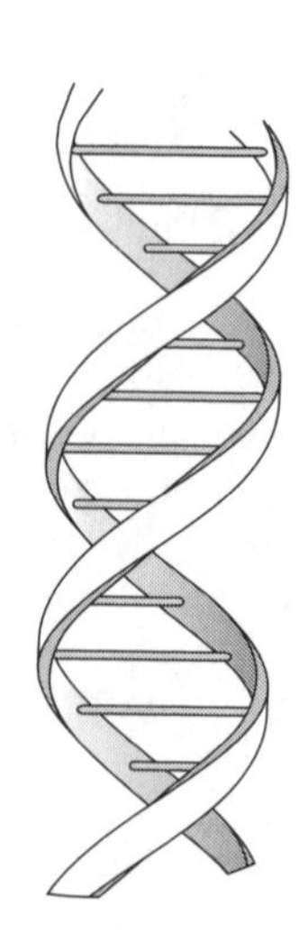

Summary

A core of the history of the NIH Guidelines is the story of their revision to modify the required experimental conditions and procedures. This daunting task for the Recombinant DNA Advisory Committee (RAC) and the director of NIH consisted of inching, through a gamut of public reviews, a series of new versions toward the desk of the HEW secretary, who must agree to the changes.

By 1977, the restraints on the scientists had become oppressive, especially because none of the dreaded risks had manifested. Relief involved a series of formal sequential steps that began in early 1976 with the presentation of the draft 1976 version of the Guidelines to the Director's Advisory Committee (DAC) on February 9–10, 1976. From the transcript of the hearing, extensive correspondence, and interaction with the RAC, the Guidelines as initially promulgated in July 1976[1] had been revised en passant as we perfected our determination to stay within the shadow of the Administrative Procedure Act (APA) to avoid chilling the Guidelines into rigid regulations. In January 1977, the RAC began to write a revision which inspired an exchange of American and European views (the EMBO meeting) and a major review of *E. coli* K-12 as a possible pathogen by experts in infectious diseases (Falmouth conference). A major proposed revision of September 1977[2] was published and presented to the DAC in December. At the meeting, participants cited insufficient documentation of the Falmouth conference and raised questions about containment of viral research. The director and the RAC held a critique of major issues and underwrote an international meeting (Astoria, England, January 1978) which led to rationalization of experimenting with viruses. A new proposed revision was thereafter published for public comment. In July 1978[3] the HEW general counsel Peter Libassi was asked by the secretary to chair a public hearing on this proposed revision. A voluminous affirmative correspondence was ready for Mr. Libassi as he opened the hearing on September 15, 1978.[4] Afterward, Libassi read all the commentary, separately interviewed scientists, and met with a national meeting of institutional biohazards committee (IBC) members. Three months later, the final version of the revision (December 22, 1978)[5] was waiting on Secretary Califano's desk, as he ordered a change in composition of the RAC and a provision for voluntary compliance, enabling the NIH Guidelines to cover all recombinant DNA research in the United States. Approval of this revision also provided a mechanism for future Guidelines revision. ("Major actions" could be proposed by anyone, would undergo public review and comment, and would be discussed and acted upon by the RAC in open session, with final approval by the NIH director.) Major Guidelines revisions occurred in January 1980,[6] November 1980,[7] and April 1982.[8]

Taking New Measure of the Barriers

It was far easier to set up the NIH Guidelines than to take them down. At Asilomar, the restraints had been set purposely high, with the expectation that they would be revised when revision was warranted. No means of achieving revisions was built into the proposed Guidelines presented to NIH in January 1976.[9] Within the next few months, however, the first step in revision was accomplished with the insertion of a section entitled "roles and responsibilities" in the Guidelines issued in July.[1] Here, a primary responsibility of the RAC was designated:

> The NIH Recombinant DNA Molecule Program Advisory Committee has responsibility for (1) Revising and updating guidelines . . .

While the RAC had the nominal responsibility for revision, as a federal advisory committee its power extended only to making recommendations. Moreover,

> The Recombinant DNA Molecule Program Advisory Committee *advises* [italics mine] the Secretary HEW, the Assistant Secretary for Health and the Director, NIH on a program for the evaluation of potential biological and ecological hazards of recombinant DNAs.[1]

As derived from the terms of the charter of the RAC, the HEW secretary was the ranking advisee. No delegation or process of revision was described.

Other than the quotation above, the NIH director was unmentioned in the text of the 1976 Guidelines. More specified roles for the director were added, little by little, such as approval of recommendations for exemptions or certification of approval of new host-vector systems. (The first full description of the much-expanded director's roles and responsibilities appeared in the 1978 [Libassi] revision, where they suddenly blossomed to 23.5 column inches in the 6-point type of the *Federal Register.* [5]) An Olympian order (Section IV-E-1-b of the Guidelines) was also added: "the Director shall weigh each proposed action through appropriate analysis and consultation, to determine that it complies with the Guidelines *and presents no significant risk to health or the environment*" [italics mine]. As for "weighing" each of its "proposed actions" which might bear on public safety, I could but keep myself as fully informed as possible about what the RAC was doing. For this purpose, those who composed the Kitchen RAC were careful to carry back what the issues were and how the voting went, particularly keeping me abreast of the areas of dissent. Thus, from the nature of the voting of the RAC arose many of the director's acceptances of recommendations or their return for further RAC consideration. At first, I tried to keep away from the meetings of the RAC, in order to maintain an appellate role. I found it useful, however, to appear from time to time, if only to deliver a brief homily. Later, we might have some tension over duties of the RAC, but the speeches were few. Because its meetings were attended by anyone who wished, it was sometimes an opportunity for me to deliver a message meant for the public. The committee and I never had a discussion of the protocol of working out Guidelines revisions. We took up our separate roles from the start.

A Daunting Continuous Process

As 1977 arrived, 6 months had transpired since the release of the Guidelines, and more than 100 laboratories were receiving NIH funds for research on the increasingly exciting course of opening some of the mysteries of genes. Inevitably, they were also discovering the limitations of the Guidelines under which they were made to work. In the fall of 1976, the RAC was already considering a major rewriting to begin early the next year.

Revision Begins

In February 1977, a subcommittee appointed by Hans Stetten that consisted of Littlefield, Barkley, Gottesman, Helinski, Rowe, Walters, and Day met to discuss the approaches to revision and assigned portions for redrafting.[10] They met again in March and April and planned to circulate their draft to the full RAC in May. However, the beginning of revision was already stirring other groups of scientists into action. Some were scientists who warned that the international playing field must be leveled by adjusting American rules that were considered oppressive. As these comments circulated, a meeting was held to provide an opportunity for American scientists to discuss with leaders in other countries the differences between the two sets of guidelines governing recombinant DNA research in the world.

NIH/EMBO Meeting, March 21–23

The two major forms of DNA guidelines, the Williams report in the United Kingdom and the NIH Guidelines, differed significantly in their definitions and details of physical containment. The British did not adopt biological containment in their constellation of barriers. You will recall that it was Sydney Brenner who had led an enthusiastic discussion of biological containment at Asilomar, but, in his role as a participant in development of the British rules, he abandoned this advocacy and at this time was developing a new means of risk assessment, which he had not yet revealed.

EMBO, including John Tooze, its executive secretary, in conjunction with NIH's Emmett Barkley, and with NIH funds, organized an NIH/EMBO workshop on "Parameters of Physical Containment."[11] It was held at the Ariel Hotel in London's Heathrow Airport in March. Participants included 12 scientists from the United Kingdom, Germany, the Netherlands, France, and Switzerland and 8 Americans from NIH, CDC, and the Naval Environmental Biology Laboratory. Illustrating that elite scientific leaders of recombinant DNA research in both worlds could find a way to common standards, the participants extensively examined the issues and definitions to create more uniform practices and equipment. The results were then discussed by the RAC, which moved that its substance be incorporated into the first revision of the Guidelines with a target date of publication in September 1977.

Falmouth, June 20–21

Among the most celebrated (and maligned) events of the entire controversy over recombinant DNA research was the Workshop on Studies for Assessment of Potential Risks Associated with Recombinant DNA Experimentation held at Falmouth, Mass., in June 1977. The idea of such a meeting had first come up in discussions between consultants informally gathered with the RAC at NIH in 1976 to consider experiments for risk assessment. The belief that Asilomar had badly neglected the subject of the safety or hazards of *E. coli* K-12 and made it the object of excessive restraints in the use of this host in recombinant DNA experiments had led to a call for a scientific reevaluation of the data on this organism. NIH scientists Wallace Rowe and Malcolm Martin therefore called on Sherwood Gorbach, Professor of Medicine and Microbiology at Tufts University School of Medicine, to convene infectious disease specialists and experts in enteric bacteriology to separate more clearly the myths from the realities associated with this organism. NIH agreed to pay the costs. The idea was discussed at RAC meetings, and RAC enthusiastically recommended such a conference. A steering committee, with Gorbach as chairman, was appointed.

About 50 participants and observers from the United States and abroad were invited to attend a 2-day conference in June.[12] As Gorbach later commented:

> They were not selected for their interest in molecular biology or recombinant DNA but for their expertise in clinical infectious diseases, epidemiology, gastroenterology, endocrinology, immunology, bacterial genetics and animal virology, mainly representing disciplines from whom little or no input had yet emerged during the recombinant DNA debate.

The researchers who gave presentations notably included E. S. Anderson, from the Central Public Health Laboratory in London, and H. Williams Smith, from Huntingdon, England. Both had presented their work before the Asilomar meeting (see chapter 1), but at Falmouth achieved greater attention.

Gorbach wrote me a five-page letter dated July 14, containing a summary report shortly after the meeting:

> The participants arrived at unanimous agreement that *E. coli* K-12 . . . cannot be converted into an epidemic pathogen by laboratory manipulations with DNA inserts and does not implant in the intestinal tract in man . . . with no evidence that non-transmissible plasmids can be spread from *E. coli* K-12 to other host bacteria in the gut . . . [and] cannot be induced to virulence. In summary, concerns about *E. coli* K-12 as an epidemic pathogen had been overstated at Asilomar and often thereafter. All the scientific evidence that has been accumulated provided assurance that *E. coli* K-12 is inherently enfeebled and not capable of pathogenic transformation by DNA insertion.[13]

The Lasting Pain from Falmouth

At the next RAC meeting in Bethesda, the RAC pursuit of revision was much influenced by fresh reports of attendees at the Falmouth meeting. I remember that awareness of these same reports led me to demand—perhaps better said, to *entreat*—Rowe and Martin to convince Gorbach to produce a detailed report of the proceedings as soon as possible, in anticipation of the need for documents to buttress the forthcoming issuance of the proposed revision of the Guidelines.

Gorbach's letter was clear and convincing to me, but as the weeks piled up, no more extensive report was forthcoming. I slowly realized that, of course, Sherwood Gorbach was a professional. As such, he could not be induced to risk sending us a manuscript condemned to printing and distribution for political purposes before it could legitimately appear in a top-flight peer-reviewed journal. The definitive publication appeared in May 1978,[14] long after the publication of the proposed revision of the Guidelines in the September 27, 1977, *Federal Register* and their review by the DAC 3 months later. As a firm adherent to the unbending protocol of scientific publication, I could understand but nevertheless feared that this could prove to be a costly charge of error in the coming days of public appraisal of the revision.

No one should expect a consensus judgment over the results of scientific experiments ever to be unanimous. A report of the absolution of *E. coli* K-12 that appeared in the *Washington Post* of July 19 quickly prompted a letter of rebuttal to NIH from another conference member at Falmouth:

> In absolutely no way do I consider the data presented at Falmouth as evidence in support of a relaxation of the current DNA guidelines on recombinant DNA research. . . . There remains the possibility that the recombinant DNA carried by these "safe" hosts can be transferred to more invasive strains of *E. coli*, other enteric bacterial species, or even the somatic cells of their metazoan hosts. . . . There was absolutely no consensus reached which suggested that the probability of transfer of chimeric DNA by plasmids was sufficiently low to be disregarded. . . .[15]

This letter was faithfully preserved as an exhibit for the meeting of the DAC now planned for December. Also preserved was the telegram received from Jonathan King, who attended the meeting but also did not share its consensus.

> As a participant in the Falmouth Conference on Risk Assessment of recombinant DNA research I am deeply concerned over incomplete and possibly misleading report of proceedings. Though conversion of *E. coli* K12 to pathogenic strains is not likely, there exists major problem of transfer of foreign genes to wild strains of *E. coli*. Presentations at Conference made clear that strains of *E coli* cause human disease. More detailed statements with others are in preparation.[16]

To the growing accumulation of comment upon this arguably "watershed" event in the recombinant DNA controversy was added a response to Jonathan King from Wallace Rowe, a member of the RAC and one of the conference organizers.

> Dear Jon,
>
> . . . What we are probably both afraid of is that the inference might be made that "Recombinant DNA Research is safe." The conference did not say this, and I'm not sure that it will ever be possible to say this is a complete generalization. What the Falmouth conference did say is that the odds are overwhelming that Recombinant DNA Research with K-12 and non-mobilizable plasmids (or EK-2 lambda phages) is *safe*, and that this can be critically tested by further *in vivo* testing. This cannot be extrapolated to viruses, to other bacteria, or even to other strains of *E. coli*. Further it does not mean that we can do away with guidelines or insistence that persons working with microorganisms follow responsible sterile techniques. However, this is an exhilarating message, as much in that it signals that rationality still has a chance to function, as in the security it gives that immensely exciting and important work will be able to proceed without so much anguish, anxiety and hassle.
>
> Wally [Rowe][17]

Roy Curtiss, given to widespread distribution of influential letters when he had a change of mind about certain risks, now wrote a letter (April 12) to Paul Berg that was also distributed to members of Congress. In contrast to his famous letter declaring abstinence from recombinant research upon the declaration of the original moratorium in 1974, Curtiss had come now to regard *E. coli* K-12 as harmless. Similar commentary from many other sources indicated that the new Guidelines should feature some relaxation of containment for experiments with *E. coli* K-12, the most commonly used host permissible under the rules.

Nothing provided more tension between more persons than the energy expended over differences in the view of *E. coli* K-12 as a host-vector system in recombinant DNA experiments. Wally Rowe's letter to Jonathan King could not remove the animus felt by King and his colleagues among the Science for the People adherents.

Already mentioned in an earlier chapter was the Academy Forum held about this time in the National Academy of Sciences.[18] It epitomized the manner of behavior and beliefs of dissident factions created by this revolutionary research. On March 7–9, 1977, the Great Hall was crowded with participants. The pictures of the scenes include shots of a troupe of young women, wearing signs proclaiming "We Will Not be Cloned," escorted by Jeremy Rifkin. Sitting next to them was Professor George Wald, wearing a peace sign. The whole effect had the air of an American scientific three-ring circus. On the last day I was invited to be the penultimate voice explaining the federal government's role. I soberly explained the actions of the RAC and ORDA. An old friend, Bengt Gustafson from the Karolinska Institute, mused that when I rose for the government, "It was like the Romans had come. . . ."[19] At the end of my talk, Ms. Francine Simring of the Friends of the Earth stood up ". . . to congratulate the NIH and Dr. Fredrickson on the wonderful job of disseminating materials, testimony, and xeroxes to all interested parties." Long a target of this redoubtable critic, I am recorded as having replied, "I'm glad to meet you at last, Ms. Simring, even at this distance."[20]

An observer later captured the dissonance of the period we were all experiencing. A perceptive scientist and writer, June Goodfield, reported sadly witnessing "present hardening of the lines, the present way in which people have tended now to come to feel that this is an issue over which there are victories to be won or victories to be lost . . . points only to be gained and points only to be conceded . . . is a very, very dangerous [course] . . . what we have to do is to regard this issue not as a battlefield, but much more as an opportunity to reassess the nature of the social contract between science and society."[21]

When Basic Definitions Were Challenged

It was during this period that Stanley Cohen informed both ORDA and sources on the staff of the Senate Health Subcommittee (see chapter 7) that even the definition of recombinant DNA might be in error. This, too, the RAC mulled over and added one solution to the questions piling up about revision. The director would therefore have to keep a list of "nonnovel" recombinants as a result.

By the end of 1977, NIH was supporting recombinant DNA work in no fewer than 230 projects in 110 institutions, and many others were taking place under support of the National Science Foundation and the Department of Agriculture. Similar research was also under way in some 20 countries abroad, mainly under the U.K. Guidelines, but the NIH Guidelines were also frequently spread across work tables in European laboratories. No convention had yet been convened to strive for parity between the two sets of rules for conducting such research. Moreover, Americans became increasingly aware that work being undertaken in Europe was prevented in the United States by more restrictive rules. None of the hypothetical dangers of recombining genes in the laboratory had yet appeared, and pressure for relaxation of the NIH Guidelines was rising throughout the research communities around the world.

Time To Test the Revision

After a labor of almost a year, the RAC was now content that it had a revision of the rules adjusted to the declining evidence of hazards from the research. The proposed text was sent to me by the RAC on September 1, 1977. The decision was not mine alone. In publishing the initial version of the Guidelines in 1976, we had first informed the department to get the secretary's blessing to publish. Two weeks after getting the RAC's proposal, I moved it simultaneously to Assistant Secretary for Health Richmond and to Secretary Califano with a draft announcement that proposed revision of the Guidelines would be published forthwith in the *Federal Register*.[22] The DAC would be convened in December for a public review of the changes. I thought that a final decision on their promulgation would await review of all the comments received, but I never contemplated that we'd still be another year in limbo.

Proposed Changes in the NIH Guidelines

The Recombinant DNA Research Guidelines proposed in the September 1977 *Federal Register*[23] did not, in my view, threaten to lessen our responsibility for the safety of the public and the environment and the protection of workers engaged in this kind of research. The proposed changes were extensive, largely concerned with technical details, and not easily summarized in either general or specific terms. Comprehensive analyses of the new document against the first set of rules were specially prepared, however, for public distribution and especially to guide the DAC for its review now set for December.[24]

The introduction to the proposed new Guidelines indicated that the present revisions took into account many communications from both scientists and nonscientists since the original publication of the Guidelines. More than 100 such comments had been received by March 1977.[24] The spectrum of these comments was very broad, but the single most mentioned subject of concern was rules involving experiments with *E. coli* K-12. None of the putative hazards had been encountered.

Other Changes in the Proposed Guidelines

Numerous changes in physical containment agreed upon in the joint EMBO/NIH meeting in March were adopted, the result being a tightening of the definitions of P1-P4 in numerous details and a greater uniformity around the world. As for biological containment, a revised nomenclature for host-vector (HV) systems, from HV1 to HV4, was provided with new definitions for classification.

Experimental Guidelines

Prohibited experiments were not changed for any classes, but a provision for possible exemption of the limit on large-scale fermentations was extended, provided the RAC made this recommendation to the director for final decision.[25] Numerous detailed experiments were permitted with one-step lowering of containment. The Falmouth conference results had particularly encouraged the RAC to remove *E. coli* K-12 from the provision of the Guidelines requiring data to be submitted for approval by the director for host-vector systems.

Roles and Responsibilities

The principal investigators were assigned new responsibility for safety and compliance with the rules. The institutions were also assigned a number of heretofore unspecified responsibilities, including training of staff in standards of microbiological safety practices, instituting necessary medical procedures for protection of the health of research personnel, broadening the expertise of members of the IBCs, and designating a biological safety officer for every institution having a P3 or P4 facility. The U.K. Williams report had placed a stress on institutional responsibilities in line with our gathering recognition that the behavior and most of the decisions made by investigators and institutional committees must be implemented peripherally, under local responsibility of the institution. The roles of the RAC were further clarified in making recommendations for revision of the Guidelines, performing the first step of a two-step procedure for certifying host-vector systems, and recommending any exemptions from the Guidelines to the director of NIH. As for the roles of the director, he was now acknowledged as having ultimate responsibility for HV certification and for "interpretations of the Guidelines."

The Price of Informal Rule Making by Comment

As the proposed revisions were sped around the land and to countries abroad, flocks of letters came back within the month carrying explicit critiques, indicating that molecular biologists and many laymen were reading every word of the compressed lines of 6-point type favored by the government's favorite publication. It was clear that we had so far successfully adhered to this part of the regulatory process in our determination to lash informal rule making to the full spirit of the APA, with complete and expedited attention to every comment. The objective was to preserve the opportunity to continuously adjust constraints in tandem with new information bearing on safety as well as research opportunity.

Keeping the Balances

As much as we were constantly at the side of the scientists to hasten an orderly forward movement of knowledge, we were also conscientiously trying to play the role of protector of the public interest. Members of the Kitchen RAC still read and discussed each and every one of the criticisms, concerns, and reassurances received from the correspondents, whatever the origin and nature. Each was recognized and commented upon. If we didn't understand an apparently serious question, we searched for wisdom of those around us. The scientific community, at least, knew what the Kitchen RAC was struggling with each day. Singer, Rowe, Martin, Gottesman, and others—Talbot and Gartland, particularly—not only worked slavishly but served as valuable conduits to the outside scientific communities.

Perishable Penmanship

The preparation of the "director's decisions," ultimately published in the *Federal Register* with each revision of the Guidelines, was a meticulous process. I genuinely felt I had to be intellectually comfortable with each word in the documents before they were published. I confess that each day that the cut-and-pasted drafts passed before my tired eyes, I was buoyed by the absurd belief that these responses would sometime be preserved and reread by posterity. I thought for example, of Judge Wolsey's decision permitting the admission of James Joyce's *Ulysses* to every library, a writ glorified by publishers and featured in all of the Clifton Fadiman anthologies that I had devoured in high school. No doubt, however, the *Federal Register* commentaries would soon be dust, having achieved their honest purposes. (I confess that in working on the July 1978 revision, I impulsively made a test of the attention of the contemporary audience by embedding in the text a small barb at Jim Watson for his theatrical recantation at the December 1977 meeting of the DAC.[26] Within a week I had a crisp note from Cold Spring Harbor and gained reassurance that the scientists at the time were reviewing the decisions set in 6-point type as critically as if they were peer reviews of their own scientific papers.)

There is no question that this method of inviting comments and public display of the answers to them was an exercise in communication far more valuable than mere window dressing to satisfy the APA. It sped up the exchange of information and views on the still-arcane rules for governing this new field of scientific endeavor that was presumed to be potentially dangerous. Moreover, it exposed the NIH authority to helpful criticism and advice.

The Public Comments

One may witness the essence of a few samples of correspondence to the NIH director in the 2 months after the proposed new Guidelines were published and prior to the DAC review:[27]

> *October 11:* "I urge that representatives of contrasting points of view be included to the greatest feasible extent at the December meeting of the Director's Advisory Committee (DAC). The breadth of representation at the February, 1976 DAC meeting was perhaps the decisive factor in establishing the legitimacy of the initial set of NIH Guidelines."[28]

> *October 13:* "I am delighted that the earlier commitment to review and modify the Guidelines in the light of more current assessments of risk has moved ahead so expeditiously. [But] why is the Director of NIH asked to prepare a list of those combinations of DNAs that are *not* considered novel and therefore to be excluded from coverage by the Guidelines?"[29]

> *October 18:* "I want to strongly support the conclusion of the RAC that 'everything they have learned tends to diminish our estimate of the risk associated with recombinant DNA in *E. coli* K-12.' I also support their conclusion that this does not mean that we should drop our guard entirely but . . . begin the process of reducing the required containment levels commensurate with our reduced estimate of risks."[30]

> *October 19:* "The Duke University Biohazards Committee proposes that all experiments involving cloning of *E. coli* K-12 DNA be subject to containment at the level of P1."[31]

> *November 2:* "*DNA containment page 49603 (5).* This section [pertaining to handling recombinant DNA molecules which do not contain viable organisms] is a disaster, and if it remains in the guidelines, it could undermine their credibility. DNA is *not* an *organism;* it is a *chemical.* Handling a chemical by 'good microbiological technique' is absolutely ridiculous."[32]

> *November 8:* "There *is no real justification* to include most novel recombinant DNA activities employing . . . *E. coli* K-12 EK1 or EK2 systems in the Guidelines . . . The Introduction states that 'the revised Guidelines continue to be deliberatively restrictive, with the intent of erring on the side of caution . . .' I have grave doubts whether scientists or the U.S. Government, represented by the NIH, have a right to 'intentionally err' just

> to accommodate irrational fears, concerns and politically motivated considerations."[33]
>
> *November 9:* ". . . The institutional committee established in connection with recombinant DNA should be called an *institutional biosafety committee* rather than a 'biohazards' committee . . . Continued use of the term 'biohazards committee' helps to foster fears which, in my opinion do not appear to be warranted."[34]

The stage is set; time for changing the Guidelines.

The DAC is Rolled Out Again

Still No Hazard

Since the last meeting of the DAC in February 1976, far more laboratories around the world had been pursuing recombinant DNA experiments, and the absence of any projected hazards was becoming embarrassingly noticeable. Nevertheless, throughout 1977, many of us were also on the Hill as Congress vigorously proposed laws to regulate the research (see chapter 7). This highly polarized field had led to organization and aggregation of groups of opponents, which, having failed to prevent the research from occurring, were now determined to prevent the Guidelines from being relaxed. The current applicability of the Guidelines only to government-supported research was a genuine cause of distress.

Changes in the Opposition

By 1977, the opposition to revision was never organized in a solid front, but the character of the hard core of opposition groups had shifted significantly. We were now encountering the opponents coalescing and concentrating behind the banners of certain major environmental groups, where the team approach was developing to a high degree of hostility. Discussions would find our opponents sometimes employing the "good cop, bad cop" dualism of interrogation and critique. Scientific issues were submerging beneath issues of process and form.

Public Airing, December 15–16, 1977

New Ingredients into the Mixing Bowl

Several veteran members of the February 1976 DAC meeting were willing to return for more.[35] Freshly recruited ad hoc members, including several from abroad, broadened the mix of points of view.[36] A dozen additional witnesses were invited to participate and give short presentations.[37] We had also prepared the table to accommodate as many voices as possible of "public witnesses" who requested to be heard.[38] Six members of the RAC (Gottesman, Helinski, Littlefield, Rowe, Walters, and Zaitlin), the safety expert Barkley, and Maxine Singer were also invited to present or answer questions about the proposed revisions to the Guidelines. Among many of the spectators around the big room in Building 31 at NIH were other members of the RAC, the Federal Interagency Committee, and congressional staff. Finally, two of the molecular biologists, Anna Skalka and John Tooze, joined the chairman of the Committee on Genetic Research (COGENE), W. J. Whelan, as observers for biologists in some 21 member countries in the International Council of Scientific Unions. Skalka's report published by COGENE[39] serves as an independent record of this event.

Documents

A great deal of material was provided to invited guests and participants for their review before the meeting. The information included the "NIH Environmental Impact Statement on NIH Guidelines for Research Involving Recombinant DNA Molecules"; a "Report on International Activities of the Federal Interagency Committee on Recombinant DNA Research"; "Background on the Proposed Revisions of the NIH Guidelines"; as well as letters received with comments on the proposed revised Guidelines.

A Brief Summary of the Meeting

Per habitude I was the chair of this meeting. It was not a smooth ride. If we may take just one or two topics for example.

E. coli *K-12*

A big green book containing the five-page letter from Sheldon Gorbach with his summary of the Falmouth conference was among the large amount of materials distributed and displayed. Donald Helinski, a man of even temperament and almost cherubic appearance, was asked to open this subject:[40]

> ***Dr. Helinski:*** . . . Another type of proposed revision in the experimental section that should be noted is the deletion of certain classes of experiments, and primarily this involves experiments which are concerned with DNA from prokaryotes that naturally exchange DNA with *E. coli*. . . . I stated earlier that the proposed lowering of containment levels in certain categories of experiments has as a major basis our increased confidence in

> the use of *E. coli* K-12 in the form of EKI and EK2 systems as a host for recombinant DNA experimentation. . . . Given the central role of *E. coli* K-12 in recombinant DNA research at the present time, our Committee has been actively soliciting information and new experimental data pertaining to the biological properties of this strain of *E. coli*. . . . The results of a variety of recent experiments with *E. coli* K-12, its Chi-1776 derivative, and various plasmid and phage vectors have greatly strengthened our confidence in the EK1 and EK2 host-vector systems for recombinant DNA research. . . . [p. 321]

The response was akin to the first tremors felt by the passengers on the Titanic.

> ***Ms. Pfund:*** The Guidelines have already been revised downward. Why is this? The introduction to the Guidelines states that the Advisory Committee based its revisions on new data which diminished the estimate of risk associated with recombinant DNA in *E. coli* K-12 systems. What are these data? Where are they published? Why have they not been widely circulated and subjected to comprehensive review by experts and the public alike? [p. 227]

> ***Mr. Hutt:*** Could you give us some idea, though, why the information in the Green Book was not put as a preamble to the *Federal Register* notice? [p. 229]

> ***Dr. Sinsheimer:*** It is not merely that the Green Book wasn't available with the *Federal Register*. It is also that many of the references in the Green Book are to unpublished materials. [p. 231]

> ***Dr. Ahmed:*** We have to address this question about peer review of scientific information, particularly non-published kinds of information that circulates within the scientific community. . . . You have to develop some mechanism by which the public out there can have access to information in the early stages. [p. 237]

Comments from invited and public witnesses took expected routes. Representatives of the Sierra Club, Friends of the Earth, and the Environmental Defense Fund took this opportunity to voice their opposition "to any loosening of the Guidelines and to call for enactment of federal legislation."

> ***Mr. Dach:*** The Guidelines were marred throughout by ambiguity and lack of clarity. . . . Obviously not written by people with any experience in enforcement . . . Loose language creates loopholes . . . It is clear that there is no adequate public representation on that Committee . . . nobody there who has experience in writing regulations, or writing legislation that later has to be transformed into law. [p. 232–233]

> ***Ms. Pfund:*** The fact that the Advisory Committee gained access to new information and used it in developing the new Guidelines without securing

outside comment while decisions were being made makes a mockery of public participation in this process. . . . the background document . . . says . . . "there are categories of experiments for which the containment levels are specifically mandated in the 1976 Guidelines, but for which some discretion is permitted in the proposed revisions." . . . What does that mean? . . . The logic and justification for such discretion should be matters for discussion of a wide range of individuals and not confined to an advisory committee whose composition and scope is not representative of the various positions and perspectives on this issue. [p. 228]

Dr. Helinski: There is always a question of what is the proper point for extensive public input in the case of revisions of the guidelines . . . knowing that this public hearing would be held to give everyone an opportunity to discuss the proposed revisions . . . [p. 229]

Dr. Shaw: Perhaps Ms. Pfund could provide the Committee . . . with specific recommendations as to at what points and how she would like the public to become involved. [p. 230]

Comments from others, including scientists from the United States and abroad, stressed the necessity for revision to keep the Guidelines abreast of new knowledge.

Dr. Davis: On scientific grounds, then, I would conclude that the original Guidelines have been far too severe, and that the proposed revisions are more than justified . . . we have been victims of rather gross misunderstanding of the problem, arising with the best of intentions. [p. 225]

Dr. Chilton: I favor the idea of excluding certain kinds of recombinant DNA experiments from regulation by the Guidelines. By excluding from consideration a large number of experiments which everyone agrees are innocuous, we are able to focus . . . on the more critical issues. [p. 360]

Dr. Tooze: On our side of the Atlantic we still believe we are talking about conjectural hazards and no proven hazards. We still think we are talking about guidelines and not laws. That colors the whole tone of the discussion. . . . [p. 347]

Dr. Skalka: In my opinion, there is little use in designing experimental Guidelines for "permissible experiments" which cannot be practically met. . . . Since we have heard that *E. coli* K-12 carrying a DNA insert cannot become an epidemic pathogen . . . the basis for special concern and conservatism in this area of research . . . seems to have disappeared. [p. 342]

Dr. Tooze: We believe that in all countries, America and all the European countries, the discussion of containment of viruses has been skipped over. [p. 347]

Dr. Ginsberg: I am completely unclear as to why polyoma has assumed such a high-risk status. Why are the Guidelines written like this? [p. 340]

Dr. Rowe: It's very hard to say, because that's over into the political area. [p. 340]

Dr. Ginsberg: Well, let's talk as scientists now. Let Don worry about the politics. [p. 340]

Dr. Rowe: Historically it is just the picture of viral genomes being delivered in new host-range systems that had many people concerned. . . . We very deliberately said let's placate the fears that were very clear at this time. . . . This was an over-political decision. You can't divorce the scientific from the political. We said this is an unpalatable type of scenario, and we had better . . . have some more data. [p. 340]

Dr. Rosenblith: We are faced with a draft and we are now hearing one testimony after another that this is a political draft. . . . What are we faced with, Mr. Chairman . . . would you care to explain . . . ? [p. 344]

Mr. Helms: I think that one of the things that would be a great shame here is to turn the scientific process into the political process. For one thing, lawyers would be out of jobs, but beyond that . . . [p. 345]

Dr. Fredrickson: Dr. Rowe, one of the ranking viral experts at NIH, is extremely conservative. In his role on the RAC in dealing with the containment of viruses he may have erred on the side of extreme conservatism, perhaps to avoid any appearance of conflict of interest. [p. 344]

Dr. Gustafson: I worry . . . about the use of the word "political" . . . It can mean finding the lowest common denominator between various pressure groups to which . . . the NIH is responsible. [However] NIH, as a public institution, has to be concerned about the political in another sense . . . namely what is the public good or the public well-being? . . . We have got to recognize that there are margins beyond the purely scientific, that it is the responsibility of an advisory committee to the Director of the NIH to take into account. From some perspectives this is finding compromises among interest groups . . . taking into account not merely the legitimate interests of the scientific community . . . but other legitimate values that people are concerned about in our society. [p. 345]

Mr. Dach: I think . . . [the need] to bring out the range of scientific opinion . . . is very crucial. . . . What we learn from medical ethics [is] that there really is no such thing as pure science, and it does us no good to . . . ask what is the scientific opinion here, and then let us overlay politics . . . on top. It is clear that values affect your interpretation of scientific information, and that should be made explicit rather than trying to be buried under the table. [p. 368]

Before the meeting adjourned, each member of the DAC was solicited for his or her opinions or final comment. In an article a month later, Nicholas Wade gave his impressions of "glamorous" comments made during this portion of the program.[41] Wade expressed the view that most biologists agreed that the rules should be relaxed, but he concentrated on the criticisms of how the NIH had gone about the revision. He emphasized Peter Hutt's negative views on the procedural basis of the proposed revisions and the obstinacy of the agency and its director in refusing to turn to Section 361 of the Public Health Service Act with its sweeping powers to control communicable diseases to regulate gene-splicing research.

As the meeting came to an end, the chairman did not struggle to offer a commentary on the likely fate of this revision of the Guidelines.

A Bad Dream: Post-Session Analysis

Immediately after the meeting of the DAC, I was exhausted and left for a few days of skiing. I left a record of that short vacation in an essay in my diary.[42]

> I had carried here much mental baggage. In the early morning of the second day I woke suddenly at a quarter to two. The sea of unconsciousness had rolled back and a vivid and complex dream was laid out before me just before I had fully awakened. . . . a morality play adopted in the wildly farcical style of Tom Stoppard appeared.[43]
>
> In the brief light of that Alpine moon I knew that I had, and lost again, a fleeting glimpse of the truth and folly of our recent groping for The Revision during an exhausting hearing . . . With firm belief that "we" were right and "they" were obviously wrong, the roles were starkly drawn. The dissenters . . . were clinging to Process, struggling to strangle us in a bog. . . . Drowsily I fought indulgence in psychology, dissecting the anatomy of dissent, a slippery and dangerous subject. . . . And I knew I was prime for similar scrutiny. . . . Once in the dream . . . I remembered feeling of pride in the deportment of the members of the RAC . . . glowing with well-scrubbed candor, sane . . . good manners even if appalled at the antagonism they encountered. . . .
>
> Which are the true Concerned? . . . A few nobles . . . Sinsheimer, Curtiss, other academicians . . . perhaps more ingenuous than the rest of the body of scientists, . . . they state openly that they don't share the pure intuitive optimism.
>
> Who are the Non-Scientist Dissenters? . . . Most are defenders of the special interests of nature and ecology . . . Skeptical knights in a perpetual crusade against polluters, special commercial interests and highly paid legal maneuverers. I would adopt their cause, but not their manners.
>
> On this . . . somnambulistic tour . . . among the stars, Watson was arguably first in magnitude, in reputation, and . . . on this day—in foolishness. He had confessed:
>
> > As one of the signers of the original moratorium, I apologize to society. . . . I think we are totally diverted away from the true mission of the

> National Institutes of Health . . . on . . . in a sense, a witchhunt, because there is no evidence that any danger exists.[44]
>
> The dramatic apology for "starting it all" was too late, much too late now to dry up all the anxiety flowing from Asilomar. Watson was highly effective theater, however. . . . With the piercing eyes, the wild hair, the childish stammer. He struck exposed nerve ends, because deep, deep down, nearly everyone said to himself that all the truly awful things like flu, Lassa virus, the killing Legionnaire bacilli, the rickettsia of tularemia, the plague . . . had learned hundreds of years ago that they could not have the full run of the earth, and were bound to strictly limited adaptations. Now kept mainly sleeping in tubes, they are not smarter than their human conquerors, even though these latter are themselves far from comprehending most of Nature. Nature has erected barriers from experimenting with genomic combinations for millennia. The ability to survive is the arbiter of what is, what once was, and what will be.

The Final Comments

Most participants sent a letter after arriving home. There were a few compliments on the quality of the meeting. I will add only a sample of the afterthoughts:[45]

> ***Dr. Neel:*** One of my . . . concerns is . . . the whole sweep of investigation concerning recombinant DNA. Be aware . . . of certain tensions within the genetic community, stemming from the intellectual arrogance of the molecular geneticist, flushed with the outstanding successes of reductionist biology, as contrasted (I hope) with a greater measure of humility of those of us who are man-oriented. [p. 137]
>
> ***Dr. Chilton:*** . . . was impressed by the clear and forceful presentation of dissenting views, made possible by the diverse interests represented. . . . The revised Guidelines appear to steer a cautious middle course. [p. 148–149]
>
> ***Mr. Dach:*** I enjoyed the opportunity to testify before the Advisory Committee and feel we have made some progress in addressing the failures of the proposed revised guidelines . . . I urge you not to finalize revisions of the Guidelines until the substantial modifications outlined herein. [p. 172–174]
>
> ***Ms. King:*** The NIH is to be congratulated. . . . Your efforts have certainly helped to allay fears and concerns. . . . However if the public and the scientific community are to continue to have faith in NIH policy . . . your policies and processes must be improved . . . particularly the process by which these particular revisions were generated. [p. 177–178]
>
> ***Mr. Thatcher:*** The NIH should take nominations for . . . [RAC] from the following groups: . . . labor unions . . . environmental groups, the knowl-

edgeable community governments such as Cambridge . . . and Ann Arbor, and groups of scientist activists such as Science for the People. . . . [p. 209]

Drs. Cavalieri/Rosenberg: Periodic updating and reconsideration of the Guidelines . . . is a necessary activity and we are glad to see that the process is taking place. . . . However . . . many members of the scientific community feel that there are broader issues to be considered . . . No controls have been set up governing the future non-research usages. . . . the potential dangers of every proposed industrial application of the technology . . . [p. 229–230]

Mr. Hutt: I wish to reiterate my general support both for the concept of the Guidelines and for the specific intention of the proposed modifications to reduce or eliminate current requirements wherever such reduction or elimination is scientifically justified. NIH was right when it concluded to issue the Guidelines, in order to reassure the public about the safety of this experimentation, and it is equally right in modifying the Guidelines to reflect new scientific information. . . . I have enormous admiration for the great amount of work that NIH and outside scientists have put into this effort, and believe that it deserves the praise and support of the entire public. [p. 239–256]

Hutt's message, inter alia, was that "Greater care must be taken in the future to reflect due process of law in proposing and adopting amendments." It was his opinion that "the regulatory or procedural aspects of the Guidelines do not meet the same high standards as the scientific aspects . . . understandable, since they were developed by scientists rather than by experts in administrative law." Hutt also was unquestionably correct in criticizing the feebleness of the preamble to the proposed revised Guidelines, both in general and for each specific revision, explaining the impact of changing each provision, the evidence for the change, and assurance of the decisional process. In short, we were acting under the penumbra of the APA and sometimes at the periphery. "By attempting to short-cut this process [of regulation], NIH has unfortunately acted in unseemly and unnecessary haste, and has given an impression that can only intensify the concern and suspicion of those who believe that inadequate controls exist for this type of this research."

Was our fight to keep these "rules" as guidelines and not regulations a quixotic one? Surely, Peter Hutt felt it was. Bruised from this obvious fall over procedural carelessness, I was determined that we should simply have to do better and stay the course.

DAC Dividend

The contentious talk about "political motivations" behind the prohibitive barriers for experiments with viral DNA led to swift action by Rowe and

Martin, in conjunction with Tooze of EMBO, to organize an international meeting on viruses and recombinant DNA. It took place in Astoria, England, January 26–28, 1978. The results were reviewed by a committee chaired by Harold Ginsberg in the first week in April. The outcome was a significant rationalization of lower viral containment levels and a removal of the "politics" stigma.

Cycling Back to the RAC

I was now stepping up communications with the RAC. A letter was sent to each RAC member from the director's office on April 12, 1978.

> . . . Over the past three months an analysis has been completed of all public testimony and correspondence received . . . [on the proposed revision] . . . I have met regularly with your chairman and other members of our staff to consider these comments. . . . I would very much appreciate your judgment on the possible revisions proposed in the enclosed document.[46]

Working from the Kitchen RAC analyses, Perpich and Talbot had compiled about 60 specific points for the RAC to consider. Perhaps just one of these can convey the flavor.

> Compliance. It has been suggested that a section on penalties be included in Part IV. Would the Committee object to a compliance section that would state that violation of the NIH Guidelines may result in suspension, limitation, or termination of NIH grants or contracts?[46]

In the same way, we queried the RAC over the whole range of changes that arose from public criticism. These included detailed clarification of roles and responsibilities, gradual but increasing shifting of central to local decision making at the institutions, especially in the IBCs. As suggested in commentary, the latter would now be designated "biosafety committees" to eliminate the "biohazard" connotation. Moreover, ascertainment by IBCs that Guidelines standards were being met was deemed desirable, for the pressures in Congress to increase national inspection and monitoring were ominous even as the perception of hazards diminished and experimentation was expanding the use of the technology. The RAC was also asked for their reaction to the suggestion that at least one public (outside) member be among the membership on the IBCs.

Director Meets with RAC

My next meeting with the RAC was held on April 27–28.[47] I had intended to leave a few remarks for the public record. Actually, I spoke for nearly an hour, with a summary of where we were and what we had learned from the last revision, including:

> There is no evidence to date that manipulations of recombinant DNA molecules yield any harmful products from research . . .
>
> . . . Although it is not possible to reduce the risks to zero, *the burden of proof should now shift to the opponents of the research*. . . . Guidelines in other countries place fewer restrictions on scientists. . . . Ours are too complex and must be revised on the basis of new evidence.
>
> There need to be . . . shifts in the locus of responsibility toward the institutions, necessary competence within the IBCs and mandated public participation for more local decision-making. . . . A compliance section with penalties for all violators of the Guidelines that could extend to cessation of all grant support. . . . The need to permit voluntary registration of non-NIH recombinant DNA research, including private sector research.
> . . .

A Sensitive Question

We were being pressed by Congress on the possibility of universal coverage under the Guidelines, including the question of voluntary compliance by industry. I asked the RAC to consider whether it would be willing to review industrial projects, keeping in mind "penalties for the unauthorized release of proprietary information. . . ."

The question of the review of projects involving proprietary information—with portions of the meeting to be closed to the public—met with various emotional responses from RAC members. Many expressed willingness to review requests from the private sector. Other members did not wish to be involved. We would be returning to this issue in earnest in 1979.

On the Writing of Guidelines for Science: Hans Stetten Tells a Luciferase Tale

There are numerous examples of the hobbling of important experiments by the now aging and sclerotic Guidelines. A particularly exasperating example of the imperative needs for revision can be drawn from the dilemma of William McElroy. He was famous for his ingenuity in recruiting fledgling biologists to bring fireflies to his laboratory in Johns Hopkins for some ingenious experiments. From these insects McElroy discovered the enzyme luciferase and learned how the enzyme lighted the lantern of the insect to its gentle glow in the summer evening. Now moved to a laboratory in San Diego, McElroy and his coworker Marlene DeLuca applied to NIH for permission to clone the gene coding for luciferase in a suitable host-vector system to obtain this valuable laboratory tool inexpensively. The 1976 Guidelines provided that insect DNA must be handled in a P3 laboratory. Alternatively, P2 conditions could be substituted in work with insects not harboring any pathogenic agents that had been *bred for at least 10 generations within the laboratory* to ensure that they were virus free. The firefly was considered harmless, but no one had ever tried to get the insect (with a 2-year life cycle) to breed in a laboratory for 10 generations. The experimentalists filed for a reduction to P2 conditions. The RAC was unable to devise any adverse scenarios and was unanimously in favor.

Stetten's comment:

> Since the Guidelines did not state that exceptions could be granted, the NIH Director ruled that exceptions would not be granted until the guidelines were formally revised. Despite the fact that a number of legally trained persons had been overseeing the activities of our Committee, no one had warned us of this complication. . . . A few weeks later . . . I learned that the . . . concept that breeding insects for ten generations in the laboratory should dilute out viral contaminants was itself false. . . . So we found ourselves trapped in an absurd quandary. . . . Because of a provision written into our guidelines, built on a premise proven untrue for a number of insect-virus associations, we were depriving the scientific community of the benefits . . . the experiment promised. . . .[48]

Dr. Stetten Steps Down

Although Hans Stetten ended his tale of the luciferase embargo with the statement that it provoked his resignation from the RAC, he was more extensive in his eloquent valedictory at the end of the RAC meeting on April 28, 1978.[49] His eyesight was failing, and he had grown fatigued and unhappy with the bureaucratic tendencies in the administration of the Guidelines. At the end of my discussion with the RAC, I closed with the sad announcement that he, who had been invaluable in starting up the RAC in 1975 and steering it through its tumultuous and demanding course, had asked to step down. We agreed with his suggestion of Dr. Jane Setlow as his replacement. It was the end of an era in the RAC.

A New Proposed Revision Enters

Over the next several months, involving the work of the RAC, its Kitchen variant, and others of the director's staff, there had been prepared a distinctly new revision of proposed Guidelines.[3] It was accompanied by an 82-page environmental impact assessment and introductory pages of a

288-page laboratory safety monograph (a monumentally revised and expanded replacement for the "Supplementary Information on Physical Containment" that had been part of the first Guidelines issued in 1976 and prepared by a special committee of safety and health experts working with Emmett Barkley).

The environmental impact assessment contained the following judgment: "As best can be determined from all evidence compiled to date and analyzed in numerous scientific and public forums, there will be no adverse environmental impact from Recombinant DNA Research conducted under the Director's proposed revisions." To this collection of documents were added the lengthy decision of the NIH director, my responses to all comments received on the previous draft, and, at the beginning, a draft release statement for the secretary.

Prior to their publication in the *Federal Register*, all these materials were passed from NIH to the department through the assistant secretary for health about the first of June. Two weeks later, I visited Richmond and his deputy, Charles Miller, to urge them to move this precious package along. They had no objection to my assisting the delivery upward, and I scheduled a meeting with the secretary on June 9.

Joe Perpich and I found the secretary in an affable and discursive mood, seemingly not affected by the bulk of the papers in our possession. When Rick Cotton arrived, we delivered the packet into his hands—Guidelines in yellow, drafts of the secretary's and NIH director's decisions in red, and the environmental impact assessment in green. Relieved that this enormous task had at least proceeded this far, we passed on to a long discussion with the secretary on the course of the second session of the Congress.

Recombinant Odyssey: Beijing

A personal invitation for an excursion to China during July 5–10, 1978, was the result of advice to President Carter from his assistant for national security affairs, Mr. Zbigniew Brzezinski.[50] Upon returning from a trip to China in May of that year, he was encouraged to take another careful step to continue a move toward full normalization of diplomatic relations with China, begun with President Nixon's historic visit with Chou En-Lai in the People's Republic in 1972. Recognizing the high interest of China in science and technology, Mr. Brzezinski had chosen this subject for the first official delegation to visit China since the Nixon visit. Our United States Science and Technology Delegation to the People's Republic of China would consist of a 4-day visit by the heads of the major U.S. federal science agencies with counterparts in China.[51]

Guidelines as Fellow Traveler?

The news of the impending trip was soon in the Washington papers.[52] On the morning of June 29, I had a call from Secretary Califano to wish me a good trip. William Foege, head of CDC, and George Lythcott of the Health Services Administration were already in China on separate missions, and he observed glumly that everybody gets to China but him. I promised to bring him some chopsticks and pick out a good hotel for his trip. Then I was inspired to say, "Joe, I'd like to take the proposed revised DNA guidelines to China, if you'd let me release them." He replied, "Take them and release if we send you a cable." I had hoped that this might move the revision further down the tracks in my absence.

This apparent coup had an immediate effect on the upper staff of the department. Within 30 minutes, a call came from Rick Cotton, executive secretary, who wanted me to get the (latest) revision to him that weekend. One hour later, Peter Libassi, on a conference call with fellow attorney Jim Hinchman, screamed in mock rage that "You're not going to China; [it's an] outrageous ploy to get the Guidelines through; the summit of a mountain of unbearable devices to get your way!"

Beijing

Upon our arrival, Air Force One reassuringly took a berth at the airport. We were greeted by Leonard Woodcock, the U.S. liaison officer, and his wife, we not yet having an ambassador to the People's Republic. Placed in a long line of limousines, we were driven to the sumptuous Friendship House that had been the accommodations for President Nixon. We had several meetings with our Chinese counterparts, mine being primarily Huang-Chia-Szu, president of the Chinese Academy of Medical Sciences and Kuo Hsing-Hsein, director of the First Bureau, Chinese Academy of Sciences. We had an extensive meeting on July 8, followed by the official toasts and traditional dinner in the People's Palace as the guests of Vice Premier Fang-Yi, ninth in rank in the government and a principal spokesman for science in China.

The last day we returned to the palace for a rare treat, an hour's audience with Vice Premier Deng Xsiao Ping, nominally "Number Two," but judged to be the most important man in China. This self-proclaimed "Small

Bottle" looked fit for his 73 years. We sat in a circle, each in a comfortable deep-upholstered chair with an adjacent white-enameled cuspidor. Deng was relaxed in our presence, chain-smoking, his right foot moving rapidly to and fro. When we were introduced to him individually and our functions described by Frank Press, Deng responded with a grave nod and an expressionless appraisal of each of us. Press, in metaphor fitting the grandeur of the setting, remarked, "Mr. Vice Premier, these are the most powerful leaders of American science; and their being away means that all the laboratories and science establishments in America are closed down for 5 days." Vice Premier Deng replied without a drop in cadence, "You could stop for 5 years and we would still be 50 years behind you."

Shortly after we arrived in Beijing, a cable arrived, addressed to me.

> E.O. 11632: N/8
> Tags: OEXC CH—US
> SUBJECT: DNA GUIDELINE REVISIONS
>
> 1. The Department has been asked to pass the following message to Dr. Fredrickson, Director, NIH, member of the Frank Press delegation.
> 2. Quote: I have further reviewed the materials you sent on the proposed revisions of the NIH Guidelines of Recombinant DNA Research.
> 3. Further discussion of the proposed revisions are needed before final approval. We need to discuss the proposed exemptions of research under the guidelines, the standards and procedures that should govern such exemptions, and the HPE [*sic*] that I should give to the revisions.
> 4. Do not, repeat, do not release the proposed revisions to the Chinese, so that we will have full opportunity to discuss the proposals thoroughly on your return before they are issued. Joseph A. Califano, Jr.
>
> End Quote. Vance.

At Mr. Woodcock's office, the duty officer felt that my proposed reply might be considered as too obvious code and inappropriate for official cable traffic. Thus, on my return I used a teletype at NIH to simulate a "telegram" directed to the secretary. It was printed on a rice paper temple rubbing, containing the following message that was carried to the department:

> Quote: Per Instructions, Paper Tiger Still in Cage. Propose Earliest Discussions of Revisions. D.S. Fredrickson. End Quote

The secretary's cable arrived just before the joint meeting with the Chinese on biomedical research at which I was to present the U.S. side. After a long description of NIH, I turned to the subject of recombinant DNA, alluding first to the interest of the Chinese in nitrogen fixation to improve agriculture. Then moving briefly to the subject of Asilomar and to the U.S. Guidelines, I handed out copies of the 1976 version. I said that the rules were presently being revised and that I had hoped to bring a copy of the new Guidelines. "However, there are technical details which have to be reviewed," and I promised to send them when published "in several weeks." During the Chinese presentation, Professor Huang said they were hoping to begin using these techniques in China, but such laboratories had not yet started. The representative from the main Chinese academy had nothing to add except that they were eager for exchange of such knowledge with America.

Home, Appointment with Secretary

On the return flight, I called on July 11 from Fairbanks, Alaska, and asked my secretary, Bel Ceja, to schedule an appointment with Secretary Califano on the 13th. I stopped first by Rick Cotton's office to talk about the secretary's hang-up over showing the revised Guidelines to the Chinese. Cotton felt it was the secretary's recent experience with the release of the swine flu vaccine[53] and a need to signal his concern and insistence on procedure and process before issuing the proposed revised Guidelines. He then told me Califano now wanted another hearing on the proposed revision, this time chaired by general counsel as a formal departmental review. Rick sensitively suggested that I might wish to be cochair of the committee. He had called Julius Richmond to learn whether he would object. Richmond, in turn, called me and graciously withdrew his hierarchical claim. We agreed to help write a secretarial memo to precede the release of the proposed Guidelines.

I then met with Secretary Califano alone to tell him about my impressions of China. I presented him with a silk banner with Mao's picture along with one of the latter's little red books.

Notes of an Impatient Nuisance

My diary reveals evidence of a developing neurosis during the next 2-week period.[54]

> By the next afternoon, July 14 (Friday), there is nothing from Cotton on the essential Secretarial preface. An ominous sign, because he is a superb draftsman and very fast. His secretary says Rick "is terribly busy." I ask her to tell him that I'll be down the next morning (Saturday) to assist him. Rick himself comes on the phone, perhaps taken aback by my aggressiveness. He agrees we meet at 11 o'clock tomorrow. Joe Perpich readily agrees to come with me.

Saturday morning at HEW, Rick's secretary says she is just finishing typing a draft of the preface. . . . It is perfectly and fairly done. We agree on all terms, there will be a hearing on October 12 and final comments will be closed within 45 days thereafter. Thus all sides are now committed and only JAC has to sign on Monday and we will at last go to the *Federal Register.*

There was an interesting rumor in the [Washington] *Star* yesterday that Attorney General Griffen Bell was to leave, Califano to become Attorney General and Paul Rogers will take HEW. Could this be going to happen? Doc Dazansky [fabled political sage and owner of the Georgetown Pharmacy] does say that JAC would like eventually to go to the Supreme Court. Can you imagine the Senate confirmation hearing if something went wrong with rDNA?

> Senator Kennedy asks the first question: "Now Mr. Califano, there is this matter of the revision of the Guidelines, your releasing them "radically revised" just before the *E. coli* mutant spread eastward from Palo Alto. Were you exercising proper procedural caution there?"

Joe had to be on the defensive, I understand that, and find his manner of doing so at once graceful and annoying. Will the scientific community understand?

Maxine Singer calls me from the Gordon Conference on Nucleic Acids.

I say to her: "Maxine be careful what you say this year. I remember what you did in 1973!"

She replied: "I'm saying nothing this year, being especially careful." She is sending a summary of the conference regarding the unrest at the endless delay in the release of the guidelines. . . . Maxine has already spoken to the scientists to comment on the revision proposed.

July 20, 1978. Yesterday at 3:35 we were notified, after a day of cliff-hanging, JAC had finally signed off on the Guidelines, thus moving them into the *Register.* The Secretary's initial statement was an appeal for comments, especially on the sections that establish the mechanisms for administering and revising the Guidelines. And then he defined the review he had in mind:

> To review the comments on the proposed revision, I am establishing a departmental review committee, consisting of Mr. Peter Libassi, the Department's General Counsel (Chairperson); Dr. Donald Fredrickson, the Director of NIH (Vice Chairperson); Dr. Julius Richmond, Assistant Secretary for Health, and Dr. Henry Aaron, Assistant Secretary for Planning and Evaluation. I have asked this committee to hold a public hearing to insure full and complete opportunity for comment. In order to hold this hearing and to issue the revised guidelines on a reasonably prompt schedule, no extension of the 60-day limit for public comment will be possible.[3]

At last, we might be within sight of land.

The "Libassi Hearing"

> Conference Room 529
> Hubert H. Humphrey Building
> 200 Independence Avenue, S.W.
> Washington D.C.
> 9:20 a.m., September 15, 1978
>
> My name is Peter Libassi. I am the General Counsel of HEW and the chairman of this Task Force. On my immediate right is Dr. Fredrickson, Director of the National Institutes of Health, who is Vice Chairman of the Task Force with me. On my left is Dr. Julius Richmond, the Assistant Secretary for Health and the Surgeon General. And on my far right is Dr. Henry Aaron, the Assistant Secretary for Planning and Evaluation.
>
> The purpose of this hearing this morning is to receive public comments on the National Institutes of Health proposed revisions in the guidelines for the conducting of recombinant DNA research.
>
> Secretary Califano has asked that we hold this hearing in connection with NIH's solicitation of public comments on these revisions [that] reflect NIH's experience with the current guidelines since they were published in 1976 . . . the mechanism for administering and revising the guidelines . . . procedures . . . for otherwise prohibited experiments . . . for exempting certain classes of research . . . standards for the exercise of administrative discretion under the guidelines by the Director of NIH, and for the composition of the various public, scientific and advisory bodies mentioned in the guidelines, particularly the Recombinant DNA Advisory Committee, and Institutional Biosafety Committees.[4]

Thus opened a novel reception of public comments on proposed revisions of the NIH Guidelines. Such hearings had previously all taken place at the campus of NIH, with the director in the chair and acting as editor-in-chief of whatever conclusions were reached and subsequently disseminated. But here was an opportunity for critics of all stripes to get the direct attention of the HEW secretary through the first ranks of his staff, his chief attorney, the surgeon general of the United States, and another high official who was a social scientist.

The reader has already met Libassi (graduate of Yale Law School), one of the attorneys appointed by Secretary Califano to key staff positions in 1977. Julius Richmond, M.D., was a distinguished child psychiatrist and head of the Judge Baker Guidance Clinic at Harvard, when he was appointed surgeon general and assistant secretary for health in 1977. Earlier we were fellow members of the Institute of Medicine, and his gracious willingness to take over my post as president temporarily had enabled me to fill the vacant chair of the NIH director. Although my superior, Richmond wisely left the curatorship of the recombinant DNA problems in my hands, recognizing that the task had become virtually a full-time job. Henry

Aaron (Ph.D.) was an economist whom Secretary Califano had appointed assistant secretary for planning and evaluation. I knew him less well, but his previous post at the Brookings Institution and his general reputation recommended him as a successful third leg of a table at which law, medicine, and social science could hear petitions for and against changes in the Guidelines.

The charge to the task force was threefold. First, members would take testimony from the witnesses who came to the Humphrey Building in person on this day.[55] Second, they would read the approximately 170 letters sent to NIH from August through October, containing comments on the proposed revised Guidelines which had been published in the *Federal Register* of July 28. Third, the task force would then analyze the criticisms and comments and edit the Guidelines accordingly for presentation of a final revision to the secretary for his approval. The target date for completion was mid-November.

Records

The transcript of the testimony at this September 15 hearing[4] and extensions of the remarks of witnesses and the voluminous text of the letters containing comments[56] are today preserved in separate volumes of *Recombinant DNA Research.* A separate record of the task force analysis of testimony at the hearing also survives.[57] (Excerpts selected from these documents are displayed in Appendixes 9.1 and 9.2 at the end of the book.)

Testimony before the Task Force

Opponents and proponents of the proposed revised Guidelines appeared in about equal numbers and were heard in no particular order. Each was given 10 minutes and the privilege of leaving extensions of their views. No repartee was permitted by the chair, but many of the participants were already more than familiar with the positions of their fellow speakers.

Hearing Records

The analyses of the testimony conducted by the task force appear in part in Appendix 9.1. Some of the flavor of the hearing and letters received can be captured in the following four cameos:

- Witness Congressman Richard Ottinger appeared distraught and angry:

> . . . Quite simply, the whole matter is rapidly getting out of control, and it is at this juncture appropriate for HEW to take the reins and protect the public from any further loss of control until Congress acts. The failure of Congress . . . is not one upon which HEW should now base any excuses for inaction. . . .

. . . The failure of Congress resulted from some of the most vigorous, and in many instances distasteful, lobbying I have ever seen on any issue. When the scientific community establishment gangs up on Congress saying that its expertise is absolute, and that there is no other recourse for protecting the public, it seems clear that Congress is hard pressed to resist. . . . On the other hand we heard from some of the most eminent scientists in the country and in the world, who testified that there were potential dangers of very grave magnitude. . . ."[4] [p. 120–122]

- Witness T. Blessing, of Washtenaw County, Mich., displays a protest of preemption:

Resolution of the Ann Arbor Michigan Democratic Party, April 1978.
. . . It is the responsibility of the Ann Arbor City Council to ensure the health, welfare, and safety of its citizens. To fulfill this responsibility, the Ann Arbor Democrats believe that private and public recombinant DNA research should be subject to local review and approval."[4] [p. s359–s361]

- Witness Carl E. Kline, of Western Maryland Clergy and Laity Concerned, declaims a broader protest:

. . . We are convinced that ultimate value lies in 'life itself'; all of it is good, not simply life that is human, and we are not advancing if we increase the quality of human life and further threaten life itself. Deeper perhaps than that, as we move closer to genetic engineering, we are convinced there is something irreparably human about human genetic mistakes.[4] [p. 371]

- My diary note of September 15, 1978:

The environmentalists prefer to arrive in teams . . . Scientists and friends of science come usually alone. Donald Brown, molecular biologist from the Carnegie Institution of Washington, bitterly opposed to the Guidelines in the beginning, acknowledged surprise at the sensitivity of the draft revisions. In the afternoon, a tall man rose to speak quietly and firmly about biology, the promise and the responsibility. By the time Frank Ruddle of New Haven had put up his gold reading glasses and made his short statement, my companions at table stirred and asked who he was. Informed he was Professor of Biology at Yale and a foremost somatic cell geneticist, Libassi murmured, "I am impressed."[58]

Adjudication of the Issues

Under Libassi's direction, a decision document was created from the hearing record. (See Appendix 9.1.) Issues were highlighted and analyzed, and recommendations were made as to possible alteration of the Guidelines. In technical areas of changing containment standards for specific experiments, the environmental impact assessment prepared for this revision was often used for reference. Excerpts from this decision document illustrate the application of a formal legal analysis to recombinant DNA matters.

Comments by Letter

Beyond the publication in the *Federal Register* of the proposal for revision of the Guidelines, no campaign was conducted to elicit comments. But the letters poured in. Comments were predominantly from scientists in academic institutions, but others came from government agencies, environmental organizations, pharmaceutical interests, scientific organizations in the United States and Europe, such as the International Council of Scientific Unions, INSERM (France), and EMBO, and from private citizens. Each of these letters was considered by the Kitchen RAC as well as the HEW task force, and responses were prepared for the director's decision document. Excerpts from letters commenting on these topics are located in Appendix 9.2, coded by page number in the full collection arrayed in the indicated volume of *Recombinant DNA Research.*

The Libassi hearing completed 3 full years of conversations and debates over restraints placed against experiments. We had arrived at a display of excess that was harming the advance of the science. There were (and there still are) unknowns remaining, but the awkward excess in the Guidelines needed now to be peeled away. The next moves would take place largely in the office of the secretary of HEW. The endgame included some surprises.

10

A Major Action

1978–79

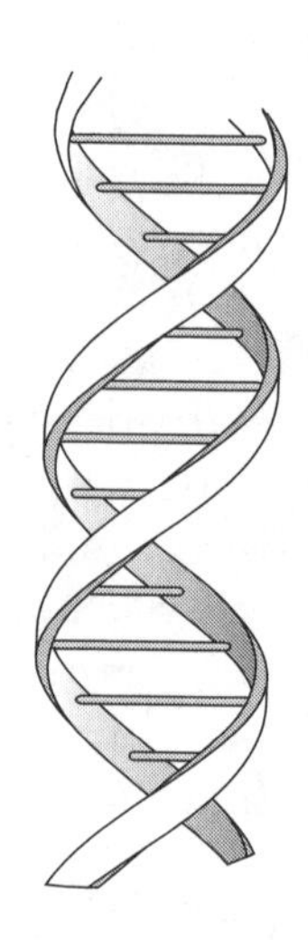

Habemus Regimentum Recombinatum![1]

Federal Interagency Committee (FIC) Unrest

The letter from Senator Adlai Stevenson commenting on the proposed revision of the Guidelines (see Appendix 9.2) was not neglected. Included was a suggestion that the days of the recombinant DNA hearings in the Congress might not yet be ended. More important, it was evident that some of our fellow agencies were conveying complaints of neglect to the senator. NIH had indeed had its hands full the last half year and needed to meet again with its sister agencies. Comments from the director of the National Science Foundation (NSF) (see Appendix 9.2) and the Department of Agriculture indicated areas particularly needing our attention to hold our coalition together.

Department of Agriculture

The agricultural scientists were represented in the commentary harvest. After years of making new species by their own early kinds of manipulation of genes without the oppression of rules or guidelines, these workers found themselves threatened. During the imposition of the Guidelines, the plant scientists were always cooperative but ever wary of what NIH might do to complicate their research. Now they were particularly concerned about the problems of release into the environment and wanted to avoid administrative encumbrances they never had before. The EPA and OSHA, as well as the FDA, also were anxious of failing their own mandates as revision proceeded.

National Science Foundation

NIH's most significant partner in the support of recombinant DNA research was the National Science Foundation. Herman Lewis had been their representative at the meetings preceding Asilomar, and his cadre of molecular biologists supported by NSF was steadily growing. It was not easy for NSF to have to depend upon NIH for more and more decisions affecting its scientists. In a letter commenting on the revisions (see Appendix 9.2), Richard Atkinson, the NSF director, revealed his discomfort with some areas of the NIH-NSF interface in the rules for recombinant DNA research.[2] The FIC had not met for nearly a year, and it was clearly time for another plenary session.

Ninth Meeting of the FIC, October 12, 1978

As the FIC gathered in one of the large conference rooms in Building 31 at NIH,[3] news from Stockholm rippled through the conversation of the attendees. Nobel Prizes for 1978 had been awarded to Werner Arber of the University of Basel, for discovery of restriction enzymes; Hamilton Smith of Johns Hopkins, for confirmation of their existence; and Dan Nathans, also of Hopkins, for using the enzymes to study SV40 DNA. Nothing could serve better to highlight how the business of the committee dealt with the most exciting new paradigm in science.

Review

Because the FIC had not met for so long, I felt constrained to open with a reminder to the members of their collective achievements during the 2 years they had been a unique consortium of federal agencies. As they knew, in the spring of 1976, well before our charter had been obtained from the president, we—all the potential federal supporters and regulators of recombinant DNA research—had met at NIH and taken the first steps to avert any chaotic struggling over regulation of this new science. Our first meeting had concluded with a general agreement that all agencies would abide by the NIH Guidelines. Since then our general counsels had spent hours in analysis of the enabling statutes of each agency. It became clear and generally accepted that none had a claim to writing regulations covering all the research. A few months later, Congress groped for statutory control. As we listened to the congressional debates and became aware of the desire for the Guidelines to cover private recombinant research as well, all of us who had sought to avoid regulation by law gave up that ideal and, acting together, found agreement on the minimal requirements of such a statute.

Then we passed them to the HEW secretary as our recommendation for an administration bill. Although numerous bills were considered in Congress, it now appeared that no legislation would be passed. Thus, the FIC—if it continued to work together—would have to continue to oversee the research without a law. For some months now, input from the member agencies had been primarily concerned with congressional actions, less with the affairs of the Recombinant DNA Advisory Committee (RAC).

A new phase was beginning, with revised Guidelines proposed by the RAC in September 1977. FIC members had participated in the December meeting of the Director's Advisory Committee on the revision and, in 1978, had been commentators of the further proposed revision, recently the subject of the HEW task force hearing. Speaking then about some of the proposed changes in the Guidelines, I said the changes offered new opportunities to retain our cooperative relationship with enhancement of the roles of the FIC members under the new Guidelines.

Liaison Memberships

We had decided that all member agencies would be invited to be represented by a liaison member on the RAC. These memberships would be nonvoting because of the large dilution of participants, but such liaison members would be invited to take part in the discussions, express the concerns of their agencies, and, in a real way, represent their diverse and separate communities of interest. The agencies would also be encouraged to contribute annual nominations of replacements for voting RAC members.

Through their liaison members, all federal agencies would be allowed to participate in RAC recommendations, such as the committee's interpretations of the Guidelines, adjustment of containment levels, recommendations for waivers of prohibitions, additions to the list of exemptions, certification of new host-vector systems, and other proposed changes in the Guidelines. There was also the possibility of consensus meetings of the FIC on RAC actions, such as a recommendation to release a recombinant organism into the environment.

Appeal Mechanisms

A three-step appeal mechanism for RAC actions would be provided to federal agencies: (i) consultation with the Office of Recombinant DNA Activities; (ii) direct communication with the director, NIH; and (iii) the privilege of any agency to request a meeting of the FIC to consider a RAC recommendation of particular concern to them or their constituents.

Registration

It being generally agreed that a registry containing the type and location of recombinant DNA experiments could be valuable and perhaps vital should unforeseen events give reality to any of the hypothetical dangers against which containment measures now loomed as overkill, NIH now maintained such a registry. In institutions where NIH funded recombinant DNA research, two registration options were proposed:

- Federal agencies may elect to keep their own registries and share information with NIH.
- If an agency decides not to keep a registry, institutions will be asked to send copies of their memoranda of understanding both to NIH and to the funding agency.

The agencies were surveyed for further comments on the revision pending before the task force. Concern on specific points remained, but the proposed changes appeared acceptable to the Agriculture Department, which had formed a committee to address recombinant DNA research and also to have a liaison member on the RAC. Herman Lewis said he felt that the proposed procedures addressed most of the points in the Atkinson letter[2] concerning the needs of the NSF. Agencies agreed to study the proposed changes and send in their written comments.

The Shark

It was now mid-October. Peter Libassi would make a few more appearances but was hibernating with the proposed revised Guidelines package to make a serious go at editing. Eventually, pieces of the work and some vociferous reactions came our way. Joe Perpich and Bernie Talbot nearly always accompanied me on my frequent visits to the sixth floor halls of the Humphrey Building to be sure nothing was needed, given the urgency felt by all participants. It is during this time that I learned I had earned the sobriquet of “the Shark” for my restless patrolling offshore.

The Endgame

In mid-October the movement of documentary records into the yellow volumes of official archives (*Recombinant DNA Research*) essentially ceased for the balance of the year. A single record of the frenetic march from draft to revision appears to have been preserved in the diaries in green government journals which I had begun when I became the director of NIH. The story, heavily edited, follows.[4]

> *October 13.* The Task Force has met with the PMA [Pharmaceutical Manufacturers Association] and other representatives from the budding biotechnology houses on the application of Guidelines to industry. By the end I had decided that the only way out of a crippling delay in revision . . . was temporarily to sever the issue of NIH trying to protect proprietary rights. . . . We could not guarantee to protect trade secrets with our present authorities at NIH. To request statutory power to do so will move us backward to where we started. Libassi has accepted this severance for now, and promises "We will later explore a solution for this."
>
> *November 12.* The Secretary left last week, for a two-week swing through Israel, Egypt, and on to Rome, "to meet the latest Pope." He deftly ducked my offer to get him the revised Guidelines by pouch to Cairo. Last Thursday, I met Libassi in his office, with Richmond and Aaron. With a long face Libassi said that he'd just read part IV of the revised Guidelines and that we "had given everything away to the IBCs!" The next three hours we went over each line of part IV again, piece by piece, and dissipated much of Libassi's anxieties. The final version was essentially as before. We were able to agree that no Biological Safety Officers were necessary in P2 containment, that public notice of IBC meetings would be the rule, but the question of open meetings was at the discretion of the chair. I think we saved all the reasonable rules and got rid of those likely to produce regulatory tetany. Libassi was still skeptical of the heavy tilt of the commentary [see Appendix 9.2] toward desirability of moving more responsibility to the institutions.
>
> *November 14.* . . . Joe, Bernie and I journeyed to Humphrey *again* to meet with Libassi, at his insistence. The first half hour Pete asked me to talk to him alone. He has relapsed, is glum and depressed.
>
> "How many times have you people written regulations?" he asked.
>
> "Never, furthermore we don't want to. Regulatory activities and high science are absolutely incompatible. Guidelines, yes; regulations, no. . . ."
>
> "I've been talking to Halvorson," I said further. "Adlai Stevenson is being revved up by Stewart and Merritt on his staff to have another go. If legislation comes again," I said, "CDC can have the whole field to itself."
>
> "*You've given it all away, totally deregulated through the IBC's,*" moaned Libassi again.
>
> Each revision to ease Libassi's anxieties leaves the Guidelines sounding more and more like marine barracks orders as the verbs are tilted toward the imperatives. Perhaps tomorrow will be the last day. When Rick Cotton joins us I ask him to develop the safeguards so we can get back our lost delegations to appoint new members to the RAC. It is now down to 9 members (7 scientists). Libassi had not realized a delay was still in effect and cried, "you must get the delegations back."

November 15. Meetings were held early today with Libassi and Cotton. RAC membership was the key issue. I had settled for 16 scientists of *all* persuasions and kept the union rep out so far. (I was opposed to having to balance labor and management on the RAC.) Libassi wants to keep his "pledge of impartiality" and wishes to hear both sides. He feels he must talk to some scientists apart just as he'd spoken with the environmentalists alone. So, we have invited several to meet with him in my office next Monday. On Tuesday he will go up to brief the Hill staff. Next week we will have 150 chairmen of all IBCs in Reston for briefings.[5] And we'll arrange an appearance for Libassi, if he agrees.

November 20. It's 4 pm . . . a November dusk in my office. The lamps illuminate the Vanderveldes on the wall and the rug on the book table. In my usual chair sits Peter Libassi. He faces Ed Adelberg [former RAC member and chairman of the IBC at Yale], Hal Halvorson [ex-president of the American Society for Microbiology], and Dan Nathans, soon to go to Stockholm. Adjacent to him is Phil Leder, a mild-mannered molecular biologist *enfant terrible*, soon to leave NIH for Harvard. Around the side sits Maxine Singer, along with Kitchen RAC veterans Bill Carrigan, Joe Perpich, Bernie Talbot, Dick Riseberg, and I in the background. It is Libassi's first informal face-to-face encounter with extramural scientists on the Guidelines. . . .

The proceduralists and other post-hearing critics have already had maximum effect, causing the revisions to be loaded with imperatives, criteria, standards, procedures, checks-and-balances, and enough infra-institutional tensions to assure that the golden dove of the New Biology does not soar too high.

One-by-one the scientists chip away at the Guidelines with Libassi. In their view the worst offenders are the 'shalls' [the verb form mysteriously replacing the 'shoulds'], the requirement for prior approval of each new P2 and P3 facility. As the meeting goes on, Gartland phones in from Reston, where the 150 IBC chairmen are waiting for Libassi to address them. They are up in arms about some of the very things we're discussing and they "want to caucus." We wonder if Perpich or I will be needed. I cannot go, and Bernie, who is less flappable, will drive Peter there.

There is a phone call for Libassi, he is told he must call such-and-such a number in California *before* going to the IBC meeting. Nancy Pfund has been at the meeting in Reston and has mobilized the Sierra Club to apply "moral suasion to stiffen Libassi's resistance in his encounter with the scientific community." Libassi proved up to the challenge.

November 26. Tuesday, the 21st, I appeared before the IBC reps as the last participant on the platform. The meeting was friendly. A number of questions were asked. I expressed the hope that we could give them a 60 day period of grace to reform their committees to operate along the lines of the new flexibility and greater responsibility. I also stated that I thought the pre-approval time of their rulings by the NIH would be reduced [as it was in negotiations with Libassi after his appearance at this same meeting]. We ended on an inspirational note with my voicing my pride in them and acknowledging the essentiality of acceptance of these rules. Some of the scientist IBC members indicated little belief in the need for the Guidelines but a commendable willingness to follow them as long as we (NIH) indicated we realized the trouble they created. I urged them to be patient, confessing my belief that the guidelines would some day be looked upon as an aberration of the late 20th century with my name below them! That evening I checked with Talbot. He, Libassi, and Perpich had gone to meet with the Hill staff and there were no controversies, but *later* Libassi had insisted on a moratorium on RAC procedures *until* a new RAC was appointed. I regarded this as a potentially disastrous delay if Califano & Co. do not get their act together and accept *my* appointments speedily [presuming, of course, redelegation of the appointment authority to me]. I asked Talbot to get our list of nominees in order for our meeting with Califano tomorrow. If the appointments drag, the IBCs, the RAC and the scientists will begin to subvert the game mercilessly.

"Science depends on both optimism and rationality," I mused, "qualities placed in jeopardy by unrelieved foot-dragging by bureaucracy."

November 27. A bright November day after all. Meeting has been called by the Secretary. First met with Libassi for some *more* last-minute changes in procedures. . . . standards are now back to "no significant" instead of "no serious" risk. I have warned Libassi and Richmond in separate calls that RAC vacancies *must* be filled without *any* delay if we are to avoid a breach of faith and loss of control over the scientific community.

The Secretary arrived at his appointment 45 minutes late. Attending were the Secretary, Cotton, Talbot, Libassi, Richmond, DSF, Aaron, Warden, Perpich, Lowenstein, FDA Administrator Donald Kennedy and his chief counsel, Richard Cooper. The latter two came because FDA is to prepare an NPRM [Notice of Proposed Rule Making] designed to assure that recombinant DNA experimentation in pharmaceutical houses would adhere to the NIH Guidelines. This was the compromise agreed to by the Department to extend the coverage into a major sector of private research. FDA was to make it a condition of review and acceptance of any drug achieved by using recombinant DNA technology.

In a good mood, the Secretary reported that on his trip he was met in Poland and Israel and told the news that the Guidelines were on his desk for his signature when he returned. "So you followed me there, Don!" he chortled.

Miffed, I reminded him that the Guidelines were important to Polish and Israeli scientists, too. He was surprised I seemed so hypersensitive to "kidding." I had grown a little ragged over these trying days. (I was even tempted to ask him what the Pope had thought about the Guidelines.) JAC is now worried about 1) Stevenson threatening hearings again; and 2) that some disaster won't follow all this action on guidelines. He now wanted the FDA notice to occur simultaneously with the release of the final Guidelines . . . "all out in one whack!" (It will be fought by some in industry, may excite the EPA, but it is a good idea.) I insist that the Guidelines must not be held up for a week or more, and agreed to accept the FDA's promise to "get something together" within 72 hours. We then have a long discussion about several petty guidelines issues. One is the question of the granting of prior approval rights to the IBCs. Califano asks, "Why not wait 30 days for the NIH to approve centrally?" The second problem is the requirement of a quorum should at least one public member not be in attendance at the IBC meetings. I'm opposed to both conditions.

Finally we moved to the question of RAC appointments and I insist we *must* make them *now*. Surprisingly, JAC Jr. says "Go and get Peter Bell and let's get them done now." In the end we all agree that the *whole* package, including RAC appointments, could be ready this weekend for the Secretary to peruse and announce by the following *Wednesday!* Mirabile!

Rick goes to "torque" Bell. I waited a diplomatic 10 minutes and then begin to work on Bell and Bruce Wolff together with the material we have prepared for filling the Committee vacancies. I reminded them that I had suggested to them that one of the new appointments be Ray Thornton, the retiring chair of the Science and Technology Subcommittee in the House, who had managed the 12 days of hearings of this Subcommittee with great sensitivity.

Prior Encounters

A few words should be interjected here to assure the reader that encounters between junior department staffers and the NIH director concerning our nominations for advisory committees were traditional tests of savoir faire. The courtiers stood between us and the secretary's approval of every list.

Earlier there had been more vexing exercises. The previous July, the appointment of all HEW advisory groups (including the many peer review committees of scientists who review the scientific merit of grant applications to NIH) had come to a grinding halt when the delegations were suspended. NIH was blocked from filling the slates of 25 study sections, an act tantamount to halting all rail traffic into Chicago.[6] Earlier at FDA, Commissioner Kennedy had lost his delegation to appoint members to 12 FDA review groups. The reason given was that we were failing to appoint sufficient numbers of women, African Americans, and Hispanics. A team existed in the department, more or less under the direction of Peter Bell, that was the contact point for the expected protests from the agencies. They also had become adept at providing lists of substitute candidates meeting the criteria of gender and ethnicity but not always the technical qualifications. In this particular July episode, we protested that the pool of qualified candidates was not sufficient to meet the ethnic or gender quotas. A month later, after asking the study sections to scour harder, I met with the secretary and his team to argue our cases. By the end of the day, I had my delegations returned. When Donald Kennedy called to ask how we came out, he was chagrined that he had not pursued his delegations more aggressively. The truth, of course, was that neither we nor the office of the secretary had means to create sufficient pools of specialized cadres needed for advisory groups of the science agencies. Now we were dealing with the RAC, a target of diverging pressures for appointments. The veterans of Asilomar felt that the RAC must not lose any of its elite majority of biologists. Others wanted addition of more persons with broader political and social interests.

> *November 30.* On Wednesday I met with Sheila Pires, Bruce Wolff, Peter Bell, and Rick Cotton (with Talbot & Gartland as my spearbearers). These staff members inform us that they have decided that the new RAC shall have 8 "cloners," 6 other scientists and 6 lay or "public" members. They also have a list of their favorites for our comment. I responded with a little lecture on competencies required, reading to them the Guidelines section on duties, and spoke of holes in the present list of specialties. Bruce Wolff rejoined that I can always get expertise to review what the RAC does and that I must be sure that its appearance, at least, is "correct." I felt the RAC had behaved very correctly, and insisted that "the RAC must be the prism for adjusting the course in recombinant DNA research by using scientific results to adjust the rules."
>
> Wolff then made a shocking proposal that Susan Gottesman and Wally Rowe, two members of the NIH intramural scientific staff, must be dismissed from the RAC because they have an appearance of "conflict of interest." "*Conflict of what,*" I demanded, protesting vigorously. Bell realized that I will certainly not yield to this without cause. We leave the subject momentarily and go on to the names they suggest.
>
> First, among scientists: they want Robert Sinsheimer. I agree to try to get him, but noting the pessimism he feels about the subject, I'm unsure of his willingness. They also want Matt Meselson, and I agree readily. Patricia King and Ray Thornton are bought readily. They then raise questions about two others. . . . I struggle briefly and get them to dismiss both. Rick Cotton liked Ahmed and Novick, whom I suggested as alternatives. They want a *unionist* and I opt instead for a *lab worker*, otherwise we'd have to get an industrial representative. Sheila produces two younger suggestions, Piñon of San Diego, who has a good recommendation from Helinski and Rodriguez. Gartland says that he also has worked with Boyer and judges him suitable. The Department team has two grander suggestions, Philip Morrison, the "cosmologist" at MIT, and May, a theoretical ecologist at Princeton. We agree they should go on a list to call.
>
> Today I made some calls. I reached Bob Sinsheimer in Santa Barbara. He doesn't have the time since he is now Chancellor, and though he appreciates the offer, he cannot accept. Ray Thornton said he "would like

very much" to participate; Meselson, who was on his way to the dentist, said "no" because he is off to a sabbatical at Cal Tech. Meanwhile, Gartland has collected Novick, Piñon and Rodriguez and is pursuing Mays.

I then go off to have more talks with Pires and Wolff on some other substitute for Sinsheimer. (Should the Undersecretary call Sinsheimer?, they wonder. The Undersecretary refuses to do so.) We are very close to finishing this last candidate, if only Pat King, whom I led to the edge of acceptance this morning, agrees to be a member by Saturday.

Tonight Wolff has called me at home to renew his demands for the "skins" of the NIHers on the RAC. What a pity when the substance of science is set aside to serve political appearance. I thought, "Working scientists, who have placed their few tiles in the eternal mosaic, do not know how fortunate they are." Alex Rich asks me tonight on the phone, "who will write the book on the fiasco of '78?" He asked, obviously well-informed about the contest in progress.

December 5. . . . Because everyone in the Secretary's office, including the Secretary, was struggling with President Carter's decision to rescind pieces of the budget, I've been away from the Guidelines until this morning. First, I aroused Peter Bell who was slightly sleepy and suspicious. Ms. Pires has apparently given over another day to work on our selections for the RAC.

"All right Peter," I ask, "may we say that by 5 p.m. tonight there will be a *final* RAC?"

"Y-y-yes, by 5 p.m.," he replies.

Next is Libassi whom I raise after 3 calls, the last at 8:55. Sheepish Peter. He has not yet (!) "gotten the package" to Cotton. [Joe Perpich had placed in my hand the bulky rewrite of the Guidelines, Statement of the Director—which I had carefully edited once more—and the Task Force Decision Statement.]

"Peter, for two years I've inched this missile to the launching pad. Now it's slowly draining out its propellant while you dawdle over the 'Secretary's statement.' "

"I've promised you, I now promise again: I'll get it to Cotton today. Joe must go ahead with a press conference."

Later Cotton returns my call.

"I've not received [from Libassi] Joe Califano's statement yet, and he will want to study it carefully. He goes at dawn tomorrow to Memphis. [The Democratic midterm conference where he debates Kennedy on Health Insurance.] It will be Monday or Tuesday before he will get to the Guidelines statement!"

"I'm *very* disappointed Rick . . ."

Noon is dragging. Pires calls, she is sending her own slate to the Secretary. He can choose hers or mine. [Are they really of equal weight, I ponder?] According to Talbot she will make four changes:

1) Nathans to replace Baltimore. [I explain this to Baltimore who calls me.] He's seen a list, but is worried about a commercial interest he may get in rDNA.

I reply "You may have a conflict, for which there are procedures to cope, but you also may be bumped for Nathans. Hold on, though, I will need you like a rock if Nathans cannot serve. . . ." I explain that I have arranged for Nathans to call me at 9 a.m. because I knew he was at the Grand Hotel in Stockholm, getting his formals pressed for the Nobel dinner on Sunday;

2) Pires is getting some Agricultural soil man at Davis for our alternate. Fine, if the man is competent;

3) She suggests another acceptable switch, and lastly

4) an unacceptable suggestion.

I am glum and disconsolate. Bernie is tolerant and Joe tries to comfort me with reminders of all we have achieved. Joe has also just noted FDA's intent to go to the Federal Register, with their controversial draft NPRM to cover *all* recombinant inventions, *before* the Guidelines are released.

Maxine Singer calls me at 5:00 p.m. It's dark and the lights are down to keep the pain from my eyes.

"What do you want us to do?" she asks, "Wally and I are ready."

"Don't march yet, I don't know what you can or should do. I have to think."

Maxine is sympathetic with my agonies, but cold. She has heard all of Pires' list and Wolff's misreading of *conflict of interest.* If we should yield, our small failures will dwarf our gigantic achievements.

December 6. Tonight there was a reception at Ted Kennedy's house for the departing Paul Rogers, who will not run again and leaves a big hole in the rational forces who guard health sciences.

I am in a far better mood tonight. Libassi phoned this a.m. Maxine Singer had called him last night and convinced him that Baltimore must stay and that NIH'ers must neither be removed from the committee nor prevented from future appointments. . . . Libassi promised to convince Joe Califano that the latter *must* see us on Monday or Tuesday and clear the package. Later we heard that Cotton felt Joe could not be disturbed next week—I said another week was impossible—that I was leaving the country. Libassi agreed to arrange a meeting between Cotton, Libassi, Bell, his team and me on Friday to get the committee done. At 2 p.m., I was in session with the Dental Institute—note fluttered to my desk from heaven, "Califano on phone." He was speaking about the wife of a dear friend who was ill. . . . Then, he turned to how awful the budget hearings with OMB were going—stacks of appeals, no understanding.

"Joe, may I speak to you briefly on one more thing."

"Of course."

"The DNA papers are the victims of a distressing snafu—we must move them early next week."

"Where are they? What's happened, I'll call . . ."

"I plead no special case—only give us a meeting, no later than Tuesday."

"Absolutely, I'll call Cotton now."

[Hallelujah, visions of opening clouds]

After seven, note to call Cotton A.S.A.P.

Cotton: "Joe wants to finish the DNA."

"Great, Rick. I'm at your disposal—we must see that the package is ready by Friday c.o.b."

"O.K." he agrees.

I call Libassi. "Joe want papers out by Monday. Or Tuesday. Please set up meeting with Rick to finish all outstanding points by Friday."

Tomorrow (Thursday) we're at Libassi's office—now we may just finish the job.

I've asked Dan Nathans to call me. He's at the Grand Hotel, in Stockholm, ready to get his Nobel Prize. He desperately did not want a place on the RAC . . . I could not press him; he's earned a choice. . . . Now we must keep Baltimore. Tomorrow's session will be very interesting—and I do not feel constrained to be too kind.

December 7. It was another long morning in Libassi's office. Wolff, Pires, Libassi and I met, with Cotton occasionally present. The debate was tough.

First, Wolff wished to air the principle that RAC was a public body advising me, different from our purely technical study sections on which the NIH staff has served as experts. He acknowledged that RAC was a hybrid, requiring complex technical advice. Weary, I was close to conceding that NIH experts might not need to serve, but I swerved back to reality and rejected the false concept of these scientists being different from their external peers. Libassi, smarting from his chat with Maxine, saved me from concessions on this point and proposed that the issue of NIH scientists, as members of advisory groups, be taken up with Califano separately. At the least, the current NIH members would serve out their terms.

Second, David Baltimore, seen as a "hawk" by the "other side," was the subject of continuing contest. I racked my brain for someone who might be an alternate for him. Baltimore would be a big loss in my view.

Peter Libassi, still impressed by Ruddle from his performance at the public hearing, suggested him. Ruddle is not strictly a "cloner" but such a first class person that I suggested Talbot call and ask him if he would serve on the RAC. Krimsky of Tufts, a member of the Coalition for Responsible Genetic Research, viewed the scene clearly, even if through slightly tinted lenses. He was judged agreeable to all.

The afternoon was marred by an angry call from Peter Bell. Jane Setlow, chairman of the RAC, had called to berate him for "what political people, your people, have forced on the RAC." Since Jane revealed she had a list of the candidates under discussion, including Krimsky, it would appear that Bell was right in assuming that she'd gotten the list from NIH. I was angry at Setlow myself. [She had called me when I was busy and had obviously not waited for me to get back to her.] Bell is right, he should not be beset by scientists who have gotten lists from all the calls made by Pires, Gartland or Talbot. I apologized. However, Chris Russell of the *Washington Star* and Harold Schmeck of the *New York Times* are now on the phone compiling a story of "NIH v. HEW" on the new RAC. We are scheduled to finish the package with Joe Califano at 2:45 on Monday. It should be in the Federal Register on Tuesday the 19th, *if* we aren't torpedoed. The siege is on, and the Departmental juniors may yet have to learn the wrath of the scientists.

December 8. A climax began at 8:15 in the morning when Peter Bell called me in the car on my way to the Department. He was rewriting a memo from Pires to the Secretary on the RAC. Ruddle was in for the "Baltimore/ Nathans seat," and the NIHers are *on.* Peter thinks at the end of each battle that I'm completely satisfied with the outcome, when in fact, I'm licking my wounds and planning how to recover my losses by some other means. I explain why Baltimore's loss is still serious. Bell reads to me bits of the Pires memo on the NIHers over the phone All the jagged edges of the adversarial clash are still there, a broken puzzle for the Secretary to *choose* to piece together in 20 minutes. I protest hard. There must be a consensus memo prepared. I want one; he wants one and we must strive for it. I warn him of the reporters, who will seek to fan the embers of an NIH-HEW split. We tussle and then finally agree on wording. He goes back to the typewriter. We will have at least 10 more incremental skirmishes this day.

Next Libassi calls. He wants me to "call off the running dogs." I explain that I've not started any attack. I, in turn, complain of the latest Pires memo to the Secretary and explain how both Ruddle and Baltimore would contribute to an excellent committee.

Bell is back and he reads a proposed cover note to the Secretary, how "[DSF] agrees we now have an *excellent* committee." I reply, "*good*, except for the 'adjectival excess.' " Bell takes umbrage and charges me with being fickle. I say, "Peter, I mean this! With Ruddle we have a good committee, with Baltimore a good committee, and with Ruddle/Baltimore together, an excellent committee." He has now delivered Baltimore back to me and is justifiably miffed that I seem never to be satisfied.

Rick Cotton, the inexhaustible workhorse, calls to complain. It seems OSTP and Frank Press and everyone else are calling in about the "HEW staff attempting to load the RAC with creeps and thus to cripple it." Rick is transparently accusing me of orchestrating this. I deny this but warn him of the possibilities that the journalists have had a hand in sharing names, but he is not to be mollified. I insist that, no matter, the package be done tonight. Bell calls to say, "If Joe sees something bad in the news, he may junk the whole thing." I warn him, as I warned Cotton, of the rebellion that is going on among the scientists. "Once the genie is out of the bottle, I cannot force it back."

Peter Libassi calls to ask me to alert our Press office for Joe's statement. He is working to finish the whole package for JAC, Jr. and to meet him at the airport with it tonight.

I sat down with Bernie to tote up the RAC that we now appeared to have. We have completed it (except for the addition of Parkinson at Pitt, an industrial hygienist, vouched for by several knowledgeable scientists). I am happy with this compromise and have changed the totals on my sheet to read *16* scientists and 7 others.

Don Kennedy also called early this morning. The scientists at Stanford have been getting on him. He asked for an update of the struggle, most of which was news to him. I declined to give him the tentative list of the new RAC members. However, I said that I thought we were not losing the game, and winning most of the scrimmages.

Late in the afternoon, Libassi, drowning in more calls, asked me to "calm things down a bit." I expect that he meant Maxine, and said we were in agreement on the Committee as it now stands. I later called Maxine and told her that Baltimore would almost certainly be on.

She agreed and said that David Hamburg had been helpful during the last 8 hours. I suggested that everyone should rest, lest we upset a now desirable detente. She said, "no one plans to move, as of now." These silent armies! We wished each other a quiet weekend.

At home at 6:00, Peter Libassi once more called to ask if Schmeck had talked to me. I said "No," and that "all the lions have gone back into their caves to rest." His chuckle is reassuring. There will be a package to Joe tonight and the Monday meeting is on.

I shake hands with Bernie Talbot on the way out tonight. Could this really be the last weekend?

December 13. Herewith the previous three days of the saga:

Monday, the meeting with Califano began with an ominous one hour delay. It seemed that JAC had arrived at 3:30 and had decided to go jogging for 10 minutes with Ben Heinemann. Joe Perpich and Bernie Talbot were with me, bearing multiple copies of the final documents. We stared at each other as the clock ticked. At last, the Secretary and Heinemann showed up at the Conference Room in sweaty jogging suits. JAC was visibly upset, nervously drumming on the table and obviously unbriefed. He gestured to me to summarize a set of memos that were proposed in final by his local staff. I suggested that Libassi give a better precis of them. Peter started to read his earlier transmittal memo, but Califano shifts quickly to the "final committee RAC memo." Bell starts to describe the RAC memoranda, including Pires' memo. "Baltimore is on and the RAC is now as we have all agreed . . ."

"How many Blacks?" JAC asked suddenly.

"One."

"Not enough! We've got to have at least four!!"

All of us stared incredulously. Slowly I suggested we might be able to find a total of 3. The nervous drumming increased.

"Jesus, Don, we've got to have at least 4; we've got to increase Blacks on *all* these committees of yours."

The situation is evidently going askew due to some recent tensions. The reports of the weekend trip to Memphis ran through my mind. There, I heard later, President Carter had received tepid applause and Ted Kennedy had stolen the convention by his remarks on health.

Ways to reach a possible solution began to flood through our minds. One switch could be made fairly easily, leaving two appointments to go. We decide what to do, while JAC is biting his nails on the other side of the table. I suggest we all agree to sever announcement of the committee appointments from the other materials to be submitted to the Federal Register. This was accepted.

Now a second bombshell explodes. It is an FDA memo to JAC. The Secretary takes out the memo and turns to Commissioner Kennedy and me:

"How many private concerns will this [memo] leave uncovered?"

"Maybe one or two," Kennedy replied.

"I must call the EPA to propose the same rule as FDA has here."

I choked for a moment, thinking of Doug Costle and his meager EPA scientific staff. What sort of internecine struggle are we headed for now, after 2 years of avoiding the turf wars? (Costle agrees to have something appropriate drafted, but is anxious about his authorities over industry.) I turn to JAC to say that we *must* announce to the press this weekend. He is angry—but not too much.

"Shall it be Friday released for the Sunday papers?" the Secretary wonders aloud, staying in command of his political instincts. "No," he decides, "that's Vocational Ed; should be Sunday or Monday?" Califano tells Donald Kennedy and me to prepare page of implementation in the Guidelines on voluntary compliance. He then leaves abruptly, promising me that "it will be done this week!"

Bell, Talbot and I huddle, judging how best to meet the new demand . . . Talbot stays on to help. I go home with Perpich, exhausted with frustration.

Tuesday, Talbot started early with Pires to clean up the wreckage of the RAC and pursue the Secretary's demand. Bell tells me afterward, somewhat apologetically, that yesterday, just before the disastrous "final" DNA meeting we had with him, JAC Jr. had met at the White House with President Carter and the NAACP. There was enormous pressure placed on him by the President. Certainly this could explain the decisions Joe made yesterday while steaming in his running suit. He has been made the scapegoat of Carter austerity and Kennedy hubris. I don't envy him.

Great fiddling all afternoon in order to get signatures accepting the Guidelines. JAC is now going to sign them!

Wednesday, Cotton promises me to take the DNA press release to Califano at 8:30 am. I arrange to edit the memo on the newly expanded RAC. Talbot has found several very satisfactory alternates. Wolff says, "Congressional checks are not complete. They will be ready by tomorrow noon (Thursday)."

Thinking of my own private obligations, I fear disaster. "Doubtless I'll leave for Europe on Saturday with those poor Guidelines stuck in Califano's pear tree, along with two turtle doves, DNA science in the U.S., and the last shred of my reputation."

December 14. . . . No news all day. . .

At 8:10 p.m. Libassi calls me at home [with a burst of gallows humor], "We have got to get one Japanese on that Committee."

"There are no Arabs, either—not even one Egyptian," I retort glumly.

Libassi continues, "*Califano has signed off on it all!!* Copies go to press tomorrow with embargo until Sunday's newspapers—the names of RAC members are not yet through Congressional checks—be released a day later."

I call Joe, Bernie, at once. We've got to release two years of tension. No way to expel a puff of white smoke from Building One, after all these months of deep black mental emissions signaling disappointments. We are exulted, and, with the aid of the former priest, Charles McCarthy, Director of OPRR, a simple message is composed: "Habemus Regimentum Recombinatum . . . ," a mock telegram to the Secretary on that handy Western Union machine in Building One. (Joe Califano never commented on my sophomoric releases.) It didn't matter. My fealty carried me through more than the long recombinant war. The high jinks were enough to buffer the stresses and disappointments on the way.

Denouement

The elation was quickly denied. Two days earlier I had learned that Maxine Singer had just written a very critical editorial for *Science* on the state of the Guidelines revisions. (She had earlier told me, upon receiving an invitation from Phil Abelson to write the editorial, that she would yield to me, if I wanted to write it.)

After perusing the draft which she brought to me, I brusquely, with no soft soap, stated three concerns that I had with it.

"There were the errors in *fact*. Two NIHers have *not* been asked to resign."

"Wally said he *thought* he was being asked to resign."

"I deny that. Sue certainly was *not* asked." Maxine admitted that the editorial gave the wrong impression.

"Secondly, no outstanding scientists were *denied* membership by the Department."

"And what about Baltimore?" Singer asked.

"Baltimore is *on* the new RAC list. Third, the ending of the editorial is destructively pessimistic. It suggests that the Guidelines will never work now."

I urged her to use her First Amendment rights. She had, after all, been indispensable in carrying NIH through the war she had helped begin.

The Singer editorial appeared on January 5, 1979.[7]

Editorial. Since 1973 substantial effort has been made to deal prudently with the concern that recombinant DNA experiments might prove hazardous. Unprecedented restriction of experiments was asked of a community bred for independence of mind and spirit. Despite widespread skepticism that hazards actually existed, the effort to minimize risk was a remarkable success, primarily because of the respect accorded to those scientists who called for prudence and later to the National Institutes of Health (NIH) as a scientific institution. To produce standards of laboratory safety, NIH formed the Recombinant DNA Advisory Committee (RAC) made of up prominent scientists, who developed the Guidelines for recombinant DNA research published by NIH in June 1976. Under the Guidelines work has proceeded safely and research accomplishments have been spectacular.

The confidence of the scientific community in the wisdom behind these efforts is rapidly eroding. Responsibility for the erosion lies with the Secretary of Health, Education, and Welfare (HEW) and his staff, who have now assumed direct supervision of recombinant DNA policy-making, and who have adopted procedures unsuitable to the complex problem of controlling creative activities. Two developments that have already had unnecessary adverse consequences are particularly disturbing.

First, there has been a long delay in promulgating amply justified revisions of the Guidelines. These were produced in July 1978 by NIH and its advisory committees, who gave unprecedented attention to comment by the scientific and general public and produced a document with broad support. Yet HEW imposed still another round of review; assurances that it would be expeditious proved empty. Experiments critical for realizing the practical and intellectual promise of recombinant DNA and for making risk assessments have been held up for months.

Second, the authority to appoint RAC members has been transferred from NIH to HEW. Contradicting its own definition of good process, HEW considered new members without adequately consulting the scientific public and in disregard of much of NIH's advice. In response to NIH urging and intervention by an alerted community, the most misguided inclinations were finally corrected. These had included questioning of the qualifications of an eminent molecular biologist, himself one of the first to call for caution; questioning the independence of NIH scientists serving as RAC members; and consideration of individuals known to be intractably opposed to the research. Nevertheless, the insensitivity of HEW to vital issues is still apparent in the makeup of the new RAC. The RAC's job will be to advise the director of NIH on highly technical matters. The revised Guidelines require that 20 percent of its members be non-scientists and that major actions be published for comment before adoption. This should have been sufficient to protect the public interest and still allow for the expertise required to deal with scientific matters. But under recent proposals the new RAC (about 25 members) will have seven to nine (depending on how one counts) non-scientists and only token representation of the molecular biology of eukaryotes. In the absence of adequate expertise in relevant areas and with lack of sufficient distinguished scientific leadership, it will be difficult for the reconstituted RAC to win respect for the Guidelines.

Lincoln Kirstein wrote: "Despite the populist politicians, certain crafts must live by elitist criteria." He included science as one of those crafts. When an egalitarian and humane society decides to support such a craft, public officials have the delicate task of nurturing elitist criteria while protecting the general interest. In the present case the two are interdependent, since safety depends on the diligence and therefore the confidence of individual investigators. Neither scientific criteria nor the public interest has been well served by HEW's actions.

Rebuttal

Shortly after the appearance of the editorial, I wrote a response but decided not to send it for publication. I sent only one copy, to Maxine.[8]

Notes on an Editorial. In an editorial in *Science* (January 5, 1979) Singer has described her perceptions of the outcome of the two-year effort culminating last month in new NIH Guidelines for Recombinant DNA Research. Under the title "Spectacular Science and Ponderous Process," she gives her account in the tones of a requiem for lost independence and time, and implies that NIH (acting for science) has yielded its sword to HEW bureaucracy. My memories of the ordeal cannot be very different from hers, but I have assessed the outcome differently. I think we have won a significant victory over a dangerously excessive reaction first set in motion by scientists. NIH will shortly issue the fourth volume of its public record on "the DNA controversy." From such raw material contemporary scientists or future historians must draw their own conclusions about the setting in which the 1976 rules were revised and whether the changes introduced in the final review process were necessary or even worthwhile.

Singer states that the HEW Secretary and his staff "have now assumed direct supervision of recombinant DNA policy-making." The new guidelines, however, continue the full delegation of responsibility for administration to the NIH Director and now give him the authorities for revision that the old guidelines so seriously lacked. The director must exercise his discretion under demanding procedures and standards. These have been created, however, to reassure the public and Federal agencies that they may participate, and are not designed to lever the Secretary and his staff into the chain of decision-making. The detailed analysis led by the HEW General Counsel undoubtedly added a few weeks to the long period of revision. A modest additional burden of procedural safeguards was one result. Another was the stripping away of any grounds for further complaint from the most fervent dissident that the public had not been exhaustively consulted or that adequate provisions had now not been made for external surveillance of what scientists are doing with techniques for recombinant genes.

As part of the latter concession, the Recombinant DNA Advisory Committee (RAC) was changed. Generally a two-level review has been accepted as the model for scientific advisory apparatus, the first a technical review by the experts (the study section), followed by the policy review involving both scientists and laymen (the Advisory Council). The Director's Advisory Committee (DAC) previously was used to fill this bicameral requirement under the old NIH guidelines. The 1976 rules, however, were devoid of most discretionary authority. Our experience with the DAC also suggested that DNA technology was too complex a subject for non-experts to cope with under the traditional two-tier system. It emerged that, when the non-expert is unable to comprehend so much of the details, his "public policy role" must be performed in the midst of the experts. Here, at the least, the layman is in a position to see if the experts appear to be listening to each other and paying attention to the evidence. Moreover, for performance of the case-by-case analysis required of RAC, and upon which forward movement in many laboratories is dependent, two separate stages of advice preceding the director's review and decision will create intolerable delays and confusion.

Singer is right, though, to sense that compression of RAC into a single, mixed advisory group is dangerous. . . .

In the anxiety created by the reconstitution of the RAC, some of the major achievements of revision have gone without notice. The new guidelines retain every important substantive change proposed by NIH and its scientific advisors. There is provision for continuous and orderly evolution of the rules—even to their eventual elimination when the need disappears. Many experiments now judged to be harmless are exempted, and reductions in containment for many important kinds of experiments have been decided upon. The discretion and responsibility for following the rules has begun to return to the research institutions, where they properly belong.

Probably the exemplary cooperation and patience of the scientists whose work has been at stake during this long evaluation have contributed most to achieve restoration of public confidence on the side of science. This being so, it is disappointing that an elitist image of science, including a sacred right of self-determination in technical matters, should be cast as the ultimate argument against the new use of the guidelines. It seems to me that the survival of the elite in our time must depend more on the ability to accept popular challenge to self-determination and to win the arguments in the open on the merits of the case.

The "Despoiling" of the RAC

Now that the moment had finally arrived, bitter fruit in the harvest of good news threatened to defile the victory. Before turning to the sweep of the revision in the Guidelines, I wish to more completely discuss the reactions of disillusionment of a sector of the scientific community in the "capitulation" of the NIH director and the alterations in the RAC revealed by its definition in the new Guidelines.

> IV-E-2. The NIH Recombinant DNA Advisory Committee (RAC) is responsible for carrying out specified functions cited below as well as others assigned under its charter or by the Secretary HEW, the Assistant Secretary for Health, and the Director, NIH.
>
> The members of the committee shall be chosen to provide, collectively, expertise in scientific fields relevant to recombinant DNA technology and biological safety—e.g., microbiology, molecular biology, virology, genetics, epidemiology, infectious diseases, the biology of enteric organisms, botany, plant pathology, ecology and tissue culture. At least 20 percent of the members shall be persons knowledgeable in applicable law, standards of professional conduct and practice, public attitudes, the environment, public health, occupational health or related fields.[9]

Predatory Tactics?

We were unaware of how widely, or with what tactics, members of the HEW staff had solicited suggestions for RAC membership during the early winter. In December, one of them called Philip Handler, president of the National Academy of Sciences. Could Dr. Handler suggest for possible appointment to the RAC the names of "distinguished scientists, knowledgeable or active in the field who espouse a *relatively restrictive course with respect to research on recombinant DNA?*"

Those of us who knew Phil Handler can imagine the sulfurous mood in which he dictated a letter to Secretary Califano. "Such new appointments," said Handler, "inject confrontational politics into the process of scientific evaluation and decision-making at the earliest level." He urged that this and other controversial or potentially controversial technical subjects be processed as a layered structure. "The first phase of review should be conducted by a committee consisting exclusively of persons either working or highly knowledgeable in the area in question, persons of impeccable scientific credentials, without regard to their political or ideological positions. . . . When deemed desirable, the report should immediately be sent to a balanced second body. . . ." Handler concluded that "Such an arrangement is decidedly to be preferred to a single review body in which the findings and judgements formed by the technically qualified members might clash with and be submerged by the ideological or political considerations of others, thereby depriving the Department and the public of the opportunity to benefit by both fully informed and duly debated deliberations."[10]

In the matter of the staff choosing among the ideology or politics of candidates for the RAC, Handler was dead right. His rigid hold on the canonical two-tier review firmly established at NIH for awarding of funding of research was to be expected. Nevertheless, the subject matter of this controversy lent merit to a try for a unicameral body in which laity were accorded seats among the expert scientists. This was a particularly advisable time, for the rules for experiments once considered potentially dangerous to the public health would now be expected to be under pressure for downward revision. My defense of this strategy is laid out in my response to the Singer editorial as presented earlier in this chapter. It also was derived from observation of the lay members of groups such as the NIH DAC struggling to come up to speed in first encountering the arcane complexity of molecular genetics. It was my hope that we could build a better bridge across the gap between science and the public when safety was an issue.

The manner of the transformation of the RAC, however, was profoundly threatening to more than one of the pioneers of Asilomar. A zealous department staff, not comprehending the hallowed nature of peer review to scientists, was seen as usurping the proper roles of scientists, NIH, and its director, and likely to tilt the RAC with a few people judged to be incompetent or too radical.

Other Negative Reactions

A diary note dated January 29, 1979, reminds me:

> Twice this week, Dick Atkinson has called to tell me that a few scientists at Stanford were urging NSF to refuse to follow the revised Guidelines, as a protest against the manner in which the RAC had been handled. In the evening I discussed this proposed insurrection with Alex Rich (from the NSF Board), George Pimental of NSF and Gil Omenn (OSTP). I urged them (1) not to 'secede from the Union,' for the result could be a Presidential order and a law; and it was far better (2) to save their weapons as a resource in case the RAC did fail to function.[11]

Positive Reactions

Shortly, however, more positive voices were heard. Jim Watson weighed in with a widely read article:

> . . . We must not fight back, however, by saying (as Maxine Singer did in a recent issue of *Science*) that only we scientists can judge unquantifiable conceptual risks. We are as unqualified as everyone else to assess such situations and we are stewing in this mess because we said otherwise. If there were firm facts showing that recombinant DNA was leading to dangerous new bugs, then it would be proper for the public to help us decide what we should do next. But since no evidence exists to let us decide rationally whether we or the public should worry at all, no Recombinant Advisory Committee should exist at all. . . .[12]

An editorial comment, from a source usually friendly to NIH, echoed Watson's theme:

> . . . The scientific community should get rid of its antiquated notion of being something apart from the society that pays its bills, reaps its benefactions and stands to suffer from its excesses and mistakes. Contemporary research is too complex to be performed by anyone but scientists; on the other hand, it is too powerful and important to be governed only by scientists.[13]

The New RAC

Having been delayed while the ritual "reactions from the Hill at the new appointments" were sought, a press release from the secretary on December 29 announced the RAC appointments. He noted "the new appointments enlarge the committee from 11 to 25 members, and significantly enhance its capacity to deal with the legal, ethical, and other non-scientific issues which surround recombinant DNA research." Of the 14 new members[14] (see Appendix 10.1), 6 were practicing scientists, broadening the medical expertise of the RAC; one was a scientist with a national environmental organization; and one was a social scientist who had written a book on Asilomar. The two lawyers included a member of the National Commission for the Protection of Human Subjects of Biomedical and Behavioral Research and a retiring member of Congress who was already arguably a ranking nonpracticing expert on recombinant DNA research. The new RAC had been given a slight ideological cant to the left but had gained valuable expertise and met the secretary's demand for broader ethnic and gender representation.

Universal Relief

Regardless of the flap over the RAC, everyone had a feeling of liberation. Frozen for two and a half years, the NIH Guidelines had placed unprecedented constraints on the construction of the scientific framework on which the medicine of the next millennium would surely rise. Although the rules had legally bound only the federally supported recombinant DNA research in the United States, there was a sense of relief in most of the laboratories in the world. The Guidelines had an authoritative international position, even in those countries where the U.K. rules were in place. The interest in the struggle for revisions in the last 2 years by the Genetic Manipulation Advisory Group (GMAG) and EMBO was genuine, for the rules everywhere could not move too far from where the NIH Guidelines were standing.

Few of the participating scientists had ever expected the rules that took shape at Asilomar to endure unchanged for three and one-half years (an eternity in laboratory work). It was assumed from the start that the hypothetical risks of proceeding would soon be clarified as experimentation advanced, and the rules would be correspondingly relaxed. The paradoxical paralysis had temporarily stranded this revolutionary opportunity in molecular biology. An excessive sensitivity of political processes in America to fear of the unknown was partly to blame, but a major reason lay in the original Guidelines themselves, which were silent on means for their timely modification. Our struggles to preserve the NIH Guidelines from conversion to statutory regulations had diverted attention from this major deficiency.

New Roles and Responsibilities

In the last proposed revision prepared before the Libassi hearing and published in July 1978, the responsibilities of the director and of the RAC had been substantially unchanged.[15] The greatest single achievement of this proposed rewrite was the insertion of language to permit continuing revision by the NIH in the future. But the instructions to the RAC and the director were feeble and mundane:

> IV-B-1. The Office of the Director shall be responsible for:
> IV-B-1-a. Final interpretation of the Guidelines
> IV-B-1-b. Revision and amendment of the guidelines after appropriate notice and opportunity for public comment.

In the revision promulgated in December 1978,[16] the tireless Peter Libassi had converted those toneless preludes into a masterly composition of remedies.

Redeeming the Director

First, Libassi fleshed out the shadowy figure of the director, NIH, to the proportions of a multilimbed Hindu deity presiding over a wonderland of duties, focusing upon a new Section IV-E-1-b-1, which detailed the responsibilities of the director, NIH.[16]

RAC

The duties and roles of the RAC were not greatly changed, but its partnership status was acknowledged in new emphases that complemented the highlighted *major* and *lesser action* roles of the director.

IV-E-2-a. The RAC shall be responsible for advising the Director, NIH, on the actions listed in Section IV-E-1-b-(1) and -(2).

Future Revision

Libassi had capitalized on three major features of the operation of the Guidelines that NIH had already practiced and polished to an effective level: a triangular partnership of RAC, director, and public, which we had effectively developed to swiftly and extensively collect public comments and carry through analysis and final action, with every step being publicly recorded. Future revisions would emanate from RAC recommendations to the director. There was no mention of a requirement for a decision point higher than NIH in the department. With good behavior, we were at last home free to practice our form of "rule making" in the ideal frame which we had initially envisioned.

New Players

Voluntary Registration and Compliance

This epoch-making revision also contained the following announcement:

IV-F-3. Voluntary Registration and Certification. Any institution that is not required to comply with the Guidelines may nevertheless register recombinant DNA research projects with NIH by submitting the appropriate information to ORDA. NIH will accept requests for certification of host-vector systems proposed by the institution. The submitter must agree to abide by the physical and biological containment standards of the NIH Guidelines.

The subsequent section reminded institutions that they should consider applying for a patent before submitting information to HEW which they regard as potentially proprietary, and indicated that "provisions for protection of [such] information will be proposed in a future supplement to these Guidelines."

In the *Federal Register* of December 22, 1978, immediately following the revised NIH Guidelines, there appeared a notice of intent to propose regulations signed by Donald Kennedy, the FDA commissioner. This notice stated that:

The Commissioner intends to propose regulations to require that any firm seeking approval of a product requiring the use of recombinant DNA methods in its development or manufacture demonstrate the firm's compliance with the requirements of the NIH guidelines . . .

The FDA notice included reference to the need for secrecy in the handling of proprietary data and the prohibition of disclosure under the general federal confidentiality statute (18 U.S.C. §1905). Thus, nearly all recombinant DNA research in the United States—and much abroad—would comply with the NIH Guidelines. Sizable sources of congressional pressure had been released by these moves. The RAC, however, did not find the adjustment to "partnership" with industry an easy one (see chapter 11).

First Meeting of the New RAC

The new expanded RAC met together for the first time on February 15–16, 1979. The largest meeting room, Room 10 in Building 31, was now cramped for space, the numerous additional ad hoc participants crowding the ring of seats around the table. All but one new member of the RAC was present at this first meeting. It was like old times to listen to Maxine Singer as she guided the newer members from the first experiments, to the Berg letter, to Asilomar and on up to the revision of the Guidelines.

I then gave a few remarks to remind the old and new members of the following:

> The compression of RAC into a single, mixed advisory group is an experiment. The new committee shares an enormous responsibility to suppress both partisan and theosophical tendencies, for an excess of either could lead to delays or even paralysis. The new RAC is the most visible of current attempts to increase lay participation in scientific process. We may all have to travel across the "uncertainty gap" on the bridge that you will be attempting to construct.[17]

Initial Success

"Tonight Bernie [Talbot] called to say that Ms. Simring had addressed the RAC briefly in the manner of launching a new ship, which I judged was not inappropriate for so veteran an observer. The RAC had worked until 8:00 p.m. 'Smooth as satin' . . . The RAC also made it through its second day with few snags. The experiment has succeeded so far . . . ," I noted in my diary.[18] And a wonder, too, if one considers the menu of the RAC on these two days. A small part of the minutes of a future RAC meeting appears at the beginning of chapter 7, indicative of the massive demand on the previous members of this committee that would now be a test of the adaptability of the intrepid acolytes who had come to join us.

Recombinant Odyssey (Eurasia) (1978–79)

Russia, 1978

Shortly after the Libassi hearing in October, two members of the HEW task force packed their bags for a change of venue. Julius Richmond and I flew to Moscow as members of a delegation renewing a U.S.-Russian Medical Research Cooperation which the cardiologist Paul Dudley White had initiated during his celebrated American penetration of Russia 5 years after the death of Stalin. This time, Richmond and I were reacquainted with a treasured part of travel in the Soviet Union, the ride on the Red Arrow *(Krasnya Strella)*, the overnight sleeper between Moscow and Leningrad. Accommodations were four bunks to a cabin, down the aisle from where the samovar for the hot tea swung back and forth over a flame. As we swept by the wooden houses amid the birches lighted by the moon, Richmond and I were carried far and away from recombinant DNA. On the 25th of October, however, Joe Quinn of NIH and I found our way to the Institute of Molecular Biology of the Soviet Academy run by A. A. Bayev, the nucleic acid chemist who had led the Russian group at Asilomar in 1975 (see chapter 1). Bayev was responsible for the Soviet Guidelines, and we gave him a copy of the still unrevised NIH Guidelines. Quinn took a handsome picture of Bayev and assistants George P. Georgiev and Constantine Scriabin. In the laboratory, Scriabin told me how he had been working in Wally Gilbert's laboratory in Cambridge, Mass., in 1976 at the time of the manifestation. When Mayor Vellucci paid a visit, Gilbert had warned him not to mention that he was Russian.

Tokyo

Later, I joined a group of leaders of U.S. science agencies under Frank Press and Ambassador Pickering for a brief trip on Air Force One to Tokyo. Our purpose was to persuade the Japanese to share more funding toward cooperative research. Molecular biology was one of the hot topics, and Jinsaku Kanamori, Science Counselor of the Planning Bureau in the Science and Technology Agency took John Bryant of HEW and me to a restaurant where we were quickly engaged in *en-kai*, the unique Japanese system for getting down to serious business. Suddenly, Kanamori flipped across the table a copy of *Nature* of September 22 with David Dickson's report of the most recent action of the RAC. Thus, it was evident that the Japanese were au courant with the ins and outs of recombinant DNA. There were about 70 projects in the country, a couple of P3 labs, nothing like P4 in the plans. Moreover, there was more than one official set of guidelines, one from the Ministry of Education, Science and Culture, another promulgated by the Science and Technology Council, and a third from the Science and Technology Agency for the nonuniversity sector. This was illustrative of trouble in countries like Japan, where universities are nearly all financed by government, and there is a tendency of support for health science to be fragmented between separate ministries of education, health, and science to the detriment of attempts to set uniform rules for precautions. At the last cocktail party before departure, an elderly nuclear physicist suggested to me that NIH was "a most interesting invention" and that "Japan should have thought of this." I suggested to him that another interesting American invention for these times had been the Federal Interagency Committee on Recombinant DNA.

Europe and the Director General, 1979

Earlier, I described a 1976 visit to the European Medical Research Council and several countries of western Europe to observe the choices between the NIH and U.K. guidelines. At that time, most countries were going to adopt the U.K. rules and set up national committees on the lines of the British GMAG. The Federal Republic of Germany early on had issued its own guidelines based on the NIH Guidelines, adjusting to any changes in the latter. The environmental lobby, and the political party they formed (die Groene), desired strongly to formulate a new law to control rDNA work. Their draft legislation offered in 1979 was never adopted by the Bundestag. Had it been, Keith Gibson observes that Germany would have been the only country in Europe that had a special law for this work.[19] By 1979, many countries were inclined to follow the NIH Guidelines tempered by their own GMAG-like rulings.

The European Community, 1979

Europe as a community had only recently undertaken any effort to achieve uniform controls on recombinant DNA. Before 1989, with the implementation of the Single European Act, the founding treaties of the European Communities laid no legal basis for dealing with research and development programs in the main.[20] In 1978, however, the European Commission's Directorate-General for Science, Research and Development (DG XII XII) formulated a "Proposal for a Council Directive Establishing Safety Measures against the Conjectural Risks Associated with Recombinant DNA Work."[21] The proposal was submitted to the council in December 1978.

In 1979, I received an invitation from the Commission of the European Communities to meet with its Committee on Medical Research. With the invitation came a copy of that draft directive of the European Community on Recombinant DNA. I read it carefully.

In the iterative tone of documents by the European councils, a litany of arguments was laid out to explain the need for a network of national laws on recombinant DNA: the gravity of the hazards; overall risks increasing as the work expanded; the transnational nature of the risks, with no barriers at national borders, making it necessary to reach agreement with neighboring countries and, further, guarantees to respect the agreements "through legal dispositions, taken in each country based on a core of principles adopted in common"; the "intolerable nature" of all laboratories not observing the same rules defeating a voluntary system; and a need for national legislation, harmonized by community principles, to avoid ". . . concentration of research activities in the most permissive states." The drafters argued that "this was an exemplary value of legislation . . . the opportunity should not be missed."

In essence, the draft directive (i) required prior notification and prior authorization for all recombinant DNA work in Europe; (ii) left to the member states the categorization and containment assignments (implicitly, a GMAG-like committee in every country); (iii) specified registration, encouraged on-site examinations and other procedures to detect and sanction against breaches. The invitation came with the staff's opinion that the director-general (XII) intended to push the acceptance of this directive.

I passed the draft about the Kitchen RAC and asked if anyone else felt the chill I was having over the possibility that if this became law in Europe, the members of the Congress contemplating further DNA legislation would likely be off and running again.

Brussels

Quickly, I accepted the invitation of the Committee on Medical Research and also sought and obtained an appointment with Dr. Gunter Schuster, the highly respected and powerful head of the Directorate-General for Science, Research and Development. Meanwhile, "our draft directive" was on the secretary's bedtime reading table (see chapter 11). There would be time to go to Brussels and maybe stop in London to visit GMAG.

I left Washington on the evening of December 5, accompanied by Dr. Joseph Quinn, travel arranger from the NIH Fogarty International Center. We arrived in London the next morning and boarded the plane for Brussels, landing there at noon. We then set out in a chauffeured car for the Argenteuil residence of Princess Liliane of Belgium and her husband King Leopold.[22] We had been invited by the princess for lunch to provide us a preintroduction to Dr. Schuster before our meeting scheduled later in the afternoon.

Arriving a bare 10 minutes late, we changed clothes and entered the library at the north end of the chateau, where, warmed by a fireplace and surrounded by the floor-to-ceiling Moroccan bindings of the Graaf de Flamande (Leopold's grandfather), the princess was gathered with her daughter Daphne, Professor Luigi Donato, president of the Committee on Medical Research, and Professor C. Van Apersele of the University of Brussels. Dr. Schuster, however, was en route and unable to be present. I was able to chat with him later that afternoon.

Meeting with the Director-General

Dr. Schuster had returned 2 hours earlier from India but was fit and vigorous when we met at 2:30 in the afternoon. We exchanged apologies, he for having missed our luncheon and I for forcing him to return to his office. We were each interested in meeting, however, and came quickly to pleasant and informative discussion. Dr. Schuster inquired of my views about mandatory compliance. I told him of my continuing desire for uniform rules and their observance, of my desire for a simple national statute to achieve this, and of my disappointment at the seeming impossibility of obtaining a law unencumbered by an excess of procedures and penalties. In sum, I preferred a simple rule and chose volunteerism rather than inflexible regulation. We also discussed our recent NIH proposals to change the Guidelines, including the provision for voluntary compliance; he wished to be informed of the range of reaction to it. He was grateful for a copy of the *Federal Register* release.

At the end of our conversation, I had a distinct impression that the director-general's convictions about his plan were less certain—leaning toward seeking something less than a directive, perhaps, rather, a recommendation of the council. I am certain it was not our conversation alone but a concern about the implications of passage of a mandatory directive that was leading his opinions.

We are indebted to Mark Cantley for a definitive follow-up:

> As this proposed Directive moved into debate in the European Parliament, the staff and scientists and administrators in the Directorate were aware that some of the initial expressions . . . [in favor of the directive] . . . had been exaggerated. NIH Director Don Fredrickson visited DG XII XII Director-General Gunther Schuster in 1979, to convey to him the lessons of the U.S. experience; and emphasized the desirability of avoiding fixed statutory controls.[23]
>
> The Parliament had commenced its scrutiny, and was adding amendments more rigidly specifying containment requirements . . . but the Commission, on the advice of Schuster, decided in 1980 to withdraw their proposed directive and replace it with a Council Recommendation.[24] This (non-binding) proposed recommendation was that Member States adopt laws, regulations and administrative provisions requiring notification—not authorization—of recombinant DNA work.

In 1994, I had a chance encounter with Cantley, who suggested that my meeting with Schuster and its outcome were a footnote in the history of Organization for Economic Cooperation and Development activities in biotechnology and acknowledged "an OECD debt to the intellectual inputs of the NIH and FDA."

11

Winding Down
1979–81

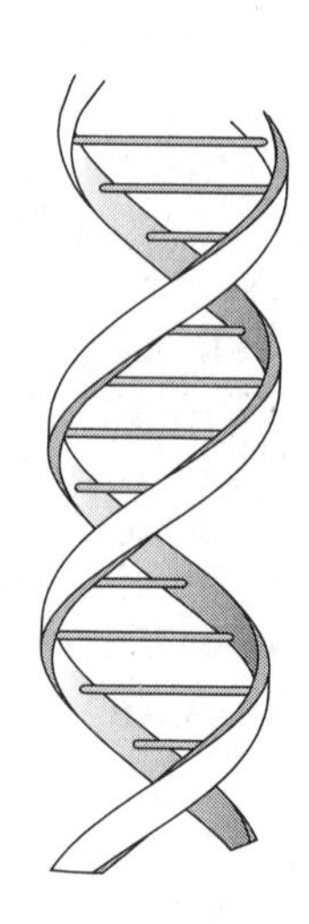

The Cross of Voluntary Compliance

From the first, it had been generally recognized by scientists and demanded by interested citizens that rules on genetic manipulations based on recombinant DNA should be extended to all users of these techniques in the United States. I have recited earlier and often our resistance to attempts to achieve this objective by law, because we were fearful of interrupting the present successful mode of single-agency management of the Guidelines.

In the last weeks of 1978, before the revision was consummated, the HEW secretary and general counsel developed a solution to the dilemma of extending the rules to universal application. They had stood with us in our objection to legislative orders, particularly the use of Section 361 to create mandatory compliance of industry with the Guidelines. A few weeks after the Libassi hearing, Libassi and the secretary had met with the Pharmaceutical Manufacturers Association and numerous commercial interests and gained from them confirmation of our conversations and the conclusions of the FIC members that industry would prefer some kind of voluntary compliance. As we sat in the last negotiations over the revision within the department, the secretary's finely tuned political instincts moved him to summon FDA commissioner Donald Kennedy to join us and draft a notice of intent to propose regulations that any products of recombinant DNA research could be approved by the FDA[1] only if the preparation or manufacture had complied with the NIH Guidelines. The EPA had also been invited to join in this uncertain play for industrial cooperation.

Later, Kennedy and I prepared a description of the details for voluntary compliance and brought the issue before a meeting of the FIC.[2] The secretary asked me to query the other agencies if they agreed that NIH was the best agency to provide a mechanism for voluntary compliance. NSF, the Department of Agriculture, the EPA, and the State Department said yes. The Department of Energy doubted that NIH had sufficient experience with industry. Our supplement regarding voluntary compliance appeared as Section VI in the January 1980 revision of the Guidelines.[3] Almost from the first, most industries intending to engage in recombinant DNA for profit seemed to realize that, in the long haul, it was prudent to comply.

Reactions to Industry's Entry under Guidelines

Anticipation

At one of the numerous symposia on recombinant DNA in 1977–78, a social scientist, Hans Jonas, had used rDNA research as a case of action in regard to moral issues in today's science:

> . . . much of science now lives on the intellectual feedback from precisely its technological application . . . It receives from them its assignments (in what direction to search, what problems to solve . . . for solving them and generally for its own advance). . . . [Science] uses advanced technology itself . . . the cost of these [uses of technology] must be underwritten from outside . . . And . . . [whether publicly or privately supported] . . . this cooperation entails the expectation of some return in the practical sense.
>
> The acceptance of this functional role [by science] has . . . destroyed the alibi of pure, disinterested theory . . . And put science squarely in the realm of social action, where every agent is responsible for his deeds."[4]

This was a statement clearly defining a pronounced shift that was beginning in the (entrepreneurial) relationship of science to commerce, altering the institutions of academic science, and, as Jonas saw it, the moral conflicts that cannot be ignored. To paraphrase Jonas's views of the conduct of recombinant DNA research, any rules to be drawn up should rest on scientific expertise in the first place (a position that was in accord with our own beliefs), but the social scientists were dubious of our faith that, under the radiance of so much new knowledge, the lion would quietly lie down with the lamb:

> . . . the stakes being what they are, the public interest impels public control of this "hazardous field" and the only coercive instrument of control, imperfect though it is, is the law. . . . The "scientific community," in spite of its hitherto blameless credentials, is not the kind of body that can vouch for members or wield effective sanctions against trespassers. . . . To rely on voluntary compliance . . . is unsound. . . . The race for breakthroughs is unstopped. Academic institutions can probably be trusted with conscientious self-policing. The pharmaceutical industry surely less so; peer re-

> view does not reach there and anyhow carries little weight. . . . Here then is an obvious case for public regulations with teeth in them.[5]

Reactions of the RAC

There were members of the RAC who were convinced of this doctrine of suspicion. During this period, RAC meetings were characterized by uncommon disharmony or culture shock in reaction to the special requirements of industry, especially secrecy of review of proprietary material. Members who were disturbed by the penalties for revealing trade secrets sought consolation from RAC member Ray Thornton but received an attorney's warning that "it is the responsibility of the authorized individual to maintain confidentiality"; failure to do so was a crime under that forbidding rule, 18 U.S.C. §1905. Those RAC members concerned how to maintain confidentiality of the papers they were entrusted with had little comfort from Mr. Riseberg's suggestion that a safe-deposit box at a bank would be a suitable option.[6]

Later in the meeting, a lengthy colloquy with representatives of the Pharmaceutical Manufacturers Association (speaking for about 95% of possible industrial users of the technology) and with individual members, Eli Lilly and Pfizer, was arranged with the RAC. The sum of the exchanges was that industry would prefer voluntary rather than mandatory compliance. If they were refused permission on a given project, would they continue anyway? The answer from industry was that they would negotiate the decision just as they would if a Freedom of Information Act demand sought to make the government give up proprietary information. Several of the RAC members still doubted that voluntary compliance would work, and exceptions to the limits on sizes of fermentation reactions made them unquiet.

When the expected proposals from industry for exceptions to the 10-liter limit on fermentation actually arrived, the suggestion of a site visit was begrudgingly met and refused by some members. As a compromise, Emmett Barkley, who had the expertise and the requisite temperament, agreed to represent the RAC in formation of a new Working Group on Large-Scale Experiments, including consultants.

In May, new member Sheldon Krimsky moved that the RAC go on record as favoring mandatory compliance for industry; a split vote favored this idea, but it went no further at the time.[7] At a later RAC meeting on December 6–7, 1979, member Richard Novick persisted with the proposal that voluntary (as opposed to mandatory) compliance for non-federally funded research be given a trial period of only 6 months.[8] Ray Thornton said industry would wish to comply with the NIH Guidelines for several reasons: (i) industry would like to have the approval of the NIH and RAC for this type of research; (ii) there were strong indications from other agencies that a failure to follow the NIH Guidelines would result in a failure to be licensed; and (iii) industry would wish to protect itself against civil sanctions under the umbrella of compliance.

After discussion, a motion by Krimsky, as amended by Richard Goldstein, emerged, stating that the RAC opposed continuation of NIH's voluntary compliance program. This motion failed by a vote of five in favor, nine opposed, with four abstentions. A subsequent motion was passed that stated that the voluntary compliance was experimental and that a review of the voluntary compliance program of the RAC would be conducted at its June 1980 meeting.[9]

In September 1979, the RAC adopted procedures to be followed for large-scale experiments by which the NIH director could authorize the carrying out of experiments involving more than 10 liters of culture. Months later, after a period of commentary, the NIH published "Physical Containment Recommendations for Large-Scale Uses of Organisms Containing Recombinant DNA Molecules."[10] With its ingenuity for creating "regulatory structure," the concept of a Working Group on Large-Scale Experiments was broadened to cover most industrial-scale work, and movement crept toward a workable relationship with commercial laboratories in the sense of voluntary compliance.

Proprietary Information

The notice of the first RAC meeting in 1980 contained specification of a time for a closed portion of the meeting to protect proprietary information.[11] During the open portion of this meeting, a report of a visit by a working party from the RAC to Eli Lilly to examine the large-scale containment facility drew questions from RAC members. Does the NIH director have the right to ask RAC members to inspect industry facilities? If the RAC looks in, should not the company union also have input? Should the RAC not interact with the industrial practice subcommittee of the FIC? (The FIC had created an Industrial Practices Subcommittee. It was chaired by Gilbert Omenn from OSTP, and its members included representatives of USDA, Department of Commerce, NIOSH, CDC, FDA, OSHA, EPA, and NSF. This committee met a number of times in 1980 and offered an exchange of views between agencies whose constituents were also dealing with the RAC.)[12]

Minicrisis

As noted above, in May 1979 Krimsky had urged that the RAC go on record as favoring mandatory compliance for industry, and a split vote slightly favored this idea.[13] I opposed opening up the sores of Congress. By midsummer 1980, I felt that some members of the RAC were experiencing a minicrisis in dealing with their repugnance for the potentially profitable exploitation of recombinant DNA technology. Part of this was the ambiguity of profit taking in a field of such basic biology. Reports had also circulated that some companies were not registering with ORDA, or even if they did, the RAC had no way to gauge their compliance. A virtue of RAC that emerged during the period when molecular biology was suddenly becoming increasingly entrepreneurial was the conscientious vigil of the committee to at least keep all participants to the letter of the Guidelines. Before their June 1980 meeting, two members of the RAC had circulated a letter to all the others again encouraging that mandatory requirements govern all such nonfederal operations. The threat that Congress would be persuaded that new regulatory structure must be sought was anathema to me. I was moved to give the RAC my views of the looming threat that we might slide backwards into hard-fought battles previously won in the Congress.

Director's Statement to the RAC

At its meeting of June 5–6, 1980, I addressed a statement on voluntary compliance to the RAC:

> . . . I am aware that some members of the RAC have chafed under the need for examination and approval of scale-up operations in industry . . . and . . . find distasteful the scientific review of information which is subject to safeguards, including criminal penalties for violation. . . . Some members of the RAC also are of the opinion . . . that such belies the insistence of the NIH and its Director that this agency eschews a regulatory role.
>
> However, in doing these tasks, the members of RAC have made it possible for all users of recombinant technology in the U.S. to do so under two important conditions that obtain in nearly all other advanced countries in the world. These conditions are (1) The determination of appropriate procedures by a single national body, responsible for interpreting and developing uniform national guidelines, and (2) The regulation of a rapidly evolving science without passage of special statutes, avoiding inflexibility often attendant upon such legislation.
>
> These important conditions are among those necessary for maintenance of parity among the nations for the safe use of this technology and access

to its benefits. It is essential that the RAC, whose deliberations affect the course of this field of science throughout so much of the world, not upset these conditions by precipitous changes in its procedures.

Nevertheless, I am sympathetic to this *malaise* of the RAC . . .

. . . it should be stated here with candor that no federal regulatory agency *presently* has anything like the in-house competence to perform—for industrial applications—all of the present tasks of the RAC. Development of such competence will require time and expense, and it would not arise suddenly upon the passage of any statute. Furthermore we are still at a stage where the creation of more than one RAC in this country—one for industry, one for non-profit research—would be inefficient and destroy the unity of information-sharing and decision-making . . . that facilitate appropriate evolution of guidelines. . . .

. . . From these considerations, I believe that the RAC should consider at least two options as it deliberates upon the procedures for handling industrial or private requests:

1). To continue as it is until more experience dictates some new mode. Here keep it in mind that an Industrial Practices Subcommittee of the FIC is actively studying the problem and will report to the FIC in the fall; or

2). To restrict its review and decisions to questions concerning biology and related issues of laboratory containment; and when containment for large scale use is the issue, to judge only whether the specified P-LS level is appropriate. The regulatory agencies will be fully apprised of such actions and they can make on-site evaluations if they deem them appropriate.

. . .

. . . Finally, let me emphasize that it would not be appropriate for the RAC to decide, in exasperation, that it would like to avoid handling *any* proprietary data. From time-to-time such material will have to be evaluated by the RAC in pursuit of its duties. . . ."[14]

The seductive scent of profit in commerce of molecular biology was rising as we entered the 1980s, and prescient members of the RAC had been viscerally responding to a threat of "de-idealization" of what they perceived as a search for fundamental knowledge. The majority of members, however, preserved the unity of the committee and turned their attention to relieving what had become the most burdensome of the duties of the RAC.

The "*E. coli* K-12/P1 Recommendation"

The First Major Action

In early 1979, Gartland estimated that about 85% of rDNA experiments were using as the host *E. coli* K-12, the *bête noire supérieure* of the history of the recombinant DNA controversy. It was no surprise that the first test of the new provision for major revision of the Guidelines would be an attempt to exempt much of the use of *E. coli* K-12 from the rules. A study of Section IV-E-1-b-(1) in the December 1978 Guidelines reveals that such a "major action" was neither simple nor swift. The proposal must first be published in the *Federal Register* for at least 30 days of public comment prior to a meeting of the RAC. The proposed action, together with all the written comments received, was to be taken to the RAC in open session, and members of the public were given opportunity to speak on the subject. A recommendation on the proposal was then made by the RAC to the director, NIH, and the decision of the director was to be published in the *Federal Register.* If the move proved controversial enough, a further provision permitted the director to publish his proposed decision for a second period of public comment prior to the final decision. As events were soon to prove, it might take at least 7 months for the culmination of the first of these major steps forward in the renovation of the "new" NIH Guidelines.

Proposed Exemption of Certain E. coli *K-12 Experiments*

On April 13, 1979, RAC members Wallace Rowe and Allan Campbell caused notice of a proposed major action to appear in the *Federal Register* to add certain exempted categories of recombinant DNA molecules to those already in Sections I-E-1 to -4. It read as follows:

> Those recombinant DNA molecules that are propagated in *E. coli* K-12 hosts not containing conjugation-proficient plasmids or generalized transducing phages, when lambda or lambdoid bacteriophages or nonconjugative plasmids are used as vectors, are exempt from the guidelines. . . . This action is being proposed because of the large body of information that has accumulated concerning the *E. coli* K-12 host-vector systems, all of which points to the safety of such systems. . . ."[15]

This proposal was taken up in the RAC meeting of May 21–23, 1979, and discussed at great length, exciting an array of opinions. Several RAC members opposed the proposal outright; others would not exempt but would classify such experiments as P1. A majority favored the exemption, arguing that the RAC and the IBCs must become more selective in their attentions, and this exemption would be a welcome relief. A new member of the RAC, Luther Williams, suggested that a working group conduct a rigorous scientific analysis of the *E. coli* K-12 systems, to which Rowe added that the working group should explore new administrative mechanisms, such as certifying IBCs to make such decisions. The motion to launch the working group carried an amendment proposed by David Baltimore that the experiments using an *E. coli* K-12 system be permitted under P1 conditions, registered with the local IBC, and that a copy of that document be forwarded to ORDA.[13] The Williams motion, as amended, passed by a vote of 18 to 3 with one abstention.[16] The *Federal Register* of July 31 carried the notice that the above amended proposal by the RAC would be considered in the September 6–7 meeting of the RAC.

In Washington, however, the interval between May and September can be long and disruptive. Thus, some intervening events, as recorded in my diary, must be recounted before this story is resumed.

Comings and Goings

Peter Libassi. In the Spring of '79, the federal government was heavily involved in a number of high-level studies of the biological effects of ionizing radiation. In January the Secretary of DHEW notified me that the Department had been directed by Congress to carry out new activities in the area of radiation research, and that he was designating me . . . project manager to carry out new responsibilities of: 1) establishing a program of research into the biological effects of ionizing radiation: and (2) a comprehensive review of all Federal programs of research on this subject.[17] At this time Peter Libassi was also already getting his "graduate course in radiation" by appointment as chair of an Interagency Task Force established by presidential order in May, 1978. On April 3, 1979, a reactor "scam" occurred in Unit Two at Three Mile Island in Middletown, Pennsylvania. America's most serious threat of a nuclear accident stepped up excitement over nuclear dangers in government agencies and in the Congress.

It was in the light of these events that I found myself with Libassi, on April 20, seated under bright television lights in the grandiose Salt Palace in Salt Lake City. We had been invited to appear before Congressman Bob Eckhardt (D. Tex) and Senator Edward Kennedy, co-chairs of a joint committee of Congress on the subject of the atom bomb tests over Utah and adjoining states in 1953. Scattered about the stage were open cartons of AEC files acquired under the Freedom of Information Act by an enterprising reporter. On superficial scanning, one could read minutes of meetings among some of our most eminent physicists who appeared to be discounting possible health effects of testing. For our part, we were shown pictures of sheep bearing huge blisters, and of some of their deformed lambs, followed by testimony from ranchers who had seen the Geiger counters held by AEC inspectors go off-scale when dragged across the animals' necks. I had a single question, from Senator Kennedy: "Well, Dr. Fredrickson, as the chief scientist of HEW, what is your view of the sheep incident?" I answered to the effect that "it isn't possible to conclude that radiation did not have at least a contributory effect to the death of the sheep." Before the assembled governor of Utah, Congressmen from Utah and Nevada, and one other Senator, I listened as Peter Libassi "gave Senator Kennedy his headline." The "cover-up" story in Libassi's and the

Senator's terms made page 1 at home with generous headlines in the next morning's *Washington Post* and *New York Times*.[18]

National politics is theater and no star can tolerate the understudy stealing the house with an arch-rival. I had the distinct impression that the Secretary was not pleased. The master-draftsman of the "Major Action" steps we were now pursuing, who had earned campaign ribbons side by side with molecular scientists, soon dropped from sight. Before he departed on June 20 there was a small farewell party. I presented him with a white coat such as we wore at NIH. Stitched across the pocket in red thread was "Dr. Libassi."[19] From the community of molecular biology he was overdue for recognition.

Joseph A. Califano, Jr. On July 19, 1979, my wife and I arrived at National Airport after interviews at a major university. James Carter, chauffeur to NIH directors since Shannon, was talking on the phone in the car. It was my secretary Bel Ceja, telling me I was to go directly to the Humphrey Building. At 12:00 Secretary Califano would tell his senior staff that he had resigned.[20] It was 11:40 and I went directly to Richmond's office on the 7th floor. We retreated to his inner office and looked at each other a moment. "It may be easier now, but it will never be so free again," I said. We silently mused about a country whose President fired the most competent of his ministers (the Health Minister!) to cure a crisis in energy, the dollar, and other "malaise."

At noon, Joe Califano entered his conference room in a white sport shirt. He spoke from notes, with some choking. It was a graceful farewell: to the "best HEW since the Great Society; the greatest people he'd ever worked with; the most remarkable experience a boy from Brooklyn could have had; and to the President who was the pilot and must choose the team he wants; and to all of you who had given him our loyalty and service."[21]

Back at the office I wrote a short letter from the heart.[22]

Dear Joe,

. . . Last Thursday, as I stood with the others in the dining room, I admired the grace with which you stepped from the wreckage of the splendid balloon in which we have been privileged to soar with you.

We will not again fly so high, so far, or in such bold directions. I have learned much, and enjoyed immensely these few years with you, Joe; and I admire in like proportions your courage, quick perceptions and grand impatience.

With affection and gratitude, I remain always

Your friend

July 24. This morning began in the Secretary's conference room.[23] On one side of the big oval table sat Rick Cotton, Dick Beattie, Dick Warden, Peter Hamilton, Patricia Roberts Harris, Joe Califano, Ben Heinemann, Peter Bell, Julius Richmond, Fred Bohen, Tom McFee, Tom Morris; and at the bends: Anabellah Martinez, Mary Berry, Murphy (headhunter); at the other bend: Lee Kimsle, Art Upton (Dir NCI). Opposite side: Susan Foster, Fredrickson, and Jerry Klerman (ADAMHA). Behind was a transition team from HUD.

Califano began with recitation of an editorial from the *Dallas Times Herald* describing "liberal Joe Califano" as "no loss" and "liberal Pat Harris" as "no gain." He gave some gracious remarks. Harris responded, indicating she was no less liberal than Joe Califano. . . . She is poised, well-spoken, in command and is an insider, nearly as street-wise as JAC, but as she said, a "team-player" . . . At an attempt at humor, Ben Heinemann growled an observation to Peter Hamilton that after only two months as Special Assistant to the Secretary, the latter had let this happen; "You are a complete and desperate failure, Peter." . . . It struck me again that the incoming set of lawyers could not possibly measure up to the legal talent of the Califano team that we had been allowed to scuffle with in this game of government. Joe introduced each of us to Secretary-designate Harris. Of me, he joked, "He's always fighting off the attempts of Congress to add more money to the NIH budgets." We filed out with a heavy tread.

On August 2, there was a final farewell and picture-taking with Joe. As we chatted before the flags, he asked me about the Stabilization Plan, The Five Year Plan for Financing Health Research for which he had given me responsibility in the Spring of 1978.† "You'd better stay, Don." And I said, "I will, at least another year."[24]

July 24 at the White House.[25] All 200-plus presidential appointees filed into the East Room at 4:45. A platform for speaking, with the seal of the President affixed, stood before a yellow silk curtain. The Presidential flags arrived. Secret Service placed themselves on either side of the dais.

At 5:10: "Ladies and Gentlemen, the President of the United States." ["Hail to the Chief" had been temporarily banned.] President Carter re-

---

† "The Plan" (the Stabilization Initiative). Congressional hearings on DNA had just bogged down when Secretary Califano announced to the societies of clinical investigation gathered in the San Francisco Hilton on April 29, 1978, that he would develop a 5-year plan for the funding of all health research in HEW. I had brokered the secretary's appearance before this group, but his announcement was a total surprise to me, especially the part about the director of NIH being responsible for delivering such a plan.

After a stormy summer experiment in ecumenism, an HEW Steering Committee representing all the PHS agencies completed a set of "Health Research Principles" by September.

On October 3 and 4, 1978 (just before the ninth meeting of the FIC), a National Conference on Health Research Principles was convened in Bethesda, with about 700 participants, representing the elite of academic and other interests in health research.

Eleven initiatives rose out of the plan. The conference unanimously endorsed number one, the stabilization of federal support of fundamental (investigator-initiated) research, expressed in a single objective: the combined agreement of the administration and the Congress to fund a *minimum* of 5,000 new and competing research project grants in each appropriation year. Some feared that in setting a "floor" we would get a "ceiling." But such was not the case. Between 1979 and fiscal 1988, the appropriations committees moved steadily from 5,000 to 6,500 new and competing grants per year. Moreover, the report language of the appropriation bills included the following command each year: "The Committee continues to give the highest priority to the support of investigator-initiated research projects and intends that the amounts in the bill shall be used to fund approximately 6,500 new and competing projects. . . ." (1988 example: U.S. Congress Approp. House Rep. [NIH],100th Congress, 1st sess.)

ceived a long standing ovation. At its conclusion he said, "I'm glad I appointed all of you." . . . Noting that it was hot, he removed his summer jacket, which he laid on the dais, and invited everyone to remove his. A few of us—the conservative wing—kept ours on.

In shirt-sleeves and without notes, the President spoke for 25 minutes about his Camp David ordeal 10 days ago; his decision that he, we, the government had failed the American people, and that he had resolved to take steps to change the management of his government. . . . I had an excellent position to view his blue eyes with reddened sclerae, a slightly mottled face of an earnest man, pained, but in control of himself and the situation. . . . He called for questions, one had to do with problems of a Cabinet member who had a mandate that was not particularly popular. Presidential answer: "No problems with Joe Califano opposing tobacco"; he'd "support Pat Harris, if she campaigns in the same way." No ad hominem statements . . . Invitees given questionnaires . . . "Tell us if you think you might be in the wrong position" . . . At the end out the center door a pause for a picture with the President . . ."

Patricia Roberts Harris. August 5. Because of the fixed direction of his gaze, John Quincy Adams could see little of the East Room but the door. Martha and George Washington, however, from their more central position, looked down with good vantage upon the crowded scene. Many here on this day were the progeny of slaves many generations removed, gathered as first citizens to share the pride and pleasure of one of their own, Patricia Roberts Harris, who was being sworn in to head the most complex piece of the government's machinery. The oath-giver, Judge Thurgood Marshall, lent a fierce, yet avuncular dignity to the ceremony and everybody, including the President, beamed. . . . One could not help being impressed by the . . . admirable beauty, wit and cadence . . . powerful reserves . . . and patience . . . of the majority present. There was something Jacksonian, too, about the heat, the gentle jostling and the fairly superficial flavor of the remarks, including the President's. After iced tea, we returned to the campus.[26]

Major Action, Continued

At the September 1979 meeting of the RAC, there was a lengthy discussion before and after a presentation of the now familiar evidence of the safety of the *E. coli* K-12 host-vector system.[27] Numerous letters had been received, including one signed by 183 scientists at the Gordon Conference on Nucleic Acids, most supporting the exemption as originally proposed in the April 13 *Federal Register*. Talk cycled to the polyoma experiment, then to the report of experts in immunology who had never found evidence that *E. coli* could cause autoimmune disease. The fear of *E. coli* clones contaminated with genes for insulin or other hormones surviving in the bowel was aired as a possible source of harmful antigens or dangerous source of hormonal activity. The ethicist Walters reminded some of the discussants that polypeptide hormones normally do not survive passage through the gut. However, these worries were not easily purged, even by the strong convictions of the eminent immunologists who wrote that the probabilities were vanishingly small. The chair of the working group moved acceptance of the exemption as it had appeared in the July 31 *Federal Register*. Several other amendments were added. With 10 affirmative votes, four negative votes, and one abstention, the following text finally passed:

> Those recombinant DNA molecules that are propagated in *E. coli* K-12 hosts not containing conjugation-proficient plasmids or generalized transducing phages, when lambda or lambdoid bacteriophages or nonconjugative plasmids are used as vectors, are exempted from the Guidelines, subject to the prohibitions of I-D-1 through I-D-6. Prior to initiation of the experiments, investigators wishing to carry out such experiments must submit a registration document that contains (a) a description of the source(s) of DNA, (b) nature of the inserted DNA sequences and (c) the hosts and vectors to be used. These registration documents must be dated and signed by the investigator and filed only with the local IBC with no requirement for review by the IBC prior to the initiation of the experiments. P1 containment shall be used for all experiments in these categories.[28]

A May Shower of Comment Extends through November

The archives of commentary on the actions of the RAC during the consideration of the exemption of *E. coli* K-12 is replete with letters from all positions.[29] This shower fairly howled again after the September split vote. The majority were strongly positive. The negatives included further jostling from Bereano, Bross, Cavalieri, Dach, DeNike, Garb, Hartzman, Simring, and Wright. But dissent was less predictable now. Four members of the RAC conveyed their dissent, as did Roy Curtiss. Congressman Henry Waxman, successor to Mr. Rogers's chairman's seat in the House, wrote a demand for an immediate explanation from me. By this time, some of the dissident sentiments had been written to Secretary Harris, including a letter from Senator Adlai Stevenson:

> . . . I understand that . . . Dr. Fredrickson is nearing decisions on two important changes in the . . . Guidelines. The first involves the recommendation of the Department's advisory committee to remove all but minimal controls on experiments using the *E. coli* K-12 host vector system. . . .

A discussion of the *E. coli* K-12/P1 recommendation appeared in the "Proposed Actions Under Guidelines" section in the *Federal Register* of November 30, 1979, along with my proposed action.[30] It included a lengthy history of the examination of the absence of evidence that this strain of *E. coli* is hazardous. However, I did not accept the amendment as sent to me by the RAC. I felt that the potential for confusion was still likely as the recommendation was worded.

> Instead, my proposed decision is to describe the experiments covered in the *E. coli* K-12/P1 Recommendation in a new Section which is to be added to the Guidelines, called Section III-0, as follows: . . .

In this proposed section, some such experiments were prohibited, i.e., those listed in Section I-D of the Guidelines involving pathogenic organisms, toxin producers, deliberate release into the environment, drug-resistant strains, or large-scale experiments. Some such experiments were exempt from the Guidelines, i.e., those listed in Section I-E.

The rest of the rephrasing of the *E. coli* K-12/P1 recommendation set the containment level for most other experiments as P1+EK1. Also included was registration with, but not prior approval by, the IBC, except for those experiments where "there is a deliberate attempt to have the *E. coli* K-12 efficiently express any gene coding for a eukaryotic protein," in which case prior approval was required.

There were approximately 400 letters of comment on these and a few other proposed changes in the Guidelines. Some critics made much of the lack of a consensus and complained that the RAC had overridden opposition. Most of the comments favored the proposed change.

Letter from a RAC Member. One of the letters came from a member of the RAC, Susan Gottesman. Here are some of her comments, particularly those relevant to the lack of full agreement on the resolutions.

> As you know, . . . I voted against the proposal to lower required containment for *E. coli* K-12 recombinant DNA experiments. During this additional comment period, questions have been raised about the procedures used by our committee when this proposal was approved. . . .
>
> In my opinion, the RAC discussion was extensive and relevant. There was ample opportunity for expression of a variety of opinions . . . The voting procedures . . . were appropriate and in accordance with HEW regulations. . . . In the end, what we really differed about were the conclusions that we drew from the data; I suspect that no amount of future data, discussion or experience would make the committee unanimous on this issue.
>
> I voted against the proposal because I felt that the data on possible consequences of expression of a harmful product by an *E. coli* containing recombinant DNA had not been fully explored, and that therefore P1-

EK1 might not be assumed to be adequate in all cases. The split vote of the RAC at least partially reflects the range of possible conclusions from the same data . . . and the complexity of the problem more than anything else. This is as it should be: we are making our best guesses about unknown and speculative hazards, and we need the differences of opinion to be sure that we think about everything. I do not think that such a split vote should be overinterpreted; the committee will become either totally unable to act at all or forced to a false unanimity if decisions are held up simply on the basis of split votes.

I hope these comments will be helpful.[31]

A highly respected scientist, stalwart of the RAC and the Kitchen RAC, and eventually consultant to the RAC because she had become an encyclopedic expert on the Guidelines, Sue Gottesman's words reflected the blend of expertise and conscience for which the RAC was becoming known as a beacon in the murky moments of the recombinant DNA controversy.

A Perception: Director Is Waffling. In the minutes of the RAC meeting of December 6–7, 1979, Chairperson Setlow delivered an update on the *E. coli* K-12 situation, noting that "Dr. Fredrickson's proposed decision departs somewhat from the *E. coli* K-12 proposal as recommended by the RAC in September. In Dr. Fredrickson's proposal, most *E. coli* K-12 manipulations are not exempt from the Guidelines. Registration and review by NIH would not, however, be necessary for these experiments." She noted that "Dr. Fredrickson accepted P1+EK1 containment conditions and the requirement that these experiments be registered with the local IBC. Prior review and approval . . . by the IBC would be required when there is attempted efficient expression of a gene coding for a eukaryotic protein."[32]

In the January 17, 1980, Notices section of the *Federal Register*, a brief comment relative to the *E. coli* recommendation read, ". . . The NIH Director instead issued them for 30 days of additional public comment . . . on November 30, 1979. Final action on these proposed changes has not yet been taken."[33]

An Explanation: the Cost of Protocol

During the month of December, I received more than 180 letters from the members of the molecular biology community in the United States and a few from abroad.[34] All were strongly in favor of the proposed change in the Guidelines. Two perceptive commentators expressed hope that the negotiations with Secretary Harris were proceeding satisfactorily. Other commentators were writing to the secretary herself.

Approaching the New Secretary

The absence of a *requirement* in the new Guidelines for approval of a major action by the secretary was dangerously misleading. The RAC charter said that in addition to advising the NIH director, it also advised the secretary. More important, the head of a major agency also had to obey an unwritten code, which briefly ran, "Never blindside the Chief." Certainly not on something as newsworthy or as politically "sexy" as recombinant DNA. The new secretary also knew all her prerogatives. On the day of her swearing-in, she was on the phone:

"Good Morning, Madame Secretary."

"I simply wanted you to know that I'm looking to see you in a few days—probably next week."

"Thank you and welcome aboard, it will be a pleasure to work with you."[35]

> *August 17.* Randy Kinder, the new HEW executive secretary, had called and wondered when Secretary Harris should visit NIH. I suggested that we come to see her first and send briefing papers with a suggested schedule.
>
> Today we met the Secretary. With me were Tom Malone, Deputy Director of NIH, Joe Perpich and Arthur Upton, the Director of the National Cancer Institute. With Secretary Harris was Kinder, and Assistant David Caulkins. After introductions, she waited expectantly for me to begin. I am afraid that I talked non-stop for almost an hour, risking my next welcome. We covered her coming visit, then moved to the NIH Budget and its relationship to the Research Principles, which had been released on August 15. One of our first tasks would be to get her backing before we approached the Appropriations Committees. We did not even get to the subject of DNA. After our conversation, I judged her in sum, a tough-minded, smart and savvy person . . . less of a trial to brief, but just as quick a study as her predecessor.[36]
>
> *September 12.* The Secretary picks NIH for the first of her visits. A full tour was presented, during which she was very gracious and said just the right things. As I always did on such visits I wore a white coat and Mrs. Fredrickson served coffee in my office.[37]
>
> *Two weeks later.* Editorial in the *New York Times:*[38] "Less Peril in Gene-Splicing . . . Federal restrictions are about to be eased again on those once-feared recombinant DNA experiments. . . . The Director of the NIH will almost certainly concur with the recommendations of an advisory committee to free from restrictions more than 80 percent of all current gene-splicing experiments."[39]
>
> *October 29.* Nothing has been heard from the Secretary's office for over a month. Perpich has tried to make an appointment with her for us to discuss

Health Research Plan—the Principles. He was rebuffed by the appointment staff. Today I have held a full-scale Kitchen RAC meeting to prepare a decision document on the P1/EK1 recommendation of the RAC in September meeting. Appointment scheduled with the Secretary on November 2 to get this decision document signed.[40]

November 2. Secretary Harris emerged from her office trim and composed, then saw the 18 inch stack of yellow documents I'd put on the coffee table. "You're not going to bring those in here!" she said, pointing to her inner office. I had overworked my point. She began discussing the Research Plan, and I waited patiently for my opening for P1/EK1. When I found my opportunity, the Secretary said she wanted to know more about this before we publish. I suggested next week, and thought I must bring Maxine Singer back with me.[41]

November 10. Full-scale meeting with Secretary Harris, who was accompanied by Nathan Stark, the new Undersecretary, Jodie Bernstein, new General Counsel (formerly GC at EPA), Randy Kinder, Carolyn Chin, all on one side of the table. On the opposite side: Maxine Singer, Emmett Barkley (despite his cherubic appearance, I explained to the Secretary, he is the author of the safety manual now visible in labs from Leningrad to Adelaide), Bernard Talbot, Joe Perpich and I. Maxine gave a talk with demonstration cards, I emphasized the need to act before our position is weakened by impatience on both sides of the Atlantic. We conclude the afternoon with the request that the Secretary and her GC read the document. We hoped to get it into the *FR* next week. She replied that she was inclined to put it out for 30 more days of comment or a re-vote by the RAC. I left with the sinking feeling that we were re-living the days before the Revision, with a new crew without institutional memory of that grim campaign.[42]

November 12. Meet with Kitchen RAC again. Consensus that we put decision out for 30 more days. The revised Guidelines had suggested this might be necessary.[43]

November 22. Secretary Harris is reported to be upset with NIH, because we "seemed stubborn and demanding." Saw GC Bernstein, who said she would convince the Secretary to push up the decision in less than 30 more days.[44]

By December no one in the Secretary's office had spoken to her about the decision document, "for fear of a stormy reaction." "Were we pressing the Secretary too hard?" we were asked. Meanwhile the Science Advisor had been urged by OSTP to write the Secretary. The National Science Board announced its impatience, EMBO said it could possibly wait until January. David Baltimore threatened to boycott the next RAC meeting. Again—I counseled patience and left for more legs of the Recombinant Odyssey [see chapter 10]. The RAC held its calm.

January 17. Suddenly we're told the Secretary has gone to her desk, found, and signed off on the decision document. Bernie picked it up from the

> Department and spirited it to the *Federal Register.* It now hardly mattered, for the action stood as it was reported in November, but we had stayed right with the Secretary by waiting for her to give the nod. We had chivalrously passed the first of the major actions in the new mode of revisions—a small but very significant milestone in the history of the recombinant DNA controversy.

A Salute to Tempestuous Times

I pause here to make certain that the memory of Secretary Harris is properly honored. Those who knew NIH in the ebbing of the recombinant controversy might have incorrect memories of the relationship between the two of us. My diaries of 1980–81 bulge with snippets of the *zamizdat* news sheets and legitimate news articles hinting at our mythical battles. Actually, our relationship was one of mutual respect, attesting to her pride and great interest in the welfare of NIH. I respected the grace and humor with which she ran the department, newly renamed the Department of Health and Human Services, arguably the most complicated department in the executive branch. There were plenty of opportunities for tension that had little to do with recombinant DNA, a program with which she did not desire to interfere.

The Attack on Section 301

In the spring of 1980, Congressman Waxman introduced a Health Research Bill in the Health and Environment Subcommittee of the Interstate and Foreign Commerce Committee, proposing a restructuring of NIH authorities. When a copy of the initial draft of the bill was handed to me, I characterized it as a "barbarian" act, for Section 301 of the Public Health Service Act—the holy chalice of the NIH general authorization since 1944—had been rudely excised. I had copies distributed at once to the NIH leadership as well as to the Association of American Medical Colleges and other interested partisans of research funding. I also immediately made it the only subject of my next day's one-on-one monthly meeting with the secretary. Apprised of this attack on her own authorities as well as those of NIH, she rose and declared she would seek a presidential veto. When she informed William B. Welsh, her assistant secretary for legislation, of her declaration, he retorted that the entire legislative calendar for the Department of Health and Human Services would now be in serious jeopardy. So, it also turned out, was her relationship with the NIH director. At the next staff meeting, Welsh was openly critical that I had taken up such a matter with the secretary alone. Suddenly, I received a shocking upbraiding from the secretary before the staff. Our putative split was then and thereafter publicized.[45]

Eruption

The department calendar for March 21, 1980, carried the notice that Secretary Harris was to appear that day at the Lister Hill Center at NIH. She was to be introduced by Director Fredrickson. However, on the afternoon of March 20, I was called by the White House to inform me that I was to accompany President Carter's party on a visit to the eruption of Mount St. Helens. We were to leave for Andrews Air Force Base to board Air Force One without delay. The next day, while on site with the presidential party, I was called by Storm Whaley, NIH press officer, who informed me that Secretary Harris had told her audience at NIH that "It is obvious that Dr. Fredrickson prefers to meet a raging volcano than the Secretary."

I had brought home a bag of volcanic ash and poured a generous amount into a tincture bottle. I tied a blue ribbon to the neck of the bottle and attached to it a brief note with my signature:

> March 23, 1980
> Madame Secretary,
> We can't go on *not* meeting this way.
>
> Sincerely,

I left my token with the secretary's office and went off to attend a staff meeting in the North Auditorium. The secretary came to the platform last and sat in front of her agency officers. Suddenly a hand was extended backward in my direction. I warmly "gave her five." In her remarks she noted that she had received a jar of Mount St. Helens ash from Dr. Fredrickson, who had "exemplified the ubiquitous presence of the Public Health Service at all major sites of disaster."[46]

Much later, when the new president was elected, Secretary Harris told the transition team that she would not ask for my resignation or that of the director of the National Cancer Institute, Vincent DeVita. The upshot of this was that I stayed 6 more months at NIH as the director under Secretary Richard Schweiker during the first part of the Reagan presidency.[47]

A Difficult Labor Done

Another epic moment of the DNA controversy was accompanied by venting of angst from partisans of all persuasions at the outcome of the first major action. As noted in Director's Comments of the *Federal Register* of January 29, 1980, where my "final" decision and the revised Guidelines were published,[48] six members of the RAC responded to my urging that they comment. Four endorsed the change. Two who had voted against the *E. coli* K-12/P1 recommendation at the September 6–8 meeting also wrote. One urged the "exemption" not be approved. The other urged that the final decision not be delayed.[49]

Over 200 more letters were received. Some were opposed, a few violently so. The majority were in favor. In February, the majority opinion was mirrored in an editorial in the *Washington Post*.[50]

> National Institutes of Health Director Donald S. Fredrickson's decision to lift most of the guidelines under which scientists have had to perform recombinant DNA research is a major milestone in a precedent-setting attempt at self-regulation. . . .
>
> . . . The guidelines have been a source of controversy and have been studied and revised almost from the moment of publication. As scientists gained familiarity with the new techniques, some felt that the dangers had been overdrawn. Others believed exactly the opposite, always postulating new dangers that have not yet been studied. While heated disagreements persist, a new consensus has developed that many types of these new experiments are safer than had been thought—hence Dr. Fredrickson's decision to, in effect, remove the regulations from them. . . .
>
> A few scientists among those who first voiced warnings believe they made a mistake. They have been buried for years under mountains of paper work, experiments have been delayed until the necessary clearances came through and many experiments have not been done at all because clearances were not received—and because of what now appears to have been unfounded fears.
>
> We hope that will not be the prevailing view. Despite their flaws, the recombinant DNA guidelines have been the model of a responsible approach to a dangerous technology, and of cooperative action between government and the private sector. Had nuclear engineers, pesticide chemists and numerous others acted with similar caution and sense of public responsibility, everyone would have been much better off.

More "Major" Actions

After the *E. coli* K-12/P1 revision was promulgated in the Guidelines revision of January 29, 1980, consummating the year-long first major action for revising the December 1978 Guidelines, there was residual disagreement among some of the more cautious molecular biologists. Few of the enthusiasts among the scientists dared to maintain there was *never* going to be a serious accident, but no amount of arguments now could stem a rising sentiment for throwing off or reducing the apparatus for maintaining the

NIH Guidelines. Three scientists at Asilomar, two among the organizers, were ready to make proposals that brought change. The first of these dealt with rationalizing the organization of the apparatus. This begun, a more radical change was then proposed for the RAC by two of its members.

A Second Major Action

In a letter to ORDA, Maxine Singer exercised her eminent historical right to propose a major action that review, registration, and approval of all experiments assigned containment conditions in the Guidelines be simplified.

> Seven years have passed since the scientific community first raised questions about recombinant DNA experiments. The ignorance of those early years has been supplanted with a wealth of experience and information. Just as the early stringent Guidelines were promptly adopted in response to the ignorance, we must now respond just as promptly to the current more realistic appraisals.[51]

This proposal was considered by the RAC at its meeting on September 25–26, 1980. It approved the elimination of *NIH review* of IBC decisions on any experiments for which containment levels were specified in the Guidelines. The RAC judged, however, that it was not ready to go so far as to eliminate *local* prereview (by the IBCs) of such experiments. The complications of untangling the syntax of the Guidelines to achieve relatively simple changes resulted in a harvest of motions. I formed a committee of Talbot, Gartland, Singer, and Gottesman to implement the Singer proposals and accepted the simplification achieved, which was promulgated in revised Guidelines on November 21, 1980.[52]

Among the regulatory appendages that were pruned was the venerable requirement that investigators or institutions register a memorandum of understanding (MUA) with NIH proposing to experiment with recombinant DNA. (I vividly remember the confusion attending the first public awareness of a break in the MUA requirement. Through the Freedom of Information Act, a reporter had obtained the files on the grant of a molecular biologist—a former member of the RAC—who had failed to warn his institution to update his MUA. This motivated ORDA to tell the startled dean of the Harvard Medical School that ORDA would soon be visiting Harvard. I was the chair of the DAC in its December 1977 review of the proposed revisions of the Guidelines on the day this news reached me. As the meeting broke, I had the task of alerting the attendees that they might find a story in the morning paper that would interest them—the first known violation of the NIH Guidelines.)

During its time, the MUA had been an important link with all institutions in which NIH-supported recombinant DNA research was under way. Some university officers were little aware of how they were bound into the system and of the importance of the decisions their IBCs were making in their behalf. Few presidents of great research universities were conversant with the significance of P1, P2, or P3 red-on-black decals pasted on laboratory walls.

Internal Threat to the RAC

The commentary on Maxine Singer's proposal was not overwhelming in numbers and was nearly uniform in sentiment supporting its adoption.[53] It was exemplified in a letter from Paul Berg:

> . . . I believe the NIH Guidelines have served their purpose. They, and the process used to produce them, educated scientists to the anxieties that were raised and developed advice on how to perform the work safely. There is now a widespread acceptance and voluntary compliance of these recommendations and I see no reason why that would not continue without all the red tape. The time has come to reconsider the costs and benefits of the RAC-ORDA establishment.[54]

I wondered at the time if these sentiments portended a swift general shift toward a disappearance of ORDA, IBCs, and the RAC. I felt strongly that RAC was succeeding in a demonstration of how the people and science could be brought together to work out what lay on the horizon of future medical and other uses of recombinant DNA.

I had appointed former Congressman Thornton to the chair of the RAC after Jane Setlow's term expired. Not only had he previously led hearings that were useful tutorials for his fellows in the Congress, but I also found in him a staunch fellow believer in the importance of the RAC and the Guidelines:

> The purpose of the Guidelines is (1) to establish a rapid, complete means of communication, (2) to assure that the Guidelines are conservative, yet allow research to proceed, and (3) to permit public participation in the formulation of public policy.[55]

Ray Thornton was proof that a layman could ably lead an elite science committee of molecular biologists. And this was an example desired for future use.

Baltimore-Campbell Proposal

A few months after the Guidelines revision of November 1980, a new proposal for more extensive defenestration of the Guidelines was introduced by RAC members Allan Campbell and David Baltimore: "to convert the NIH Guidelines into a code of standard practice and to reduce recommended containment levels for some experiments."[56] The proposal was designed to have several major effects. First, it would revoke the mandatory nature of the Guidelines by eliminating all the sections specifying regulatory procedures and their underlying organizational machinery. "The Purpose of the Guidelines" would now read: "Adherence to these standards by all laboratories using recombinant DNA is recommended." Second, containment would essentially be P1 except where the CDC standards for pathogenicity of an organism required higher than P1. The prohibitions of the Guidelines were to remain in force.

A full dismantling of the 6-year-old RAC was in the minds of some in the full panel of RAC members when the Baltimore-Campbell proposal was first extensively discussed at the RAC meeting in April 1981.[57] A letter from Asilomar veteran Norton Zinder encapsulated the thinking of many who had been involved in the inception of that historic meeting:

> It would be an important precedent for the NIH to dismantle the unneeded regulatory structure. If scientists are ever again to cope with *potential* hazard, they must see that what they believed were temporary measures can be undone.[58]

RAC member Patricia King remarked that if the RAC undid the Guidelines, it should be done in a responsible manner, with public input solicited. Campbell replied, "The question is whether the function of the RAC is to deal with hazard or to deal with fear." A number of RAC members felt that any dismantling must be preceded by a thorough assessment of the Guidelines and development of a policy on the matter.

In accordance with a vote of the RAC, Chairman Thornton appointed nine members of the RAC, two liaison members, and two other scientists who had long involvement with the recombinant DNA issue as a working group on revision.[59] Susan Gottesman was appointed chair of this group. And she would play a major role in achieving the desirable part of this proposed revision, leaving the core machinery intact. The working group concluded that in most organisms the barriers to expression of foreign genes, the necessity for new enzyme activities to function as an integrated part of an existing pathway, and the selective disadvantage of carrying recombinant DNA would interfere with such organisms establishing themselves in the environment and thus ultimately with their potential to cause harm. Therefore, the minimum controls associated with good laboratory practices should be sufficient. Many such experiments had already been exempted from any special procedures.

Still, the residual concern that "a particular subset of experiments may pose some possibility of risk" was present.[60] The majority of the working group supported the containment provisions of the Baltimore-Campbell proposal but wished to maintain the mandatory aspect of the Guidelines. A minority preferred either a "voluntary code of practice," as stipulated in the original proposal, or an end to all specifications for working with recombinant DNA, apart from those "good laboratory practices." The mood prevailed that if some unsuspected risks existed, it would be foolhardy to do away completely with a working system consisting of ORDA, the RAC, and the IBCs.

Further thoughtful and lengthy discussion of the proposals for revision of the Guidelines took place at the RAC meetings over the next year. Eventually, the alterations of the Guidelines in terms of *containment* as proposed by Campbell and Baltimore became the rules, but the Guidelines remained mandatory for all research involving federal support. RAC and ORDA remained unchanged. The IBC review of covered experiments also remained mandatory.[61] Industry cooperation continued unchanged, as witness the statement of a prominent biotechnology representative:

> . . . In the next several years, industrial applications of recombinant DNA research will certainly increase. With its established scientific expertise, credibility, and record of demonstrated sensitivity to public interests, the RAC can . . . without preempting any regulatory authority . . . become a focal point for efficiently addressing recombinant DNA issues important to both science and commercial development. Under such a regime, industry's commitment to full compliance with RAC guidance would be expected to continue and be strengthened.[62]

This transition period of 1980–82 from which simplification of the Guidelines was achieved without loss of their mandatory nature or the essential apparatus was a stress test of the resiliency and adaptability of this "nonregulatory" construction. The RAC survived the episode and began a metamorphosis, one which would drag its attention to another corner to the end of the millennium.

The Case of Cline

On August 25, 1980, members of the Kitchen RAC, including Charles McCarthy, head of the NIH Office for Protection from Research Risks (OPRR) assembled in my office. The subject was a preliminary report sent to OPRR that a hematologist researcher from UCLA, Martin J. Cline, had attempted to replace the defective β-globin gene in the bone marrows of two women with severe thalassemia, one located in a hospital in Naples, the other in Israel.[63] It was widely known that Dr. Cline had several times tried unsuccessfully to get approval for this experiment from the institutional review board (IRB) in UCLA and at least once had appealed to the NIH to have RAC review the protocol without local approval. We had turned down this request as a serious threat to the priority of institutional action by the IRBs. Cline finally arranged to perform the human experiments abroad, carrying with him materials from his laboratory to attempt to transfer a normal globin gene into the patients' bone marrow cells by calcium phosphate treatment. He allegedly also attempted to use a recombinant vector containing the normal gene. Colleagues at UCLA did not accept his initial denial of the latter work, and it soon became impossible for him to conceal the facts. As he knew, he had prematurely jumped the gate in the race, ignoring the necessary technological and procedural preparations that would delay the performance of the first approved human gene therapy for another 10 years. The stories of the abortive first attempt and the later successful transfer of a gene have been related in extensive detail in Larry Thompson's *Correcting the Code*.[63]

Needless to say, we were acutely sensitive to the possibility that these first reports might upset the delicate balance on which our stewardship of recombinant DNA experimentation was maintained. A frowning public attitude on genetic manipulation already existed:

> . . . It is now necessary, as perhaps it was not before, for the public to weigh the social value of freedom of inquiry; before the public needed only to appreciate the benefits of inquiry.[64]
>
> . . . When it comes to genetic manipulation, in particular with changing human types or their hybridization with alien genetic material there are questions not of risk but of essential admissibility whose responsible answering will require the ultimate wisdom of the race far beyond the special competence of scientists and their right to free inquiry . . . something more transcendent.[65]

Failure to obtain approval of the UCLA IRB was a serious infraction apart from the recombinant gene aspect. Apparently, Cline was not aware that the human subjects general assurances under which UCLA researchers were bound meant that off-site, as well as on-site, experiments and use of any NIH funds for them fell under NIH rules requiring IRB (and IBC) approval from the investigator's home institution.

At that time there were no Public Health Service or departmental committees to which such serious violations of the rules would be referred. An infraction of the recombinant DNA Guidelines, if one had occurred, or of the human subject regulations was a matter to be dealt with centrally at

NIH. Therefore, I decided that a review of the case would be held from my office, excluding myself, by a carefully selected ad hoc committee of NIH scientists. I chose Richard Krause, director of the National Institute of Allergy and Infectious Diseases, who also served as chair of the rDNA executive committee; Charlie McCarthy, head of OPRR, a former priest and an ethical authority; Mortimer Lipsett, director of the Clinical Center, an experienced IRB man who could be counted upon to balance some of McCarthy's rage with a sense of what it is to work in the trenches; Sue Gottesman, as close to an ultimate expert on the rDNA Guidelines as there was in the world; Harry Keiser, a clinical investigator from the National Heart, Lung, and Blood Institute, Richard Riseberg, attorney; and Bernard Talbot, executive secretary, all of whom I judged to be "dedicated to keeping the flame of inquiry high and free of contaminating smoke."[66]

I prepared the charge to the ad hoc committee in anticipation of receipt of the reply from UCLA:

> The Office for Protection from Research Risks has received a report from the Chancellor of the University of California, Los Angeles, dated _____ concerning research activities of Dr. Martin Cline, an NIH grantee.
>
> The Committee is requested to consider this report and such other information as it deems necessary to advise me concerning the following questions:
>
> 1. Do these activities constitute violations of DHHS rules, particularly the Human Subjects Regulations (45 CFR 46) or the *NIH Guidelines for Recombinant DNA Research?* If so, what actions should NIH take with respect to such infractions?
> 2. If there has been a clear transgression of the spirit of the Guidelines or regulations, are there steps that NIH should take in joining the larger community at interest in safeguarding the intent of the rules?
>
> The committee should begin without delay, and during its deliberations, take all necessary steps to safeguard the privacy and other rights of the individuals involved.
>
> I will anticipate a preliminary report within 30 days.

It is ironic that these preparations were taking place within the space of a few weeks during which I was also compelled to visit each of the advisory councils of the institutes and the DAC to explain an important change in regulation freshly minted in the pages of the *Federal Register*.[67] These new rules had arisen in March 1979 at an unforgettable and rare rancorous meeting with Secretary Califano. I had just made the secretary aware of several cases of egregiously fraudulent experiments by NIH grantees. With Tom Morris, the inspector general of the department, sitting grim-faced in the corner, the secretary said he presumed the culprits would be forbidden any further government support. I answered that there had long been such punishment in case of government *contractors*. Throughout NIH history, however, there had never been debarment provisions in cases of scientific misuse of *research grant* monies. It was a tradition embedded in the general horror of all scientists against blacklisting of a fellow member, preventing his access to the laboratory. Not impressed by my Galilean stance, the secretary roared, "If you don't see that such scientists can be disbarred, I'll strip your delegations to award a grant to anybody." Our conversation had never before contained such animus. He had the regulations promulgated, and during the next several months it was my task to advise the leadership of the scientific community of this new and serious order. There were some scientific members of every Advisory Council who bristled at my announcement.

Thompson describes the outcome of the ad hoc committee's actions:

> On May 21, 1981 . . . the Ad Hoc Committee . . . handed down its judgement and the punishment: Cline was guilty of violating the rules and "warrants disciplinary action." The recommended punishment came in four parts: For the next three years, a requirement to get prior NIH approval for any experiment he conducted in human patients; . . . on any experiments using recombinant DNA whether or not they involved human subjects; individual institutes would review existing federal support for Cline's ongoing work; and in the future, whenever Cline submitted a request for NIH funding, he would have to send along the committee's unflattering report. The judgement . . . would remain in effect for three years. Three days later, Fredrickson publicly accepted the report, saying: "it leads me inexorably to agree with the conclusions that Dr. Cline has violated both the letter and the spirit of proper safeguards to biomedical research. I therefore accept all the committee's recommendations."[63]

A detailed report was sent to the advisory councils of the several NIH institutes who supported Cline's work. Their concurrence and the prescription of harsh penalties was announced the following November.

The Metamorphosis of the RAC

Cline's tragically premature experiment merely underlined the possibility of human gene therapy, which already had become a predictable goal of molecular genetics. More than 20 years before, a few people had already foreseen that someday such an experiment might be possible.[68] In the late 1960s, a slow crescendo of interest by molecular biologists in the eventual possibilities of experimentally replacing a defective gene that caused a human disease was evident. By the 1970s, numerous symposia dedicated to such possible future gene therapy were held. Successful experiments to make viral vectors capable of transferring DNA raised speculation about who might be the first to successfully target the human genome.

In June 1980, a letter signed by three prominent Protestant, Jewish, and Catholic leaders had been sent to President Jimmy Carter expressing concerns about the dangers inherent in genetic engineering in the context of moral, ethical, and religious questions. This event and the widely publicized Cline experiment promptly led to the formation of a President's Commission for the Study of Ethical Problems in Medicine and Biomedical and Behavioral Research, particularly to examine the moral issues in molecular biology and possible intervention in the human genome.[69] The president's commission issued a report, *Splicing Life*,[70] which was also the subject of a House hearing chaired by Congressman Albert Gore, Jr.[71]

The commission regarded splicing life as

> . . . an issue of concern to the people of this country—and of the entire globe—for the foreseeable future; indeed the results of research and development in gene splicing will be one of the major determinants of the shape of that future, with its profound social and ethical consequences retain[ed] at the very center of the conversation of mankind.

In its search for a mechanism for oversight of gene splicing in humans, the commission seriously considered numerous possibilities. Groups could be gathered for discussion to try to set limits, but the requisite knowledge of the science was crucial to any such exercises. The role NIH had played in tending the *Guidelines for Recombinant DNA Research* was noted. "Revising RAC" became a topic for consideration. The commission remembered the expansion of the RAC with nonscientific members by Secretary Califano in 1978. In a general desire to widen the public opinions and advice on the RAC, further reference was made to a suggestion I made in 1982 shortly after I left my position as the NIH director:

> The idea . . . is to make RAC more representative of both the scientific and regulatory communities, as well as the public, to be better equipped to deal with the emerging problems and be relieved of some of the detailed burden of reviewing minor administrative concerns. Such a third generation RAC should be accountable to a Cabinet member . . .[72]

After much discussion of how the RAC might be further changed, the members of the commission wandered off target to reflect on other possible creations, including a genetic engineering commission of 11 or 12 members outside of the government; some other body with an eclectic membership;

a series of panels; or, failing these, why not "assign responsibility for oversight of genetic engineering to the body that succeeds the president's commission?" Eventually, neither the commission nor the participants in Mr. Gore's hearing emerged with any definitive solution.

A year later, I was not surprised to learn from a distance that the favorable public recognition of the RAC's performance had prompted it to begin to take on this new role voluntarily.

Departure

I remained in the post of director of NIH through the spring of 1981 and finally received word that President Reagan would not ask me to leave. Not long after that, and having sent a letter to the president, on June 16, 1981, I convened a general meeting of NIH and announced that it was time to go ". . . before I forget how to be a scientist and a physician."

12
Moral

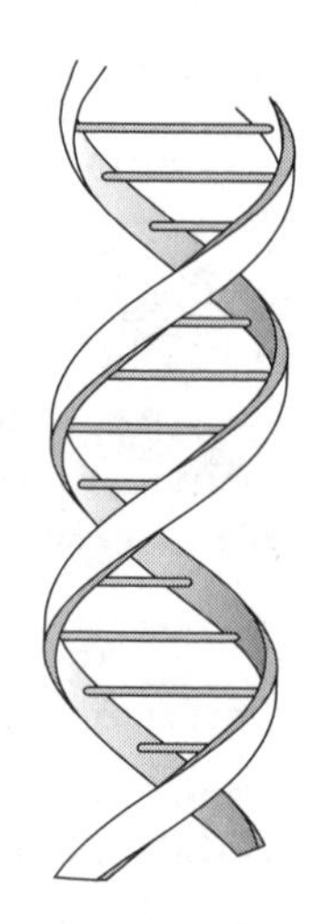

On the wall in my study hang two glass plates enclosing several documents with the heading "The Miami Winter Symposium Distinguished Service Award, 1985." One document is a replica of the *Federal Register* of Wednesday, July 7, 1976, the cover of the first issuance of the *NIH Guidelines for Recombinant DNA Research.* Adjacent is a statement:

> The NIH Guidelines became an international standard of reference and provided an unprecedented opportunity for public input into scientific matters. The use of this new technology in the ensuing decade has resulted in extraordinary scientific progress and significant practical benefits. These contributions rank amongst the most important that science has brought to society.

A third document is entitled "Moral." It is from "A History of the NIH Guidelines" prepared for delivery at a meeting in Wye, England, in 1979. When I was "grounded" with my fellow agency chiefs by Secretary Califano during the Three Mile Island nuclear threat in April 1979, I ceded my plane ticket and script to William Gartland, the director of Office of Recombinant DNA Activities, who read the paper.

This was the closing "moral":

> It is possible that the "DNA Affair" will someday be regarded as a social aberration, with the Guidelines preserved under glass. Even so, we can say the beginnings were honorable. Faced with real questions of theoretical risks, the scientists paused and then decided to proceed with caution. That decision gave way to dangerous overreaction and exploitation, which gravely obstructed the subsequent course. Uncertainty of risk, however, is a compelling reason for caution. It will occur again in some areas of scientific research, and the initial response must be the same. After that the lessons learned here should help us through the turbulence that is sure to come.[1]

Twenty years later, I have the same convictions. The initiators of the Asilomar conference, particularly Paul Berg and Maxine Singer, who had the key hand in many decisions, did the right thing by society and by science. From my standpoint, the organizing committee can also take credit for calling upon the NIH to provide an advisory committee. The RAC became the vehicle for allowing use of the technology to proceed under safety devices that were, as we know now, excessive and unnecessary. For their demonstration of prudence, patience, and dedication to both scientific principles and public welfare, the members of the RAC have earned a place in the history of science.

My "moral," however, is deficient if it leaves a pejorative connotation assigned to those who "overreacted" and were "exploitative" when the scientists took the unusual steps they did. When science makes moves that can be interpreted as threatening to the public welfare, it is proper and necessary for other citizens to provide the "turbulence" necessary to give them access to the full intent and meaning of the science. It is the "dissidents" who provided important counterweights to the assignment to the NIH director in the later revisions of Guidelines: to "weigh new proposed action, through appropriate analysis and consultation, to determine that it complies with the Guidelines and presents no significant risk to health or the environment."[2]

When the decisions of the scientists at Asilomar became a "federal case," the window was thrown open to enlighten scientists on the different powers and roles of the legislative and executive branches of the government when they become concerned with science. By virtue of its powers to appropriate federal money for science—especially for basic research—and to make laws concerning the conduct of research, Congress has the greatest power. Members of the Congress are often called upon to see beyond the short visions of many of their constituents. They must also try to weigh costs and benefits when neither is sufficiently known to provide a satisfactory calculus. Their task of making laws that weigh the value of creativity among all the other traits of human behavior is neither an easy nor an enviable one.

There are many invaluable public servants who facilitate the legislative powers of Congress. They also complement many others in the executive branch, who combine a strong sense of civic duty and respect for the law in aiding the agencies and institutions to disseminate the social benefits of science. This memoir contains many examples of such persons. By relating their deeds, it especially honors the group of people at NIH who, in their role as the "Kitchen RAC," were exemplary of the finest tradition of public service.

Since those first years of controversy over new technology for molecular genetics, our successors have been faced with the realization that Asilomar was only the first act. In the last decades leading to the millennium, new controversies over molecular genetics have appeared, including, most prominently, the attempts to supply substitute genes for humans having serious illnesses.

NIH lacks statutory powers of regulation. This characteristic of NIH stewardship was nevertheless invaluable as the agency more and more provided the forum for introducing scientists to the public and vice versa. Any change in this status that would attempt to convert NIH to a regulatory mode would be a dangerous mistake. The scientific powers of NIH, the National Science Foundation, and other science agencies, are best employed in assistance to others with regulatory authorities.

The story of the RAC demonstrates several valuable aspects of making decisions regarding arcane technology. Initially, only the cognoscenti can fashion a matrix for discussion. Later, the addition of other scientists and laypersons to the process enriches its value. The early proposals in legislation to provide "commissions," using a minority of scientists to create the matrix while setting the majority of members off on a hunt for social and ethical issues of recombinant DNA research, were patently premature. Bodies having a heavy presence of lay expertise have subsequently been constituted and given mandates to define social purposes and protection of both public welfare and scientific freedom.

Enormous power has been inherited by humankind in this new biology. It has altered the nature of older relationships between academia and commerce. The resulting entrepreneurial forces represent the only way in which new discoveries will be rapidly converted to inventions, enriching the economy and offering myriad possibilities for improved health practices in the future. At the same time, the complications in preserving objectivity in such interactions continue to be troubling. May we find the wisdom to husband this complicated new set of forces with generosity and humility.

13
Epilogue

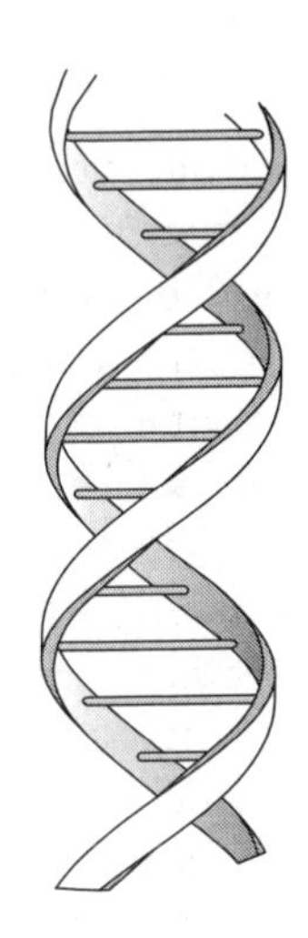

Reductionism is, without question, the most successful analytical approach available to the experimental scientist. With the advent of techniques for cloning and sequencing DNA and the development of a variety of molecular probes for localizing macromolecules in cells and tissues. . . the biologist now has available the most powerful reductionist tools ever invented.[1]

If a document marking the arrival of "panreductionism" into the community of biomedical science is desired, the minutes of the February 1982 meeting of the NIH Recombinant DNA Advisory Committee (RAC) may qualify—or, more definitively, the revised *Guidelines for Recombinant DNA Research*, recommended at that meeting and published in the *Federal Register* of April 21, 1982.[2] Absolute prohibition of all classes of recombinant DNA experiments was removed. Three of the five classes of experiments previously forbidden in the Guidelines—those involving transfer of drug resistance traits, use of toxin-producing genes, and release of organisms containing recombinant DNA products into the environment—were now allowable if the RAC could make a case for lifting the barrier by the NIH director. Although the overall structure of the Guidelines was left mandatory, additional responsibility was now shifted to the institutions or organizations where recombinant DNA research was performed; each had an institutional biosafety committee and additionally employed a biological safety officer where P3 or P4 containment levels were used. The coverage extended on a voluntary basis to nearly all of the 80-some commercial locations using recombinant DNA.

Exactly 7 years earlier, the molecular biologists had committed an extraordinary voluntary act of handing the power to a federal agency to determine how much distance had yet to be traveled before a biological revolution might be consummated. During those years, under continuous scrutiny and often against the objections of various factions of the public and members of Congress, a committee of citizens—mostly scientists—patiently steered through the storms of controversy. Molecular biologists, assiduously following the rules, progressively crafted and carefully tested the safety, range, and power of new tools based on recombinant techniques and thereby added steadily to the knowledge of their safety. As many had predicted—and for which all were grateful—no harm to people or the environment was ascertained. Thus, Pandora's box had become unlocked forever and now contained tools that would have profound effects on biological experiments and on the understanding of humankind and its evolution.

With this release, there would come profound changes in the culture of science, its relationship to society, commerce and health care, and—with many new benefits—also many new problems and dilemmas.

Technical Power

Recombinant DNA allowed scientists to obtain genes and begin to understand their structure, location, and mode of regulation in normal development and in disease.[3] For the first time, human genes could be cloned and amplified in their pure state. Specific human genes could be linked to transcriptional initiators and to translational promoters to force production of large amounts of desired human gene products, such as human insulin. The precise arrangement of genes in a complex locus, such as the arrangement of immunological genes in the human genome, could be determined. Any cloned gene could be located on its chromosome, an important step toward identification of genetic polymorphisms, greatly expanding the power to follow different phenotypes in family studies. Not the least to be mentioned is the visualization of an ultimate goal of altering the action of a disease-causing gene.

The Entrepreneurial Side

The commercial value of genomics had been sighted before the Asilomar conference, and the vision prospered under voluntary compliance with the Guidelines, spawning an important new industry.[4] A critical factor was government funding of basic research, along with a favorable institutional patent policy providing industry access to the stores of inventions arising in academic laboratories. Cetus and Genentech were among the first such firms, and by 1981, the input of venture capital had created more than 80 new biotechnology firms with budgets totaling $4.5 billion. By 1984, there were several hundred new small companies and noticeable commencement of such activities in the large pharmaceutical firms. Genetic engineering was already leading to new medicines, such as human growth hormone and insulin, and new strains of agricultural crops. Ten years later, there would be more than 1,300 biotechnology companies in the United States and an estimated 100,000 people working in this "new miracle industry." New laws fostered CRADAs (collaborative research and development agreements) whereby government scientists and profit-making laboratories could work together on joint projects. By 1992, the NIH intramural program had 145 active CRADAs.[5]

Cultural Changes

Enter the Entrepreneur

The introduction of recombinant DNA technology quickly effected a perceptible cultural change in how research was conducted by academic and government intramural scientists. Closer ties were promulgated between academic research and industrial settings. The days were numbered when *serendipity* was considered the best route to important discoveries[6] and when the highest reward for making an important contribution to the communal mosaic of knowledge was peer recognition. The reductionist approach now caused researchers to select projects with clearer objectives, including the vision of usefulness of the product, possible ownership of intellectual property, and wealth above modest standards of laboratory employment. There was new awareness of the value and cultural acceptance of entrepreneurial collaborations among scientists. The line between corporate and academic science, once a tight border in biological science—compared with the earlier accommodations achieved in engineering and the chemical and physical sciences—was softening and becoming similarly blurred. The industries were equally aware of the desirable route for filling product pipelines by arrangements for additional financial support to nonprofit partners in exchange for the first rights to ideas and inventions for development. The rise in industrial investment also meant more scientists were leaving government or academic positions for full-time industry research and development. Universities began to relax limits on outside faculty employment for gain, and institutions formed relationships with industry to keep pace with the rising costs of competing in genetic research.

It is too early to muster either a comprehensive or focused view of the ultimate reactions of government and the public to an accelerating succession of mind-boggling achievements of the "new genetics" after crossing the bridge from Asilomar into the first part of the new millennium. Pulses of excitement have included the cloning of sheep, the revelation of the apparently miraculous capabilities of embryonic stem cells, and, recently, the unveiling of a map of the human genome, a triumph undiminished by the presentation of two competitive maps to President Clinton. One was the product of a government-funded international consortium led by the National Human Genome Research Institute, and the other was from Celera Genomics Inc., a venture capital product assembled in a remarkably short time. We have yet to perceive how questions of intellectual property interests may impinge upon the fullest extension of the intelligence contained in these maps.

The Social Contract

The unwritten contract defining the profession of scientist was explicitly mentioned by one commentator (see note 19 in chapter 3) in this history of one of the greatest extended public debates on the subject. This contract will continue to be a central issue in the new millennium, not only for the agent of basic science (the scientist) and the purchaser (the federal government) but also for the public. Society will want to assert itself more vigorously in regard to the moral and ethical bases on which the costs and benefits of the gifts of genomics are calculated. Guston[7] recently provided a useful reminder of the "essential elements of the [social] contract as a unique (post-war) partnership between federal government and universities for the support of basic research, the integrity of scientists as recipients of federal largesse, the easy translation of research results into economic and other benefits, and the institutional and conceptional separation between politics and science" that it maintains. Guston believes that an element of "collaborative assurance" to justify perpetuation of such a laissez-faire system is missing. He implies that the system is not justified by the mere generation of intellectual property or, he asserts, adequately monitored by the peer review system. He proposes that the government purchaser must be able to assess the collective value of group achievement and that some of the achievement must be directed toward a larger, more collaborative target. Frodeman and Mitcham[8] similarly believe that scientists now have much broader obligations, even if increased public discourse on social objectives raises questions that lack simple answers.

Holton and Sonnert[9] argue that changes or a supplementation of the compact might pursue a "conscious combination of aspects of the Newtonian (comprehensive) search for knowledge and the Baconian (mission-oriented)" modes toward a third alternative. They would use the Lewis and Clark expedition as a first example of a "Jeffersonian mode" of research that would "motivate a specific research project by placing it in an area of basic science ignorance that seems to lie at the heart of a social problem." It is their view that the manner in which NIH manages to support basic research in the interest of health is a commendable example of a Jeffersonian institution. The successful interactions of the NIH RAC and citizens to achieve voluntary compliance and to maintain open dialogue while developing rules that do not penalize the potential users or detract from the common good proves that the compact is still a workable proposition in our society.

The erstwhile social contract has, however, become a troika, with commerce contributing its own priorities, supporting basic science, assigning proprietary value, and taking control of intellectual property. This "third party" defines the value of productivity by a calculus in which the shareholder looms larger than the common good. It remains for government and the scientific institutions (including commercial houses) to share in maintaining the integrity of basic science. The Congress will unquestionably respond to renewed concerns of their constituents generated by the speed and vitality with which genes (genomics) or their protein products (proteomics) become subjects of more intensive research and commerce. The legislature has sometimes deferred the use of federal funds to deal with dilemmas rich in ideological content. The pressure of new opportunities in research may lead to more recognition of the social contract and more rational redefinition.

Medicine and the Genes

> The science of genetics is overwhelmingly the science of disease, primarily of those "inborn errors of metabolism" that cause human suffering . . . The mighty gene is seen as medical salvation. Precisely because genetics is seen as the next big step in medicine, we start to believe it is the *only possible* step.[10]

The above opinion is from a recent book by Bryan Appleyard, a lay observer who has provided a valuable perspective on the effect of the resulting explosion of molecular genetics attending the availability of recombinant

DNA technology. It allows me to close with an update of the history of the now almost venerable RAC.

The RAC and Gene Therapy

At the end of chapter 11, the RAC in 1981 was considering whether it should take up a new mission in response to a presidential commission that had favorably considered its record and wondered if it was possibly the body to take up the leadership in the awakened interest in gene replacement in humans. As the RAC seriously considered this challenge at its meeting in April 1983,[11] recommendations were made that membership of the committee should be modified to "deal credibly with these issues" and that procedures should be developed for RAC consideration of proposals for human gene therapy experimentation. By late 1984, a new RAC Working Group on Human Gene Therapy (WGHGT), composed of laboratory scientists, clinical investigators, lawyers, and ethicists, proposed a set of "Points to Consider in the Design and Submission of Human Somatic-Cell Gene Therapy Protocols."[12] In true RAC fashion, this format was widely discussed, published for comment, resubmitted to revision for an extended period, finally approved by vote, and recommended to the NIH director in 1990. LeRoy Walters, an ethicist who was an early member of the RAC, and who returned to serve for a period as chairman of the WGHGT, has provided a useful chronology of events of the time.[13] Many contributions over years of preparation and debate occurred before gene transfer experiments were considered to be feasible.[14] Each proposal was published in the *Federal Register* and discussed in public sessions run by parliamentary procedures, and recommendations to the NIH director were determined by vote. Proposals were often deferred until more information was obtained.

The first proposals for gene therapy were submitted by NIH scientists Steven Rosenberg, R. Michael Blaese, and W. French Anderson, and the first experiment was performed at NIH in 1990.

FDA Appears. Under its investigational new drug regulations, the FDA has regulatory control over gene transfer experiments; both NIH and FDA agencies considered about 100 proposals for gene therapy in the next decade.[15] All RAC considerations were open to the public. FDA, bound to protect proprietary data, never opened its meetings to the public. The FDA had the final say independent of whether the RAC recommended approval or disapproval.

Proposal to Terminate the RAC. In July 1996, Harold Varmus, then director of NIH, published a proposal to discontinue the RAC and thereby eliminate the overlap between NIH and FDA in review of gene therapy proposals.[16]

The public commentary on this move, however, included opposition from pharmaceutical and biotechnology firms, numerous members of citizen advocacy groups, members of Congress, and scientists, including the American Society for Microbiology, the largest single life science society in the world.[17]

The NIH director was then persuaded to modify his initial proposal. Citing "the historical importance of the RAC as a public platform for discussion of the science, as well as the safe and ethical conduct of . . . research," he proposed to retain the RAC but reduce its members from 25 to 15 and withdraw its power to approve individual gene transfer experiments. Instead, the RAC would discuss novel gene transfer experiments, regularly convene gene therapy policy conferences, and maintain public access to information about human gene transfer clinical trials.[18] The changes were implemented.

Death of a Patient. The RAC was operating under these conditions in September 1999 when an 18-year-old volunteer patient on a gene transfer protocol died at the University of Pennsylvania. Investigations by the FDA and the university were aired at a 3-day public meeting of the RAC.[19] The proceedings were amplified in subsequent press coverage and revealed the potential for conflict of interest when medical scientists involved in clinical trials have corporate interests in the outcome. The debate in the academic community and the Department of Health and Human Services over new restrictions to protect the human subjects of research will undoubtedly bring changes in the rules.

Nearly a decade ago, Edward H. Ahrens, Jr., a distinguished clinical investigator of my own era, wrote wistfully of the "loss of the traditional starting point of . . . an integrationist science to a reductionist preoccupation with the technology of molecular biology and molecular genetics."[20] The preoccupation with reductionist approaches will someday revert to more patient-oriented research as the new science matures. We need not protest the brilliant radiance of this new, astounding series of additions to society's knowledge and capabilities, as long as the light also illumines the morals that ensure the distribution of the benefits of this science among humankind.

The recombinant DNA controversy, begun at Asilomar 25 years ago, is over. In the new controversies that are certain to arise, the players will change, but the stakes will not. This book has attempted to describe in detail one of the most serious endeavors in our time to address together the social compact between basic science, the government sponsor, commerce, and the public good.

Notes

Foreword

1. Maxine F. Singer, "Spectacular Science and Ponderous Process" [Editorial], *Science* **203**(4375):9, 5 January 1979.

2. The most detailed discussion of the Martin Cline case is contained in Larry Thompson's book, *Correcting the Code: Inventing the Genetic Cure for the Human Body* (New York: Simon & Schuster, 1994), chapters 6 and 7.

3. See chapter 11, p. 276. This proposal was first presented at the 1982 annual meeting of the American Association for the Advancement of Science and was published as Donald S. Fredrickson, "The Recombinant DNA Controversy: The NIH Viewpoint," in Raymond A. Zilinskas and Burke K. Zimmerman (ed.), *The Gene-Splicing Wars: Reflections on the Recombinant DNA Controversy* (New York: Macmillan Publishing Company, 1986), p. 13–26. See especially p. 24.

4. United States, President's Commission for the Study of Ethical Problems in Medicine and Biomedical and Behavioral Research, *Splicing Life: The Social and Ethical Issues of Genetic Engineering with Human Beings* (Washington, DC: U.S. Goverment Printing Office, November 1982).

5. Ibid., p. 85.

6. United States, National Institutes of Health, "Recombinant DNA Research; Request for Public Comment on 'Points to Consider in the Design and Submission of Human Somatic-Cell Gene Therapy Protocols,' " *Federal Register* **50**(14):2940–2945, 22 January 1985.

7. Chapter 13, p. 287.

8. Chapter 5, p. 94.

9. Chapter 13, p. 282–285. To his credit, during the final years of his directorship former NIH director Harold Varmus also sought to restore greater access to essential research tools developed with the aid of NIH funding. See Eliot Marshall,

"New NIH Rules Promote Greater Sharing of Tools and Materials," *Science* **286**(5449)**:**2430–2431, 24 December 1999; and Jeffrey Brainard, "NIH Issues Guidelines on Release of Research Tools Developed with Federal Funds," *Chronicle of Higher Education* **46**(19)**:**A37, 14 January 2000. The text of the guidelines can be found in *Federal Register* **64**(246)**:**72090–72096, 23 December 1999.

10. http://grants.nih.gov/grants/award/trends95/annotate.htm

11. http://grants.nih.gov/grants/award/trends95/pdfdocs/fedtabl1.pdf

12. Pharmaceutical Research and Manufacturers of America, *Pharmaceutical Industry Profile 2000* (Washington, DC: PhRMA, 2000), p. 20.

13. See, for example, chapter 2 in the recent draft report of the National Bioethics Advisory Commission entitled "Ethical and Policy Issues in Research Involving Human Participants," draft dated December 19, 2000 (http://bioethics.gov/human/humanpdf_toc.html).

14. Chapter 4, p. 72.

Preface

1. A dozen years after Asilomar, Philippe Kourilsky, today the director of the Institut Pasteur, published his recollections of this extraordinary meeting, which he attended as one of six French participants. This quotation is adapted from his book *Les Artisans de L'Hérédité* (Paris: Éditions Odile Jacob, 1987), p. 143–144. See also P. Kourilsky, "Manipulations génêtiques (in vitro): Compte-rendu de la conférence de Pacific Grove," *Biochemie* **57**(2)**:**vii (1975), and transcript of an interview with Philippe Kourilsky, March 20, 1976, MIT Archives, Recombinant DNA Controversy Oral History Collection, Box 9, Folder 113.

2. The largest single source of references for this book is *Recombinant DNA Research* (hereafter cited as *RDR*), National Institutes of Health, Department of Health, Education and Welfare, published in 20 volumes. Volumes cited are vol. 1, 1976 (DHEW publication no. [NIH] 76-1138); vol. 2, 1977 (NIH 78-1139); vol. 3, 1978 (NIH 78-1843); vol. 3 appendixes (NIH 78-1844); vol. 4, 1978 (NIH 79-1875); vol. 4 appendixes (NIH 79-1876); vol. 5, 1980 (NIH 80-2130); vol. 6, 1981 (NIH 81-2386); vol. 7, 1982 (NIH 83-2604); vol. 8, 1986 (NIH 86-2863); vol. 9, 1986 (NIH 86-2864). The archive now continues on a website (http://www.nih.gov/od/oba/). Many of the documents in that equally voluminous source, the *Federal Register*, are cross-filed in the *RDR*. All of my personal papers relevant to the recombinant DNA controversy, including sets of diaries, are being catalogued in public archives in the History of Medicine Division of the National Library of Medicine. A number of references in this book are to labeled folders in DSF Papers.

Chapter 1

1. At least four books on the early history of the recombinant DNA controversy contain detailed descriptions of the 1975 Asilomar conference: Michael Rogers, *Biohazard* (New York: Knopf, 1977); John Lear, *Recombinant DNA: the Untold Story* (New York: Crown, 1978); Nicholas Wade, *The Ultimate Experiment: Man-Made*

Evolution (New York: Walker, 1979); Sheldon Krimsky, *Genetic Alchemy* (Cambridge, Mass.: MIT Press, 1982). Additional sources include the Recombinant DNA Historical Collection at the MIT Archives; the Archives of the National Academy of Sciences, including the original tape recordings of the conference; the NIH Central Files; and the collections of the National Library of Medicine. Between November 1989 and June 1990, the following participants at the Asilomar conference—William Gartland, Leon Jacobs, Philippe Kourilsky, Arthur Levine, Andrew Lewis, Herman Lewis, Malcolm Martin, Robert Martin, Anna Marie Skalka, Waclaw Szybalski, and Pierre Toillais—were interviewed in person or by telephone.

2. The clue: F. Griffith, "The Significance of Pneumococcal Types," *J. Hyg.* **27:** 113–159 (1928). The key discovery: O. Avery, C. M. MacLeod, and M. McCarty, "Studies on the Chemical Nature of the Substance Inducing Transformation of Pneumococcal Types," *J. Exp. Med.* **79:**137–158 (1944). The confirmation: A. D. Hershey and M. Chase, "Independent Functions of Viral Protein and Nucleic Acid in Growth of Bacteriophage," *J. Gen. Physiol.* **36:**39–56 (1952).

3. A. Hunter Dupree, "The Great Instauration of 1940: the Organization of Scientific Research for War," in *The Twentieth Century Sciences* (Gerald Holten, ed.) (New York: Norton, 1970).

4. J. D. Watson and F. H. C. Crick, "A Structure for Desoxyribose Nucleic Acid," *Nature* **171:**737–738 (1953).

5. Horace Freeland Judson, *The Eighth Day of Creation: Makers of the Revolution in Biology*, p. 605–608 (New York: Simon and Schuster, 1979).

6. Ibid. Judson provides an unparalleled source of the romance of early molecular biology in his book. He introduces us to a galaxy of performers, including John Kendrew and Max Perutz, crystallographers who worked out the structures of myoglobin and hemoglobin; Erwin Schrödinger, the mathematician considering the adaptation of quantum mechanics to living organisms (*What is Life? The Physical Aspect of the Living Cell* [Cambridge: University Press, 1944]); Max Delbrück, the phage expert whose early training was with the atomic physicist Niels Bohr; Leo Szilard, who was with Fermi at the Manhattan Project before he took up the study of phage and became an ardent pilgrim among the biologists; Francis Crick, who first read physics, but fortunately for molecular biology turned to the study of living things and was the perfect complement to James D. Watson, a very young biologist who came to Cambridge after his doctorate under Salvador Luria at Indiana University. Others are described who coursed in and out of Room 103 in the Austin Wing of the Cavendish Laboratory, where the helical model of Crick and Watson was rising.

7. These classical geneticists eventually would receive Nobel Prizes in physiology or medicine—Morgan in 1933 and McClintock 50 years later. Beadle and Tatum (with Lederberg) were so honored in 1958. The increasingly rapid succession of such honors thereafter shows the surging importance of molecular genetics up to the time of the Asilomar conference. There was Arthur Kornberg in 1959; Crick, Watson, and Wilkins in 1962; Jacob, Lwoff, and Monod in 1965; Holley, Khorana, and Nirenberg in 1968; and Delbrück, Hershey, and Luria in 1969.

8. Joshua Lederberg, "Gene Recombination and Linked Segregations in *Escherichia coli*," *Genetics* **32:**505 (1947).

9. David Baltimore, "The Strategy of RNA Viruses," *Harvey Lect.* 1974–75;70 Series:57–74; Howard Temin, "On the Origin of the Genes for Neoplasia. G. H. A. Clowes Memorial Lecture," *Cancer Res.* **34:**2835–2841 (1974).

10. R. C. Parker, H. E. Varmus, and J. M. Bishop, "Cellular Homologue (c-*src*) of the Transforming Gene of Rous Sarcoma Virus: Isolation, Mapping, and Transcriptional Analysis of c-*src* and Flanking Regions," *Proc. Natl. Acad. Sci. USA* **78:**5842–5846 (1981).

11. Meeting at the Rockefeller Institute, October 2, 1966. MIT Archives, Recombinant DNA Historical Collection, Box 16, Folder 204.

12. M. Meselson and R. Yuan, "DNA Restriction Enzyme from *E. coli*," *Nature* **217:**1110–1114 (1968).

13. The international traffic to and from the Pasteur Institute and the charisma of the late Jacques Monod are poignantly revived in the recollections of his colleagues edited by André Lwoff and Agnès Ullmann: *Un hommage à Jacques Monod: Lés origines de la biologie moléculaire* (Paris: Études Vivantes, 1980).

14. Among the members of the NSF Human Cell Biology Steering Committee were Paul Berg, James Darnell, Gerald Edelman, Phillip Robbins, Harry Eagle, William Sly, Matthew Scharf, James Watson, Herbert Weissbach, Charles Yanofsky, and Norton Zinder.

15. Krimsky, *Genetic Alchemy*, p. 26–29.

16. B. H. Sweet and M. R. Hilleman, "The Vacuolating Virus SV 40," *Proc. Soc. Exp. Biol. Med.* **105:**420 (1960); R. J. Huebner, R. M. Chanock, B. A. Rubin, and M. J. Casey, "Induction by Adenovirus Type 7 of Tumors in Hamsters Having the Antigenic Characteristics of SV40 Viruses," *Proc. Natl. Acad. Sci. USA* **52:**1333–1340 (1964).

17. Lear, *Recombinant DNA*, p. 1.

18. Wade, *The Ultimate Experiment*, p. 33.

19. Lear, *Recombinant DNA*, p. 28.

20. Krimsky, *Genetic Alchemy*, p. 33.

21. Leon R. Kass, "The New Biology: What Price Relieving Man's Estate?" *Science* **174:**779–788 (1971).

22. Interview with Andrew M. Lewis at NIH, Bethesda, Md., November 17, 1989.

23. The NIH Biohazards Committee was established in 1972, with Robert Martin as the first chairman. The committee's jurisdiction was restricted to intramural operations. It approved the concept of Andrew Lewis that potentially dangerous viral cultures should be shared with outside investigators who signed and returned a memorandum of understanding.

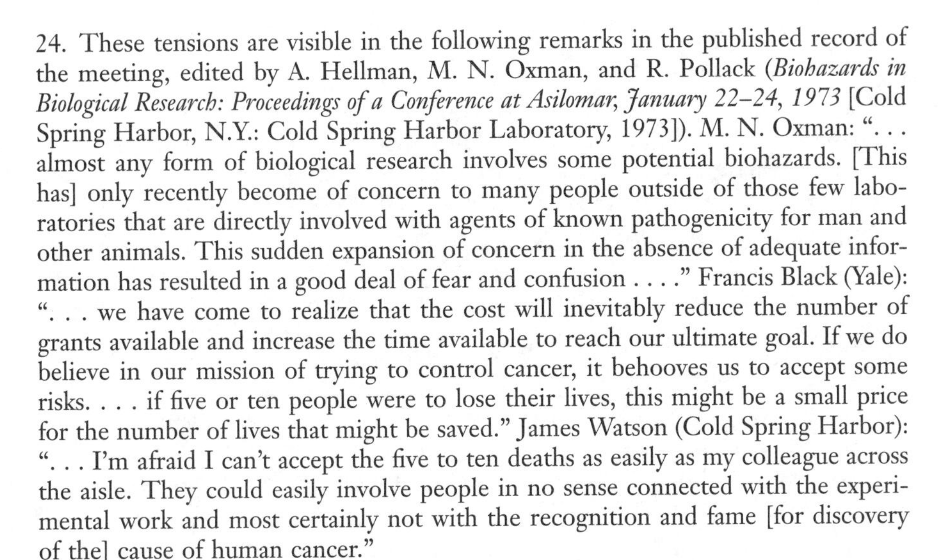

24. These tensions are visible in the following remarks in the published record of the meeting, edited by A. Hellman, M. N. Oxman, and R. Pollack (*Biohazards in Biological Research: Proceedings of a Conference at Asilomar, January 22–24, 1973* [Cold Spring Harbor, N.Y.: Cold Spring Harbor Laboratory, 1973]). M. N. Oxman: ". . . almost any form of biological research involves some potential biohazards. [This has] only recently become of concern to many people outside of those few laboratories that are directly involved with agents of known pathogenicity for man and other animals. This sudden expansion of concern in the absence of adequate information has resulted in a good deal of fear and confusion . . . ." Francis Black (Yale): ". . . we have come to realize that the cost will inevitably reduce the number of grants available and increase the time available to reach our ultimate goal. If we do believe in our mission of trying to control cancer, it behooves us to accept some risks. . . . if five or ten people were to lose their lives, this might be a small price for the number of lives that might be saved." James Watson (Cold Spring Harbor): ". . . I'm afraid I can't accept the five to ten deaths as easily as my colleague across the aisle. They could easily involve people in no sense connected with the experimental work and most certainly not with the recognition and fame [for discovery of the] cause of human cancer."

25. Janet E. Mertz and Ronald W. Davis, "Cleavage of DNA by RI Restriction Endonuclease Generates Cohesive Ends," *Proc. Natl. Acad. Sci. USA* **69:**3370–3374 (1972); A. Dugaiczyk, H. W. Boyer, and H. M. Goodman, "Ligation of EcoRI Endonuclease-Generated DNA Fragments into Linear and Circular Structures," *J. Mol. Biol.* **96:**171–184 (1975). The density of molecular biology studies at Stanford at the time is indicated by the fact that as the paper by Mertz and Davis was sent to the *Proceedings* of the Academy by sponsor Paul Berg, a report of similar findings from another department was on its way to the same journal: Vittorio Sgaramella, "Enzymatic Oligomerization of Bacteriophage p22DNA and of Linear Simian Virus 40 DNA," *Proc. Natl. Acad. Sci. USA* **69:**3389–3393 (1972).

26. EMBO Workshop on DNA Restriction and Modification, Basel, Switzerland, September 26–30, 1972. MIT Archives, MIT Recombinant DNA Historical Collection, Box 16, Folder 205. U.S.-Japan Conference on Bacterial Plasmids, Honolulu, Hawaii, 1972. MIT Archives, Recombinant DNA Historical Collection, Box 16, Folder 206.

27. Lear, *Recombinant DNA*, p. 64–65; Stanley N. Cohen, Annie C. Y. Chang, Herbert W. Boyer, and Robert B. Helling, "Construction of Biologically Functional Bacterial Plasmids In Vitro," *Proc. Natl. Acad. Sci. USA* **70:**3241–3244 (1973).

28. From the onset of its extramural grants program, the NIH protected to the utmost the autonomy and freedom of the basic researchers. The clinical investigators—whom many molecular biologists considered a foreign culture—were meanwhile feeling governmental restraints. Beginning in 1966, all institutions receiving NIH support, and later any federal support, had to have a local institutional review board (IRB) approve their clinical experiments. Soon, at least one member of the IRB was required to come from outside the institution. (After Asilomar, a similar requirement would be imposed on recombinant DNA experimentation.)

Potentially far more serious was the appearance in the early 1970s of the first proscriptions of federally funded research. First, studies of abortion were forbidden. In 1974, all fetal research was proscribed.

29. Lear, *Recombinant DNA*, p. 69. See also Krimsky, *Genetic Alchemy*, p. 72–73, citing transcript of an interview with Maxine Singer, July 31, 1975. MIT Archives, Recombinant DNA Controversy Oral History Collection.

30. The Institute of Medicine (IOM) was in the third year of its existence as a new partner of the National Academy of Sciences and the National Academy of Engineering. The president of IOM responded to the letter of Singer and Söll with the suggestion that their request should be handled by the National Research Council (NRC). (Letter from John Hogness to Maxine Singer and Dieter Söll, August 1973. NAS Archives, Folder ALS, Committee on Synthetic Nucleic Acids: Ad Hoc, Proposed, 1973.) Rather than IOM, a new organization within the academies had that year been created to oversee NRC activities in biology and health. This Assembly of Life Sciences (ALS) had yet to have its first meeting when the communications from the Gordon conference arrived.

31. Maxine Singer and Dieter Söll, "Guidelines for Hybrid DNA Molecules," *Science* **181:**1114 (1973).

32. Both Lear (*Recombinant DNA*, p. 69–74) and Krimsky (*Genetic Alchemy*, p. 73–80) cite the origins and depth of concern for the ethics of science held by the Gordon conference chairperson, Maxine Singer. MIT Archives, Recombinant DNA Controversy Oral History Collection, and transcript of an interview with Daniel Singer, July 28, 1975. Maxine recollected that when she and Söll were confronted by the two scientists, she experienced no doubt that the conference must seriously consider the concerns they raised.

33. Letter from Philip Handler to Maxine Singer and Dieter Söll, July 20, 1973. NAS Archives, Folder ALS, Committee on Synthetic Nucleic Acids: Ad Hoc, Proposed, 1973.

34. Letter from Paul Marks to Philip Handler, August 30, 1973. NAS Archives, Folder ALS, Committee on Synthetic Nucleic Acids: Ad Hoc, Proposed, 1973.

35. Letter from Paul Berg to Leonard Laster, January 2, 1974. NAS Archives, Folder ALS, Committee on Recombinant DNA Molecules: Ad Hoc, 1974.

36. Wade, *The Ultimate Experiment*, p. 36; Rogers, *Biohazard*, p. 44. Richard Roblin had met Berg at Dulbecco's laboratory in the 1960s. In the course of preparing a lecture on bioethics, he remembered the letter from the Gordon conference and wrote to the Academy inquiring what had happened to the issue. (Letter from Richard Roblin to Leonard Laster, March 20, 1974. NAS Archives, Folder ALS, Committee on Recombinant DNA Molecules: Ad Hoc, 1974.) Roblin was referred to Berg, who suggested he attend the planning committee meeting. Thereafter, Roblin served as scribe, one of his tasks being the reworking of the several drafts of the committee's report before it was released by the Academy.

37. P. Berg, D. Baltimore, H. W. Boyer, S. N. Cohen, R. W. Davis, D. S. Hogness, D. Nathans, R. Roblin, J. D. Watson, S. Weissman, and N. D. Zinder, "Potential

Biohazards of Recombinant DNA Molecules," *Science* **185:**3034 (1974). Also *Proc. Natl. Acad. Sci. USA* **71:**2593–2594 (1974). The "Berg letter," as it is often called, was signed by more scientists than the seven who met at MIT as the planning committee. Lear says that when Stanley Cohen learned Berg had been invited by the Academy to form a committee, he asked to be a member. Berg is said to have declined his offer, saying that it would consist of cancer workers. Cohen appeared at MIT one day after the planning committee meeting and learned from David Baltimore that no plasmid experts had been in attendance. Concerned that the committee's actions might selectively harm research in his area of interest, Cohen threatened to write his own letter and suggested to Boyer that he join him. Berg then called Cohen and asked him to join in endorsing the report of his committee. A number of other scientists at Stanford thereafter asked to be included, and Berg eventually invited Hogness, Davis, and Boyer to be signers as well. (Lear, *Recombinant DNA*, p. 83–84.)

38. Annie C. Y. Chang and Stanley N. Cohen, "Genome Construction between Bacterial Species In Vitro: Replication and Expression of Staphylococcus Plasmid Genes in *Escherichia coli*," *Proc. Natl. Acad. Sci. USA* **71:**1030–1034 (1974); John F. Morrow, Stanley N. Cohen, Annie C. Y. Chang, Herbert W. Boyer, and Howard Helling, "Replication and Transcription of Eukaryotic DNA in *Escherichia coli*," *Proc. Natl. Acad. Sci. USA* **71:**1743–1747 (1974); David R. Hogness, Ronald W. Davis, and Herbert W. Boyer, unpublished data cited in Berg et al., *Science* **185:**3034 (1974). See note 37 above.

39. Documents in the NAS Archives (especially Folder ALS, Committee on Recombinant DNA Molecules: Ad Hoc, 1974) supply an interesting record of efforts in the latter weeks of May 1974 to adjust to concerns on the part of President Handler. He was determined that the report of the planning committee not appear to be a private letter from the scientists to their colleagues. The committee was enjoined to agree to be constituted immediately as an ad hoc committee of the ALS and forthwith accepted this legitimation. A comparison of Richard Roblin's drafts suggests that the ALS reviewers helped stress the international importance of the follow-up and improved the stance of the document so that an impression of self-sacrifice on the part of the signers was replaced by a call to all scientists to join in the moratorium.

40. The French government also reacted quickly to the publication of the Berg letter. The Délégation Générale à la Recherche Scientifique et Technique (DGRST) set up an organization for some form of control over "research which nobody denies can be dangerous." Two committees were formed. One was to consider ethical problems arising from the research; it was chaired by Jean Bernard and included Monod, Jacob, Gros, Monier, Ebel, Chabbert, and Slonimsky. The second committee of 15 experts, researchers, doctors, and biologists later defined the safety limits using the Asilomar guidelines. DGRST reviews of research grants in the summer of 1974 imposed a moratorium along the lines proposed by the Americans. ("Asilomar and the Pasteur Institute [from La Recherche]," *Nature* **256:**5 (1975); and P. Kourilsky, personal communication.)

41. R. Roblin, "Notes on planning meeting for Asilomar conference, MIT, September 10, 1974"; Herman Lewis, "Notes on biohazard conference organizing com-

mittee meeting, MIT, September 10, 1974." MIT Archives, MIT Recombinant DNA Historical Collection, Box 16, Folder 207.

42. See Lear. *Recombinant DNA*, p. 115–118, for details of how the press coverage was arranged.

43. At the conference, Anderson gave portions of two papers: "Indiscriminant Use of Antibiotics Has Exerted More Pressure on the Bacterial Population Than Could Be Wielded by All Research Workers in the World Put Together" (*Nature* **250:** 279–280 [1974]), and "Viability of, and Transfer of a Plasmid from, *E. coli* K-12 in Human Intestine" (*Nature* **255:**502–500 [1975]). William H. Smith gave data from his paper (*Nature* **255:**500–502 [1975]). A few months after the conference, Anderson conveyed an impression of the proceedings that is reprinted with his permission. (Transcript of interview with E. S. Anderson, May 31, 1975, by Charles Weiner, MIT Archives, Recombinant DNA Controversy Oral History Collection, Box 1, Folder 2.) "In some ways, the Asilomar meeting reminds me of Bernard Shaw's definition of the English gentleman hunting the fox: the unspeakable in pursuit of the uneatable. When I say that, I am not actually decrying the people who were considering the problem. But here was a bunch of people, with no experience in the handling of pathogens, virtually, with the sole exception of a mere handful, considering hazards that were not even known to exist. There's a certain comic atmosphere about it. It's true that this is the first occasion on which such hazards have been considered possible. But, in fact, they were a bunch of innocents abroad." When I interviewed Anderson by telephone in his home in London on February 6, 1990, he emphasized that he did not intend his remarks to be unkind but that he still felt strongly that the conference was hampered by the lack of sufficient participants who had had experience in handling pathogens and infectious disease. Asked how he had voted on the final show of hands on the provisional statement, he answered, "Aye, because I hadn't time to consider all the issues, and therefore couldn't be completely negative. One had to leave the matter open at the moment."

44. Memorandum from Roy Curtiss III to Paul Berg et al., "On Potential Biohazards of Recombinant DNA Molecules," August 6, 1974. NIH Central Files, Com 4-4-7-1A.

45. K. Murray, "Alternative Experiments?" *Nature* **250:**279 (1974).

46. Harold Green. From the tape recording of the Asilomar conference, February, 1975, NAS Archives. Mr. Green later voluntarily provided valuable advice to NIH concerning its environmental impact statement.

47. The members of the Plasmid-Phage Working Group were R. C. Clowes, S. N. Cohen, R. Curtiss III, S. Falkow, and R. Novick (Chair). I am indebted to Andrew Lewis for copies of the original reports of the working groups.

48. Rogers, *Biohazard*, p. 62–65.

49. It is noteworthy that despite his enthusiasm for biological containment at Asilomar, Brenner's enthusiasm for this mode seemed to dampen as he assisted the British (Williams) committee developing U.K. Guidelines. The British rules never included biological containment.

50. Among the "lambda people" at Asilomar were D. Botstein, A. Campbell, P. Kourilsky, A. Skalka, and W. Szybalski.

51. The members of the Animal Virus Working Group were M. Bishop, D. Jackson, A. Lewis, D. Nathans, B. Roizman, J. Sambrook, and A. Shatkin (chair).

52. Open letter to the Asilomar conference on hazards of recombinant DNA from Science for the People. MIT Archives, MIT Recombinant DNA Historical Collection, Box 17, Folder 219. The signers of this letter from the Genetic Engineering Group of Science for the People were Fred Ausubel, Jon Beckwith, and Luigi Gorini (Harvard); Kostia Bergmann, Kaaren Janssen, Jonathan King, Ethan Signer, and Annamaria Torriani (MIT); and Paulo Strigini (Boston University). Although there are differences in their reports, Krimsky (*Genetic Alchemy*, p. 110–111) and Lear (*Recombinant DNA*, p. 124) agree that Berg extended an invitation to Jonathan Beckwith to attend the conference. The latter did not attend, however, and the record is not clear whether Jonathan King was invited. He was not present. The description of the initial isolation of the gene appears in J. Shapiro, L. Machettis, L. Eron, G. Ihler, K. Ippen, and J. Beckwith, "Isolation of Pure *lac* Operon DNA," *Nature* **224:**768–774 (1969).

53. The members of the Plasmid-Cell DNA Recombinant Working Group were S. Brenner, D. D. Brown (chair), R. H. Burris, D. Carroll, R. W. Davis, D. Hogness, K. Murray, and R. C. Valentine.

54. Sources of the voting tallies: Lear, *Recombinant DNA*, p. 14; Rogers, *Biohazard*, p. 100; P. Kourilsky and W. Szybalski, personal communications.

55. Five Soviet scientists attended the Asilomar conference (see Appendix 1.1). A. A. Bayev, a well-known nucleic acid chemist, spoke for the delegation. He and his colleagues were in accord with the consensus, he said. His remarks also gave the wistful impression, however, that molecular biology was lagging in the Soviet Union. As all the participants knew, research in genetics had been gravely damaged in Stalin's time, providing a stark reminder of the vulnerability of science in a totalitarian milieu. In 1978, I visited Academician Bayev in the U.S.S.R. Academy of Sciences' Institute of Molecular Biology in Moscow and delivered to him a copy of proposed revisions of the NIH Guidelines.

56. Paul Berg, David Baltimore, Sydney Brenner, Richard O. Roblin III, and Maxine F. Singer, "Summary Statement of the Asilomar Conference on Recombinant DNA Molecules," *Science* **188:**991 (1975). Also *Proc. Natl. Acad. Sci. USA* **72:**1981–1984 (1975).

57. Letter from Philip Handler to James Ebert, May 20, 1975. NAS Archives, Folder ADM: International Relations: International Conferences: Recombinant DNA Molecules: Organizing Committee: Report.

Chapter 2

1. "Introduction of the Proposed Guidelines for Research Involving Recombinant DNA Molecules, January, 1976." *RDR*, vol. 1, p. 72–73.

2. William Gartland, personal communication.

3. Clifford Grobstein, "Asilomar and the Formation of Public Policy," p. 3–10, in *The Gene-Splicing Wars: Reflections on the Recombinant DNA Controversy* (Raymond A. Zilinskas and Burke K. Zimmerman, ed.) (New York: Macmillan, 1986).

4. Harold P. Green, "The Recombinant DNA Controversy: a Model of Public Influence," *The Bulletin of the Atomic Scientists* (Nov. 1978), vol. XXXIV, p. 12–16.

5. Paul Berg, David Baltimore, Herbert W. Boyer, Stanley N. Cohen, Ronald S. Davis, David S. Hogness, Daniel Nathans, Richard Roblin, James D. Watson, Sherman Weissman, and Norton D. Zinder, "Potential Hazards of Recombinant DNA Molecules," *Proc. Natl. Acad. Sci. USA* **71:**2593–2594 (1974). Also *Science* **185:**3034 (1974).

6. Letter from James Ebert to A. Vosburg, July 30, 1974. NAS Archives, Folder ARS D Med, 1973–74.

7. The members of the Ashby committee included Lord Ashby, Master, Clare College, Cambridge (chair); P. M. Biggs, director, Houghton Poultry Research Station; Professor Douglas Black, chief scientist, Department of Health and Social Security; Professor W. F. Bodmer, professor of genetics, University of Oxford; W. M. Henderson, secretary, Agricultural Research Council; Professor H. L. Kornberg, professor of biochemistry, University of Leicester; Professor R. R. Porter, professor of biochemistry, University of Oxford, and honorary director, MRC Immunochemistry Unit; Professor J. R. Postgate, professor of microbiology, University of Sussex; Professor R. Riley, director, Plant Breeding Institute, Cambridge; M. G. P. Stoker, director, Imperial Cancer Research Fund Laboratories, London; Professor J. H. Subak-Sharpe, professor of virology, University of Glasgow, and honorary director, MRC Virology Unit; Professor M. H. F. Wilkens, professor of biophysics, Kings College, London, and director, MRC Cell Biophysics Unit; and Professor R. E. O. Williams, director, Public Health Laboratory Service, London.

8. United Kingdom, *Report of the Working Party on the Experimental Manipulation of the Genetic Composition of Micro-organisms*, Cmnd. 5880 (January 1975).

9. Two sources of the differences are available: (i) Susan Wright, *Molecular Politics* (Chicago: University of Chicago Press, 1994). Wright's text is a detailed source of information about the U.K. handling of recombinant DNA research. (ii) Keith Gibson, "European Aspects of the Recombinant DNA Debate," p. 55–71, in *The Gene-Splicing Wars*. See note 3 above. Gibson was a staff member of the MRC and a frequent liaison person between the United Kingdom and the United States.

10. Letter from Paul Berg to Philip Handler, June 24, 1974. NAS Archives, Folder ALS, Committee on Synthetic Nucleic Acids, 1974.

11. Letter from Robert Stone to Philip Handler, July 17, 1974. NAS Archives, Folder ALS, Committee on Synthetic Nucleic Acids, 1974.

12. Colin Norman, "NIH Backing for NAS Ban," *Nature* **250:**278 (1974).

13. Michael Stoker, ibid.

14. In the spring of 1977, President Jimmy Carter suddenly determined to redeem a campaign pledge to "reduce the 1,900 agencies of the executive branch to 200." NIH alone had 169 such "agencies" which were mainly study section committees, whose charters were the ball bearings of the peer review system. Months of negotiations were required to infuse the president's staff with sympathy for the essential nature of the federal advisory group structure and to achieve recovery of the charters of the committees from the demolition pens of the Office of Management and Budget (OMB).

15. Charter of the NIH Recombinant DNA Molecule Program Advisory Committee. NIH Central Files, DNA files, Box 1, Folder 1.

16. Robin Herman, *Washington Post*, February 2, 1993, Health section. DSF Papers, Folder dna\aagenl\abc.

17. Fourth draft of Senate bill no. 1217 from the Senate Subcommittee on Health and Scientific Research, June 14, 1977.

18. In the early 1960s, the educator and idealist in Hans Stetten prompted him to accept the deanship of the new medical school at Rutgers. He had asked me to consider the post of professor of medicine, and I went up to look at the plans. The promises of a new academic hospital, however, suddenly evaporated. I returned to my laboratories at NIH, but Hans stayed on as dean for several more years.

19. The best record of the early years of the RAC are those in a series of articles written by Stetten in collaboration with William Gartland and Bernard Talbot: "The Road to Asilomar: Reminiscences of the Recombinant DNA Story," *Bioessays* **1**:41, 82, 135, 185, 231 (1984).

20. D. Stetten, Jr., "Valedictory by the Chairman of the NIH Recombinant DNA Molecule Program Advisory Committee," *Gene* **3**:265–268 (1978). Delivered at the RAC meeting of April 28, 1978.

21. Letter from Paul Berg to Robert S. Stone, December 10, 1974. NAS Archives, Folder ALS, Committee on Recombinant DNA Molecules, 1974.

22. Letter from Philip Handler to Paul Berg, December 17, 1974. NAS Archives, Folder ALS, Committee on Recombinant DNA Molecules, 1974.

23. Letter from Ronald Lamont-Havers to Paul Berg, December 19, 1974. NIH Central Files, Com 4-4-7-1A.

24. Appraisal of RAC in letter from Jane Setlow to DeWitt Stetten, September 4, 1975. NIH Central Files.

25. Paul Berg, David Baltimore, Sydney Brenner, Richard O. Roblin III, and Maxine F. Singer, "Summary Statement of the Asilomar Conference on Recombinant DNA Molecules," *Proc. Natl. Acad. Sci. USA* **72**:1981–1984 (1975). Also *Science* **188:** 991 (1975).

26. The Federal Advisory Committee Act stipulates conditions under which part or all of committee meetings may be closed to the public. Thus, most NIH study section meetings are closed.

27. DeWitt Stetten, Jr., *The Early History of the Recombinant Molecule Program Advisory Committee of NIH.* NIH Central Files.

28. PPO #129 Policy, February 8, 1966, U.S. Public Health Service, Division of Research Grants. Memorandum dated December 12, 1966, from Surgeon General, U.S. Public Health Service, to "Heads of Institutions Receiving Public Health Service Grants." This rule was initially received with derision by American clinical researchers. A cameo role played by the NIH in the promulgation of the rule is described in Donald S. Fredrickson, "Values and the Advance of Medical Science," in *Integrity of Institutions* (Ruth Bulger and Stanley Reiser, ed.) (Iowa City: University of Iowa Press, 1990).

29. A uniform set of guidelines for the entire country was a necessity. The achievement of this goal is an instructive part of this story. In chapter 5 is discussed the organization of the Federal Interagency Committee on Recombinant DNA Research (FIC). The strong fiducial role of the NIH in nearly all academic and nonprofit institutions was another asset. It was not until the December 1978 revision of the Guidelines (see chapter 9) that we began to achieve the acceptance of all commercial researchers to make the coverage uniform (chapter 9) and that the RAC adjusted to voluntary compliance (chapter 11).

30. Letter from Stanley Falkow to DeWitt Stetten, August 7, 1975. NIH Central Files.

31. Letter from Richard Goldstein, Harrison Echols, et al., to DeWitt Stetten, August 27, 1975. NIH ORDA Files.

32. Letter from DeWitt Stetten to Donald Helinski, September 3, 1975. NIH Central Files.

33. Letter from Marshall Edgell to DeWitt Stetten, August 19, 1975. NIH Central Files.

34. Letter from Paul Berg to DeWitt Stetten, September 2, 1975. NIH Central Files.

35. Stetten, Gartland, and Talbot, *Bioessays*, p. 135. See note 19 above.

36. Nicholas Wade, "Recombinant DNA: NIH Sets Strict Rules to Launch New Technology," *Science* **190:**1175–1179 (1975).

37. Wright, *Molecular Politics*, p. 202.

38. Stetten, Gartland, and Talbot, *Bioessays*, p. 82. See note 19 above.

39. Also sworn in with me was Theodore Cooper as the new assistant secretary for health. He and I were old friends. In 1986 I recruited him to head the Artificial Heart Program during my 18-month stint as director of the National Heart Institute. When I left this position to return to my laboratory full-time, I recommended to James A. Shannon that Cooper would make a good replacement for me, which he did.

40. I remember vividly my first meeting with the scientific director of the National Heart Institute in 1951. James A. Shannon had an office on the second floor of

Building 3, much too small for his 6-foot, 2-inch frame. He wore his trademark bow tie, and his keen eyes took in my image and projections of hope for a research career. When the interview ended, I had no idea what he had concluded. Later, home at the Peter Bent Brigham Hospital, I heard the good news. I do know now that, if Destiny was also there, she could not have made either of us believe that someday I would sit in the first and the last of the chairs held by this powerful man at NIH.

41. A memoir of this part of my career ("Phenotyping: On Reaching Base Camp. 1950–1975," *Circulation* [Suppl. 87, no. 4, April, 1993]) also contains a reference to Garfield's *Citation Classics* that rated me as the "most cited physiologist" in the period 1961 to 1975.

42. The recognition from successful research has a variety of awards other than prizes. Just before leaving the Institute of Medicine in the spring of 1975, I was called by a voice identifying itself as the "North African desk at State." His story was hardly plausible. It seems the American ambassador had been called to the royal palace in Rabat and ordered "to produce one Dr. Fredrickson in Morocco on the following Friday." "Of course, we can't make you go," said State. "But we hope you will." I thought, "But who could turn down such an invitation?" Thus, I was introduced to King Hassan II, one of the rare sharifyian kings in the world, a man of extraordinary intelligence and royal manners. Over a period of 25 years, I attended the king semiannually until his death in 1999.

Chapter 3

1. Leon Kass, "The New Biology. What Pride Relieving Man's Estate?" *Science* **174:**779–788 (1971).

2. Senator Kennedy had Cooper and me brought to the Senate reception area before the hearing. I handed him the answers to five questions he had sent me a day before, through Lee Goldman of his staff. Senator Harrison Williams (D-NJ), who presided, also gave me a questionnaire on affirmative action to return. Senators Laxalt (R-Nev) and Randolph (D-WVa) made cameo appearances. Dr. Barbara Davis and her lawyer gave testimony against my confirmation on the grounds that she had been discriminated against in my laboratory and that of Dr. Martha Vaughan in the National Heart Institute. This matter became the subject of a trial in the federal courts (*Davis v. Weinberger et al.*, UDC-DC 75-0205) in which the court found for the defendants. Earlier, we had made numerous trips to the reception rooms around the two chambers to meet Congressmen Robert Michel (R-Ill), ranking minority member of the House Appropriations Subcommittee for Labor and Health; Paul Rogers, chairman of the Health Subcommittee of the Committee on Interstate Commerce, the ranking minority member of that committee; Dr. Tim Lee Carter (R-Ky), one of two physicians in the Congress; and Daniel Flood (D-Pa), the colorful chairman of the Labor and Health Appropriations Subcommittee, with waxed mustache, spats, and cane. On the Senate side, we had brief discussions with Health subcommittee members Senators Gaylord Nelson (D-Wis), Thomas Eagleton (D-Mo), and Jacob Javits (R-NY) and Senators Williams (D-NJ) and Ken-

nedy (D-Mass). Our tours also included members of the Senate Health and Labor Subcommittee on Appropriations, Richard Schweicker (R-Pa), Charles Mathias (R-Md), and the powerful chairman, Warren G. Magnuson (D-Wash).

3. Interview with Andrew Lewis at NIH, Bethesda, Md., November 17, 1989.

4. Stephen E. Toulmin, "The Research and the Public Interest," p. 105, in *Research with Recombinant DNA: an Academy Forum, March 7–9, 1977* (Washington, D.C.: National Academy Press, 1977).

5. Peter Hutt, "Research on Recombinant DNA Molecules: the Regulatory Issues," *Southern California Law Review* **51**(6)**:**1435–1450 (1978). Hutt, formerly the counsel of the Food and Drug Administration, later an attorney in private practice, was a most knowledgeable and unsparing critic of the NIH handling of the recombinant DNA Guidelines.

6. CDC, Center for Disease Control; EPA, Environmental Protection Agency; FDA, Food and Drug Administration; NIOSH, National Institute for Occupational Safety and Health; and OSHA, Occupational Safety and Health Administration. CDC, FDA, and NIOSH were governed by the Department of Health, Education and Welfare; OSHA was under the Department of Labor, and EPA was an independent agency.

7. Some said at my appointment as NIH director that "he already knows all the skeletons in every closet."

8. Charles Kidd, personal communication, 1998. Kidd explains that in James Shannon's early years, when NIH programs were being developed, two planners, he and Joseph Murtaugh, played important roles consonant with the job title. Kidd literally produced the seminal Bayne-Jones Committee Report that had saved NIH from the verdict of a report inspired by the first secretary of HEW. This review, chaired by C. N. H. Long, had demanded that the intramural and extramural parts of NIH be severed and the funds be sent to the institutions in block grants (NLMHMD. Records of the National Science Foundation's Special Committee on Medical Research, 1948–1949). A few hours after Congressman Fogarty dropped dead in his congressional office, the resourceful Murtaugh had a draft bill written for Shannon to send to Congress, creating the Fogarty Center for International Research at NIH within a day.

9. An appraisal of Barkley, from "Letters from director, NIH, to Kitchen RAC members." NIH Central Files. These are fragments of letters to staff members written in 1978, when victory seemed near.

10. Memorandum from Susan K. Feldman to director, NIH, "Definitions and Procedures Regarding Regulations for Research on Recombinant DNA Molecules—Information," January 13, 1976. NIH Central Files, Com 4-4-7-1A.

11. S. 3529, 92nd Cong., 2d sess., §§3(1),(2) (1972).

12. House, *Conference Report*, 92nd Cong., 2d sess., Rept. 1403 (1972).

13. H.R. 4383, 92nd Cong., 2d sess., §3(2) (1972).

14. Freedman, *Administrative Procedure and the Control of Foreign Direct Investment*, 119 U. Pa. Law Rev 1,9 (1970), cited in *Lombardo v. Handler*, 397 F. Supp. 792 (1975).

15. These reports were later confirmed to me by Professor Philippe Kourilsky, who also noted, however, that the disturbances were short-lived and involved only the scientific and academic community. Later, I also discussed the reports with Professor Jean Bernard from Paris, who considered the reports of "pandemonium at the Institut Pasteur" to be quite exaggerated.

16. In connection with the casting of the hearing, Horowitz called Andrew Lewis at NIH to see why he had evinced at Asilomar "a somewhat negative position in terms of self-regulation by the scientific community" (see chapter 1). Lewis was not called to testify. Memorandum from Andrew M. Lewis to NIH deputy director, April 21, 1975. NIH Central Files.

17. J. Lederberg, *Prism* 33–37 (1975); R. Sinsheimer, *New Scientist* **68:**148 (1975).

18. Among definitive sources of the events at the University of Michigan are J. Goodfield, *Playing God: Genetic Engineering and the Manipulation of Life* (New York: Random House, 1977), and Donald N. Michael, "Who Decides Who Decides? Some Dilemmas and Other Hopes," in *The Gene-Splicing Wars*. See chapter 2, note 3.

19. Goodfield, *Playing God*, p. 79.

20. Roger G. Noll and Paul A. Thomas, "The Economic Implications of Regulation by Expertise: the Guidelines for Recombinant DNA Research," p. 262–277, in *Research with Recombinant DNA: an Academy Forum, March 7–9, 1977* (Washington, D.C.: National Academy Press, 1977).

21. Edward Yoxen, *The Gene Business: Who Should Control Biotechnology* (New York: Harper & Row, 1984).

22. During the fall and winter of 1975, the only notable coverage was John J. Fried's short piece in the *Baltimore Sun* of October 12, 1975, under the headline "Bacteria Experiments—Promises and Dangers." Victor McElheny had described the last day of the RAC meeting for the *New York Times* of December 9, 1975, under the headline "U.S. Panel on Gene Experiments Urges Stiffer Guidelines to Lessen Hazards." Norton Zinder's "A Personal View of the Media's Role in the Recombinant DNA War," p. 109–118, in *The Gene-Splicing Wars* (see chapter 2, note 3), notes the media silence during the Michigan uprising.

Chapter 4

1. Alexis de Tocqueville, *Democracy in America*, p. 43 (New York: Century Co., 1898).

2. Institute of Medicine, *Report of the NIH Research Priority-Setting Process* (Washington, D.C.: National Academy Press, 1998).

3. Susan Wright was thereafter an attentive participant in most of the meetings sponsored by NIH and a careful gatherer of facts on the British construction and

administration of the Guidelines. Her book *Molecular Politics* (Chicago: University of Chicago Press, 1994) is also a source of criticism of NIH conduct of the recombinant DNA controversy.

4. Public interest groups notified included Federation of American Scientists; Environmental Policy Center; Friends of the Earth; The League of Women Voters of the United States; Center for Science in the Public Interest; Common Cause; Environmental Defense Fund; Environmental Study Conference; Center for Law and Public Policy; Concern, Inc.; Consumer Action; Sierra Club; Consumer Federation of America; National Consumers League; National Consumers Congress; Maryland Citizen Consumers Council; Virginia Citizens Consumer Council.

5. The participants in the Director's Advisory Committee were each supplied with (i) a description of the purpose of the meeting and the major issues that would be considered; (ii) a chronology of major events associated with the formulation of policy on rDNA molecules; (iii) reprints of articles dealing with rDNA by Stanley Cohen (*Scientific American*), Robert Sinsheimer (*New Scientist*), Joshua Lederberg (*Prism*), Charles Fried (*Baltimore Sun*), and Nicholas Wade (*Science*); (iv) a summary of efforts in countries outside the United States; and (v) a summary and the full text of the proposed Guidelines.

6. Hans had fashioned on his lathe at home a handsome gavel, which he presented to me before the meeting. I had no need to use it.

7. *Proceedings of a Conference on NIH Guidelines for Research on Recombinant DNA Molecules.* Public hearings held at a meeting of the Director's Advisory Committee, February 9–10, 1976. *RDR*, vol. 1, p. 140–349. The film records of this meeting and several other hearings were stored under the direction of Dr. Charles Weiner in the MIT Archives, Recombinant DNA Historical Collection.

8. Ibid., p. 175–201.

9. Memorandum from Maxine Singer to Donald S. Fredrickson, February 13, 1976. DSF Papers, Folder dna\dna76\corresp.

10. Nicholas Wade, "Man-Made Evolution," *Washington Star*, June 18, 1976.

11. J. Sambrook, "Adenovirus Amazes at Cold Spring Harbor," *Nature* **268:**101–104 (1977).

12. Some of the publications with commentary on rDNA by Robert L. Sinsheimer: "Troubled Dawn for Genetic Engineering," *New Scientist* **68:**148 (1975); "Recombinant DNA: On Our Own," *Bioscience* **26:**599 (1976); (with G. Piel) "Inquiring into Inquiry: Two Opposing Views," *Hastings Center Report* **6:**18 (1976); "The End of the Beginning," *Eng. Sci.*, p. 1 (December 1976); "On Coupling Inquiry and Wisdom," *Fed. Proc.* **35:**2540–2542 (1976); "The Galilean Imperative," p. 18–27, in *Recombinant DNA: Science, Ethics, and Politics* (John Richards, ed.) (New York: Academic Press, 1977); "Two Lectures on Recombinant DNA," in *The Recombinant DNA Debate*, (David A. Jackson and Stephen P. Stich, ed.) (Englewood Cliffs, N.J.: Prentice-Hall, 1979).

13. Letter from Peter Barton Hutt to Donald S. Fredrickson, February 20, 1976. *RDR*, vol. 1, p. 474–484.

14. Memorandum from Joseph Perpich to Donald S. Fredrickson, "Follow-Up to the DAC Meeting," February 13, 1976. NIH Central Files, Com 4-4-7-1-1.

15. Memorandum from Maxine Singer to Donald S. Fredrickson, "Proposed Guidelines for Research Involving Recombinant DNA Molecules," February 13, 1976. NIH Central Files, Com 4-4-7-1A.

> Finally I would like to share with you my own sense of urgency concerning your decision. My colleagues have, by and large, been patient through these several years of thought and debate, even when they have not shared the concerns that motivated some of us. Seemingly endless levels of debate have now discouraged many of them. If we want their cooperation in this, and similar problems to come, we must promptly have a definitive resolution. The "go-slow" argument expressed by several attending your meeting is, I believe, what we have had and would have, with the proposed guidelines . . . the guidelines, insofar as they require P4 or EK3 for the most worrisome experiments, will result in the continuation of a "go-slow" policy. The meeting itself, both in conception and organization, was a logical, useful, and important addition to the "process." The manner of your chairmanship was to my mind, admirable in every way. I would especially like to express my gratitude for your continuation of premises that have been the basis of my own activities in this area, namely that matters of this sort can be resolved in a rational and scientific manner, and that our community, while preserving its interests and freedom, can be both responsible and trustworthy.

16. *RDR*, vol. 1, p. 406–417.

17. Wright, *Molecular Politics*, p. 185.

18. "I admire the very fine way in which you ran the meeting on the guidelines . . . Moreover I felt that, by even calling this meeting, you did something historically significant." Daniel Callahan, February 18, 1976. *RDR*, vol. 1, p. 471. "There is my sincere admiration for your wisdom in calling this meeting and for the efficient and extremely fair manner in which it was conducted. Without your skillful direction or a less sympathetic concern that all points of view be heard, the meeting would have been a useless exercise." Paul Berg, February 17, 1976. *RDR*, vol. 1, p. 466–469. "The meeting that you ran so convincingly—both as a scientist and a responsive public policy maker at the highest level of public service—will always stay in my mind as a memorable event . . . it may be worthwhile to consider if the [DAC] should not play a role related to the broader, non-technical issues that were raised at Asilomar. . . ." Walter Rosenblith, March 29, 1976. *RDR*, vol. 1, p. 507.

19. "Decision of the Director, NIH, To Release Guidelines for Research on Recombinant DNA Molecules," *Fed. Regist.* **41:**27902–27911 (July 7, 1976). Also *RDR*, vol. 1, p. 2–11. "Guidelines for Research Involving Recombinant DNA Molecules," *Fed. Regist.* **41:**27911–27943 (July 7, 1976). Also *RDR*, vol. 1, p. 11–43.

20. Letter from Jeremy J. Stone to Donald S. Fredrickson, March 31, 1976. DSF Papers, Folder dna\dna76\corresp.

21. Letter from Arthur Schwartz, Susan Wright, Marc Ross, Robert P. Weeks, and Max Heirich to Donald Fredrickson, April 21, 1976. *RDR*, vol. 1, p. 532–533.

22. "[NEPA] requires that before undertaking any major federal action significantly affecting the quality of the human environment, the responsible federal official shall prepare and circulate to other agencies and to the public a 'detailed statement' which describes the environmental impacts of the proposed action, alternatives to it . . ." The letter continues to iterate what we had already learned, that regulations (Guidelines) could be construed as one of these "major actions." Letter from Richard N. L. Andrews to Donald S. Fredrickson, April 16, 1976. DSF Papers, Folder dna\dna76\drafteis. *RDR*, vol. 1, p. 525.

23. ". . . Some of these areas of concern which . . . [your] EIS should be addressed . . . Interfering with or re-directing evolution and natural selection . . . legal liability in the event of . . . contagions affecting the public . . . cost vs. benefit comparison . . . who benefits? who is at risk? . . . What are the dangers, economic and scientific and social . . . is genetic manipulation the proper solution to problems that have both a social and medical basis?" Letter from Lorna Salzman to Donald S. Fredrickson, May 17, 1976. *RDR*, vol. 1, p. 542–544.

24. Memorandum from deputy director for science to director, NIH. "Possible Consequences of Further Delay in Issuance of Recombinant DNA Molecule Research Guidelines," May 3, 1976. NIH Central Files. Com 4-4-7-1A.

25. Present were agencies in HEW, the Food and Drug Administration, including its Bureau of Biologics, the Center for Disease Control (CDC), the National Institute of Environmental Health Sciences, the National Institute for Occupational Safety and Health (NIOSH), Occupational Safety and Health Administration (OSHA) of the Labor Department, and the independent Environmental Protection Agency (EPA). Also represented were the Transportation Department's Office of Hazardous Material Operations and the Defense Department's U.S. Army Medical Institute of Infectious Diseases and Walter Reed Institute of Research. The National Aeronautics and Space Administration (NASA), always alert to extension of biological research in space, also had a representative there. The Commerce Department was present, as was the deputy assistant secretary for advanced and applied technology affairs of the State Department. Someone noted that the CIA had not been invited; we would be repeatedly asked if any rDNA work was going on undercover. (It wasn't, to our knowledge.) *RDR*, vol. 1, p. 420–423.

26. Memorandum from William H. Taft IV to Donald S. Fredrickson, June 7, 1976. DSF Papers, Folder dna\dna76\gl vs regs.

> From your presentation of the matter . . . the "Guidelines" for DNA Research will . . . provide guidance for DNA research . . . from HEW. Second, [they] will be imposed as condition of grant and contract support of projects by NIH. As to the second aspect . . . it may be asserted that they fall within the scope of the Departmental policy concerning public participation in rulemaking, which calls for compliance with . . . the Administrative Procedure Act, 5 U.S.C. 553 (36 F.R. 2532, February 5, 1971).

> To the extent it is sought to impose these Guidelines on unwilling applicants, without publishing them as required by section 553, however, they are subject to attack.
>
> In addition, we understand that you intend to publish the Guidelines in the Federal Register prior to the making of an environmental assessment . . . The making of a grant or contract prior to completion of such actions would also be subject to challenge in litigation.
>
> Whether the need for the proposed course of action is so strong as to justify the acceptance of the risk of legal action outlined above, is a matter for policy and administrative judgment.

27. Memorandum from Joseph Perpich to Donald S. Fredrickson, "Issues Related to Regulation—Your Meeting with General Counsel Taft," June 1, 1976. DSF Papers (Box DNAK-2), Folder dsf\nihdir\secHEW\mathews.

28. Memorandum from the chief of the Environmental Safety Branch, DRS, to Joseph Perpich, "Recombinant DNA Environmental Assessment," May 20, 1976. NIH Central Files.

29. Memorandum from Theodore Cooper to Secretary Mathews, "Response to the Memorandum of June 7 from General Counsel to Director, NIH," June 15, 1976. NIH Central Files.

30. Press conference on DNA Guidelines, June 23, 1976. *RDR*, vol. 1, p. 569–573.

31. "Guidelines Issued on Life Creation," p. 1. *Washington Post*, June 24, 1976.

32. Colin Norman, "Genetic Manipulation: Guidelines Issued," *Nature* **262**:2 (1976).

33. "Decision of the Director, NIH, To Release Guidelines for Research on Recombinant DNA Molecules," *Fed. Regist.* **41**:27902–27911 (July 7, 1976). Also *RDR*, vol. 1, p. 2–11.

34. William J. Walsh III, a staff member at the Department of State, is to be credited with the thoroughness and dispatch with which U.S. diplomats in every foreign post were alerted to this event.

35. F. R. Simring, "Folio for Folly: NIH Guidelines for rDNA Research," *Man and Medicine* **2**:110–119 (1977).

36. "A Scientific Breakthrough," *Washington Post*, July 2, 1976, Editorial page.

Chapter 5

1. Cynthia B. Cohen and Elizabeth Liebold McCloskey, "Introduction," *Kennedy Institute of Ethics Journal* **8**(2):vii–x (1998).

2. Nicholas Wade, "Man-Made Evolution," *Washington Star*, June 18, 1976.

3. Erwin Chargaff, "On the Dangers of Genetic Meddling," *Science* **192**:938–940 (1976). ("Our time is cursed with the necessity for feeble men, masquerading as experts, to make enormously far-reaching decisions.") Erwin Chargaff, "An Infor-

mal View of Science Funding in the United States," *Trends Biochem. Sci.* (March, 1976). ("That awful mess in Bethesda and Washington.") Erwin Chargaff, "Profitable Wonders," *Sciences*, p. 21 (August–September, 1975). ("At this council of Asilomar, there congregated the molecular bishops and church fathers from all over the world, in order to condemn the heresies of which they themselves had been the first and the principal perpetrators. This was probably the first time in history that the incendiaries formed their own fire brigade.")

4. Donald S. Fredrickson, "The Government's Role in Biomedical Research," *Fed. Proc.* **35**:2538–2540 (1976).

5. A quote of "telegram from Texas." David Clem, "Regulation at Cambridge," p. 241–250, in *Recombinant DNA: Science, Ethics, and Politics* (John Richards, ed.) (New York: Academic Press, 1977). In addition to Clem, valuable sources for the Cantabrigian counterpoint include James T. Watson and John Tooze, *The DNA Story*, p. 91–131 (San Francisco: W. H. Freeman, 1981). The most extensive source is the MIT Archives, MIT Recombinant DNA Historical Collection.

6. Letter from Paul Berg to Mayor A. E. Vellucci and the City Council of Cambridge, July 2, 1976. DSF Papers, Folder dna\dna76\cambridge.

7. Letter from Francis L. Comunale, M.D., acting commissioner of health, Cambridge, Mass., to Donald S. Fredrickson, January 6, 1977, including a copy of ordinance no. 874. DSF Papers, Folder dna\dna76\cambridge\jan6 cambr.rpt.

8. *Recombinant DNA Technical Bulletin*, DHEW publication no. 77-99.

9. Memorandum from James Cannon, White House, to DHEW secretary Mathews, June 24, 1976. DSF Papers, Folder dna\dna76\birthIAC. This was the first alarm we had about this early congressional attempt to corral the management of the Guidelines. Within the next year, it would become the feature struggle with Congress, particularly with the Senate Health Subcommittee.

10. Memorandum from Donald S. Fredrickson to Joseph Perpich, July 7, 1976. DSF Papers, Folder dna\dnagenl\IAC.

11. Memorandum from acting DHEW secretary Marjorie Lynch to James Cannon (White House), July 16, 1976. DSF Papers, Folder dna\dnagenl\IAC.

12. Letter from Senators Edward Kennedy and Jacob Javits to President Gerald Ford, July 19, 1976. *RDR*, vol. 2, p. 158–160.

13. Letter from Senator Edward Kennedy to Donald S. Fredrickson, July 22, 1976. *RDR*, vol. 2, p. 157.

14. Memorandum from James Cannon (White House) to DHEW secretary Mathews, August 12, 1976. dna\dnagenl\IAC.

15. Letters from President Gerald Ford to Senators Kennedy and Javits, September 22, 1976. *RDR*, vol. 2, p. 161–164.

16. Letters from President Ford to "Heads of Departments and Independent Agencies," requesting that nominations for the Interagency Committee be

sent to Secretary Mathews, September 22, 1976. DSF Papers, Folder dna\dna76\general\Sep2276Ford ltr.

17. From the charter of the Interagency Committee on Recombinant DNA Research. "Purpose: To coordinate Federal programs and activities relating to recombinant desoxyribonucleic acid (DNA) research, to assist in facilitating compliance with a uniform set of guidelines for the conduct of this research in the public and private sectors and to facilitate communication and exchange of information among Government agencies." *RDR*, vol. 2, p. 181–182.

18. "New Strains of Life—or Death," Document 5.8, in *The DNA Story: a Documentary History of Gene Cloning* (James D. Watson and John Tooze, ed.) (San Francisco: W. H. Freeman, 1981).

19. "Petition of Environmental Defense Fund and Natural Resources Defense Council to the Secretary of Health, Education and Welfare To Hold Hearings and Promulgate Regulations under the Public Health Service Act Governing Recombinant DNA Research." *RDR*, vol. 2, p. 325–345.

20. Letter from Robert Sinsheimer to Karim Ahmed, October 28, 1976. Letter from Alan MacGowan to Burke Zimmerman, November 5, 1976. *RDR*, vol. 2, p. 321–324.

21. "Petition of Environmental Defense Fund and Natural Resources Defense Council." *RDR*, vol. 2, p. 325–345.

22. Although I strongly disagreed with several of Burke Zimmerman's suggestions, I eventually hired him to join my staff because he was one of the few scientists with molecular biology training who worked for one of the environmental advocates. I felt that the Kitchen RAC and I needed his point of view, if not his petitions.

23. The member departments or agencies of the Interagency Committee on Recombinant DNA Research and their principal representatives included AGRICULTURE (Charles F. Lewis, Ph.D., staff scientist); COMMERCE (Betsy Ancker-Johnson, Ph.D., assistant secretary for science and technology); National Oceanic and Atmospheric Administration (David Wallace, assistant administrator for marine resources); DEFENSE (William R. Beisel, M.D., U.S. Army Medical Research Institute of Infectious Diseases at Fort Detrick, Md.); HEALTH, EDUCATION AND WELFARE (Marian Mlay, Office of Policy Development and Planning); Center for Disease Control (CDC) (John H. Richardson, D.V.M., Office of Biosafety); National Institute for Occupational Safety and Health (NIOSH) (John F. Finklea, M.D., director); Food and Drug Administration (FDA) (Robert L. Elder, Sc.D., deputy associate commissioner for science; Rosa Gryder, Ph.D., staff scientist, and John C. Petricciani, M.D., Bureau of Biologics); National Institutes of Health (NIH) (Donald S. Fredrickson, M.D., chairman, and Joseph Perpich, M.D., J.D., executive secretary); INTERIOR (Mariano Pinental, M.D., medical director); JUSTICE (Anthony Liotta, deputy assistant attorney general); LABOR (Morton Corn, Ph.D., assistant secretary for occupational safety and health, and Byung Kwon, OSHA); STATE (Oswald H. Ganley, Ph.D., deputy assistant secretary for advanced and applied technology affairs, and William J. Walsh, science officer);

TRANSPORTATION (William D. Owens, deputy assistant secretary for systems development and technology); EXECUTIVE OFFICE OF THE PRESIDENT (H. Guyford Stever, director of Science and Technology Policy, and Warren R. Muir, senior staff member for environmental health, Council on Environmental Quality [CEQ]); ENERGY RESEARCH AND DEVELOPMENT ADMINISTRATION (ERDA) (James L. Liverman, Ph.D., assistant administrator for environment and safety, Charles E. Carter, M.D., manager, Biomedical Programs, and Walter H. Weyzen, M.D., manager, Human Health Studies Program); ENVIRONMENTAL PROTECTION AGENCY (Delbert S. Barth, Ph.D., deputy assistant administrator for health and ecological effects, and Lawrence Plumlee, M.D., medical advisor); NATIONAL AERONAUTICS AND SPACE ADMINISTRATION (David L. Winter, Ph.D., director for life sciences); NATIONAL SCIENCE FOUNDATION (Herman W. Lewis, Ph.D., section head of cellular biology, and Laurence Berlowitz, Ph.D., special assistant to the assistant director for biological, behavioral, and social sciences); NUCLEAR REGULATORY COMMISSION (Frank Swanberg, Jr., chief of Health and Environmental Research Branch); U.S. ARMS CONTROL AND DISARMAMENT AGENCY (Robert Mikulak, Ph.D., physical science officer); VETERANS ADMINISTRATION (Jane Schultz, Ph.D., Research Service). *RDR*, vol. 2, p. 302–304.

24. Further remarks of the chair. Minutes of the FIC meeting of November 4, 1976. *RDR*, vol. 2, p. 165–175.

> Although NIH has so far been the principal site of action in the American sector of . . . governance of this provocative and controversial new research, we are not unaware of our limits . . . in either perceptions or powers to guide and effect control of recombinant research beyond our own grantees and contractors . . . and to establish for the entire nation a workable system that protects adequately the legitimate interests of all. The problem enters a new stage here—larger than the parochial interests of any agency, any department, any single sector of public interest, power or authority. If we are to serve usefully here, all of us must rise above those narrow interests of which we are often unjustly accused . . . Many demands have already been made upon Congress, and upon the Administration . . . that extraordinary means be used to determine how we proceed in this matter . . . i.e., special commissions, Congressional actions, or the courts. These may come about through failure of a rational process, involving experts representing all relevant facets of administrative government. That is the test we are to undergo. Let us begin.

25. The tasks would be registration of activity, certification of containment standards, oversight of investigators and institutions, formation of an appellate mechanism for the above, requirements for safety education and training, development of safer hosts and vectors, establishment of a mechanism to provide hosts and vectors, exchange of information, establishment of international liaison, extension of the Guidelines throughout the public and private sector, placement of ultimate authority. Ibid.

26. "Statement: The Biological Weapons Convention would prohibit any future type of warfare which employed biological agents or toxins, regardless of when the

agent was first developed or discovered. This also applies to weapons, equipment and means of delivery. In other words, the Convention prohibits not only existing means of biological and toxin warfare, but also any that might come into existence in the future." Letter from Robert Mikulak to Joseph Perpich, December 11, 1976, containing copy of statement by Ambassador Joseph Martin, Jr., of August 17, 1976. DSF Papers, Folder dna\interagency interactions.

27. Minutes of FIC meeting of November 23, 1976. *RDR*, vol. 2, p. 240–249.

28. Cohen et al. December 1980. U.S. patent 4,237,224, noting several modifications and continuations in part on May 1976, November 1978, and January 4, 1979. To the original assignees, the trustees of Stanford University, were added the University of California before the patent was awarded in 1980. DSF Papers, Folder dna\dnagenl\patents 76-78.

29. Stanley Cohen, at September 1981 molecular biology meeting in Rome, replying to a question from John Tooze of EMBO. DSF Papers, Folder dna\dnagenl\odyssey I 71-81.

30. Letter from Robert M. Rosenzweig to Donald S. Fredrickson, June 16, 1976. *RDR*, vol. 2, p. 60–61. Copies of correspondence and source citations are also found in DSF Papers, Folder dna\dnagenl\patent.

31. Memorandum with Rosenzweig letter, June 16, 1976. Ibid., p. 59.

32. This was a few years before the 1980 Bayh-Dole and Stevenson-Wydler Acts, which served to greatly increase technology transfer of government-funded inventions. Prior to 1968, mechanisms for administering patents on government-funded inventions were nonuniform and, in some instances, inconsistent. Invention rights to discoveries were normally allocated in either of two ways. Under the deferred determination policy, the patent rights were automatically vested in HEW. The inventor's sponsoring institution could petition HEW to waive its patent rights and allow the institution to own them; waivers were granted in about 90% of the cases, provided the institution showed the ability to move the invention to the marketplace.

33. This summation of the Upjohn case was specially prepared by Michael Clayton of the NIH staff. Memorandum from associate director for program planning and evaluation, "Patentability of Living Things," July 12, 1978. DSF Papers, Folder dna\dnagenl\patent.

34. Austin Scott, "Court Rules That GE Can Patent Life Created in Lab." *Washington Post*, March 3, 1978.

35. Memorandum from Joseph Perpich to Donald S. Fredrickson, "Recombinant DNA and Patents," July 26, 1976. DSF Papers, Folder dna\dnagenl\patents.

36. Letter from director, NIH, to members of the Recombinant DNA Advisory Committee, August 27, 1976. DSF Papers, Folder dna\dnagenl\patents. Also *RDR*, vol. 2, p. 48–51.

37. Letter from Donald S. Fredrickson soliciting views on patent applications, September 8, 1976. *RDR*, vol. 2, p. 21–24.

38. The responses are available. *RDR*, vol. 2, p. 62–155.

39. "I am concerned . . . Why the patent? Is it for money-making . . . I feel very strongly about avoiding any commercialism in this area and I am truly frightened. . . . I do want controls so that no scientist will be able to move into what I call a never-never land where negative results for all of society might come forth." Ibid., p. 136. "Frankly, as a scientist it strikes me as vaguely ludicrous that one could or would seek to obtain patents concerning recombinant DNA. . . . If patents can be obtained . . . I would prefer to see ownership and control vested in HEW. . . ." Ibid., p. 80.

40. ". . . It seems to me that the Institutional Patent Agreement is meant to stimulate use of new inventions made on grants from NIH. In the recombinant DNA area there is hardly need for more stimulation. The overriding concern here is that the potentialities in recombinant DNA research be realized in an orderly and safe manner. I would therefore . . . opt for the 3rd alternative. . . . Any royalties could be used to establish a research award fund for universities and/or to continue work on safety and ethical aspects of recombinant work." Letter from David Baltimore to Donald Fredrickson. Ibid., p. 80.

41. "From the particular point of industry, there are really two issues that present themselves. The first of these is the problem of confidentiality, given the Freedom of Information Act, of disclosing protocols and research results prior to the time the patent process has been able to react. This is an important issue since the patent is the legal documentation of priority of a customer-oriented effort. . . . The second concern from industry is . . . policies and purposes that are unexceptionable at their origin have a very unhappy tendency to become transformed through mindless application of regulations into Frankensteins. . . . It is my impression . . . that industry will avoid committing itself in very formal ways to policy statements out of a reasonable and deep-seated fear that this is just the first step of another wave of bureaucratic intervention into individual endeavor. . . . I think . . . it is highly probable that a state of sharp contest will continue. . . . [One] could imagine this immediate problem in genetic engineering as being an opportunity to set some new pattern whereby the Greek tradition of *paideia* could be given priority over the tradition of the Roman *lictor* bearing his *fasces*." Letter from W. N. Hubbard, Jr. (president of the Upjohn Company) to Donald S. Fredrickson, July 16, 1976. Ibid., p. 66–67.

42. "Dear Sir, I am writing in regard to an inquiry I received from a constituent . . . who said that . . . studies have been done at Stanford University and the University of California regarding patenting a technique of genetic manipulation. He believes that, since all the information has been derived through grants from the Public Health Service, it is public information and not patentable. I would appreciate your looking into this and furnishing me with information." Letter from Congressman John J. LaFalce to the assistant secretary of health, August 9, 1976. *RDR*, vol. 2, p. 68.

43. The patent notice read in part: "In view of the exceptional importance of recombinant DNA and the desirability of prompt disclosure of developments in the field, the Assistant Secretary of Commerce for Science and Technology has re-

quested that the Patent and Trademark Office accord special status to patent applications involving recombinant DNA . . . [Applications] must also include a statement that the NH [*sic*] guidelines . . . are being followed." *Fed. Regist.* **42**:12–13 (1977).

44. "HEW Seeks to Lift Patent Order on Genetic Research Procedure," *Washington Post*, February 10, 1977.

45. I received a copy of Secretary Kreps's memorandum to HEW. To clear the air about the order, the department would temporarily suspend accelerated processing of patent applications involving recombinant DNA inventions, the question should be reviewed by the FIC, and better liaison should be established in this area. Memorandum from Secretary Kreps to Secretary Califano, passed to Donald S. Fredrickson, February 13, 1977. DSF Papers, Folder dna\dnagenl\patents 76-78.

46. Copy of letter from Secretary Joseph A. Califano to Senator Dale Bumpers, February 17, 1977. DSF Papers, Folder dna\dnagenl\patent 76-78.

47. *The Patenting of Recombinant DNA Research Inventions Developed under DHEW Support, an Analysis by the Director, National Institutes of Health, November, 1977* (18 pages with three appendices). *RDR*, vol. 2, p. 2–47. Original in DSF Papers, Folder dna\dnagenl\patent.

48. Letter from Donald S. Fredrickson to Robert M. Rosenzweig, March 2, 1978. Ibid., p. 1A–1B.

49. Senate Subcommittee on Monopoly and Anticompetitive Activities, Select Committee on Small Business. *Hearings on Government Patent Policies: Institutional Patent Agreements*, 95th Cong., 2d sess., 26 June 1978. *RDR*, vol. 3, Appendix B, p. 243–252. Also in *Speeches, Articles, and Selected Papers, Donald S. Fredrickson, 1975–1981*, vol. 2, item 99.

50. "Patent on Gene Splicing, Cloning Granted to Stanford and University of California," *Wall Street Journal*, December 4, 1980. The news reports noted that Herbert Boyer, being one of the founders of Genentech, now owned 925,000 shares of that company, valued then at $40 million.

51. Memorandum from Norman Latker to Donald S. Fredrickson, March 3, 1978. DSF Papers, Folder dna\dnagenl\patents 76-78.

52. Information kindly provided by Donald Kennedy, president emeritus of Stanford University, in a personal communication, May 1, 1998.

53. Lynen was probably the most influential biochemist in Germany at the time. He had received the Nobel Prize for his work on fatty acids. Later, he and I serially received the Jimenez Diaz award, a gold medal and a lecture, and thereafter went to Madrid for a stint of 5 years to help choose the next winners.

54. Members included C. Weissman (chair), E. S. Anderson, G. Bernardi, S. Brenner, W. Bodmer, K. Murray, L. Philipson, J. Tooze, and H. Zachau.

55. *Report of the Second Meeting of the EMBO Standing Advisory Committee on Recombinant DNA. European Molecular Biology Conference, Seventh Ordinary Session*, September 28, 1976. CEBM/76/12E.

56. It would take about 25 years before gene-directed political action came to the fore again. In June 1998, a vigorous campaign led by Greens, members of Greenpeace and WWF, and other ecologists managed to place before the Swiss electorate an *initiative pour la protection génétique (IPG)* proposing to ban in Switzerland research involving transgenic animals and the patenting of genetically modified animals ("La recherche suisse menacée," *L'Express*, June 4, 1998). Scientists and the pharmaceutical industry involved themselves heavily in the electioneering, and the initiative lost 2 to 1, with 41% of the electorate turning out to vote ("Voters Reject Antigenetics Initiative," *Science* **280:**1685 [1998]).

Chapter 6

1. National Institutes of Health, *Final Environmental Impact Statement on NIH Guidelines for Research Involving Recombinant DNA Molecules*, Part I, p. 25. DHEW publication no. (NIH) 77-1489 (October 1977).

2. On January 1, 1970, President Nixon signed into law the National Environmental Policy Act (NEPA) requiring each federal agency to prepare a statement of environmental impact in advance of a major action that may significantly affect the quality of the human environment. The statement is to include a description of the proposed action, discussion of any probable impact on the environment, and alternatives to the proposed action. This draft environmental impact statement (DEIS) is then to be published in the *Federal Register* and circulated for comment for at least 90 days. A final EIS that includes a discussion of problems and objections raised during the review of the draft may be issued after revision. NEPA also provides for establishment of a Council on Environmental Quality (CEQ) in the executive office of the president. By 1973, CEQ had been established and issued guidelines. It also established a structure within the government for implementation of the law. CEQ thereafter received EISs and arranged for their publication but eschewed comments of approval or disapproval about them.

3. Memorandum from Rudolph Wanner to Joseph Perpich, "Analysis of NEPA Requirements Relative to DNA Recombinant Molecule Research," April 22, 1976. DSF Papers, Folder dna\dna76\drafteis.

4. Two years later, the history of the 1974 filing was recounted in a memorandum from Gala Fuller to Richard Riseberg, "Applicability of the National Environmental Policy Act of 1969 to Programs Implemented by DHEW (or Its Components—NIH and DNA Recombinant Research)," April 19, 1976. The attempt to obtain a generic EIS status was continued by a contractor, Dalton, Little and Newport. Their efforts had proved to be fruitless by 1977. Memorandum from Richard J. Reisberg to Joseph G. Perpich, "Recombinant DNA Research," April 6, 1977, and staff memorandum on "NIH Generic Analysis, May 6, 1976, to April 11, 1977." NIH Central Files, Com 4-4-7-1A, and DSF Papers, Folder dna\dna76\drafteis.

5. "Chronology of Events in the Preparation of Environmental Impact Statement: Guidelines for Research Involving Recombinant DNA Molecules," prepared on March 31, 1977, has been a valuable resource for this discussion. DSF Papers, Folder dna\dna76\drafteis.

6. *General Administration Manual*, Department of HEW, chapter 30-15-30B(6).

7. Memorandum from Joseph Perpich to Donald Fredrickson, "Summary of the Singer Analysis," May 11, 1976. DSF Papers, Folder dna\dna76\drafteis.

8. Memorandum from director, Office of Environmental Affairs, to principal environmental officer (HEW), "Comments on Memorandum by Gala [*sic*] Fuller on the Applicability of NEPA to NIH Actions Such as DNA Recombinant Research," May 7, 1976. Ibid.

9. Memorandum of meeting with Green attached to memorandum from Joseph Perpich to Donald Fredrickson, "Your Meeting with Secretary Mathews on the DNA Guidelines," May 25, 1976. Ibid.

10. *Draft Outline Environmental Document for Proposed Guidelines for Recombinant DNA Molecule Research*, May 27, 1976. Ibid.

11. "Recombinant DNA Research Guidelines. Draft Environmental Impact Statement," *Fed. Regist.* **41**:38426–38483 (September 9, 1976).

12. Memorandum from deputy director, Office of Environmental Affairs, to deputy director for science, NIH, June 28, 1976. DSF Papers, Folder dna\dna76\drafteis.

13. Letter from deputy director, Office of Environmental Affairs, to principal environmental officer (H). Comments on DNA draft #3, August 4, 1976. NIH Central Files, Com 4-4-7-1A; DSF Papers, Folder dna\dna76\drafteis.

14. Roughly translated: "What is simple is faulty, but what is complicated is unusable."

15. Letter from Paul Cromwell to principal environmental officer, "Draft #4 of Recombinant DNA EIS," August 19, 1976. NIH Central Files, Box Comm. 4-4-1-1A; also DSF Papers, Folder dna\dna76\drafteis.

16. Letter from director, Office of Environmental Affairs (Custard), to principal environmental officer (HEW) (Osheroff), "Draft EIS for Research Involving Recombinant DNA Molecules," August 23, 1976. Ibid.

17. "Public Comments on Draft Environmental Impact Statement of September 9, 1976," in National Institutes of Health, *Final Environmental Impact Statement on NIH Guidelines for Research Involving Recombinant DNA Molecules*, Part II, Appendix K. DHEW publication no. (NIH) 77-1490 (October 1977).

18. Marcia Cleveland (Natural Resources Defense Council), "Comments of the Natural Resources Defense Council, Inc. on The National Institutes of Health's Guidelines for Research Involving Recombinant DNA Molecules and the Draft Environmental Impact Statement on Recombinant DNA Research Guidelines," November 1, 1976. Ibid., p. 124–153.

19. Letter from Marc Lappé, Institute of Society, Ethics and the Life Sciences, to Donald S. Fredrickson, September 8, 1976. Ibid., p. 8–20.

20. Letter from Lewis J. Lefkowitz, Richard G. Berger, and Deborah Feinberg, Department of Law, State of New York, to Donald S. Fredrickson, October 19, 1976. Ibid., p. 80–110.

21. Letter from Susan Wright, John Wright, Marc Ross, Arthur Schwartz, and Richard N. L. Andrews, to Donald S. Fredrickson, October 12, 1976. Ibid., p. 30–39A.

22. Letter from Burke Zimmerman, Environmental Defense Fund, to Donald S. Fredrickson, October 18, 1976. Ibid., p. 54–76.

23. Letter from Robert L. Sinsheimer to Donald S. Fredrickson, September 7, 1976. Ibid., p. 6–7.

24. Loretta Leive, "Comments on the Draft Environmental Impact Statement," November 1, 1976. Ibid., p. 123.

25. Letter from Solomon Garb to Rudolph Wanner, October 8, 1976. Ibid., p. 28–29. This commentator, a member of the National Cancer Advisory Board, wrote many vitriolic comments on the NIH Guidelines. In the August 1976 *York (Pa.) Dispatch*, he was quoted: "Many scientists have been seriously worried that an escaped virus from a laboratory doing recombinant DNA studies could cause an epidemic like the Legionnaires' epidemic—or much worse."

26. Letter from Lansing M. Prescott to Donald Fredrickson, October 19, 1976. Ibid., p. 77.

27. Letter from Margaret Puls to the director, NIH, October 14, 1976. Ibid., p. 49.

28. Letter from Marshall Hall Edgell to Donald Fredrickson, October 15, 1976. Ibid., p. 50.

29. Letter from Philip Handler to Donald Fredrickson, October 18, 1976. Ibid., p. 53.

30. National Institutes of Health, *Final Environmental Impact Statement on NIH Guidelines for Research Involving Recombinant DNA Molecules*, Part I, p. 113–136. DHEW publication no. (NIH) 77-1489 (October 1977).

31. Green Diary 1, p. 25 et seq. This is an early citation from the diaries I kept irregularly as director of NIH. These diaries being green notebooks, they have become known as the Green Diaries, numbered sequentially, with entries identified by page.

32. Richard A. Rettig, *Cancer Crusade: the Story of the National Cancer Act of 1971*, (Princeton: Princeton University Press, 1977).

33. Letter from the undersigned NIH officials to Secretary-Designate Joseph A. Califano, Jr., January 7, 1977. The letter, never sent, was signed by DeWitt Stetten, Jr., Leon Jacobs, Joseph G. Perpich, Thomas E. Malone, Mortimer B. Lipsett, Leon Schwartz, Storm Whaley, Robert S. Gordon, and Seymour Perry. DSF Papers, Folder SecHHA\JACJr\advent.

34. At noon on the 14th, I had lunch with Ted Cooper and learned that he had been summarily relieved of his position as assistant secretary. In the hall outside his office, Roger Egeberg, deputy assistant secretary of health, wandered about with a red-stained towel around his neck, saying "I've just been guillotined." Kenneth

Endicott, then head of the Health Resources Administration, had been given 24 hours to clear his office. Ken had been a longtime director of the National Cancer Institute and merited more respect than this brusque dismissal. Lou Hellman, head of the Health Services Administration (HSA), was given a month to find another position. Jim Isbister was allowed to stay in the HSA and thus be eligible to compete for return to his old job as head of ADAMHA, once a part of NIH but then a separate agency including the new units for Alcohol, Drug Abuse, and the former Mental Health Institute. The FDA commissioner's office was already vacant, Max Schmitt having departed to be a vice president at the University of Illinois. There was no word on the status of David Sencer, head of CDC.

35. "Califano to Depoliticize NIH," p. A4. *Washington Post*, January 27, 1977.

36. I fulfilled this latter mission handily, for Rogers was in the Navy Medical Center across the street with a fractured left humerus incurred on a nearby ski slope. Later he called me, saying, "Congratulations, Don." "For what?" I asked. "Oh, didn't you know? Don't say I told you."

37. Green Diary 1, p. 25 et seq.

38. "Secretary Califano Lauds NIH; Cites Dr. Fredrickson as 'Best,' " *NIH Record*, February 23, 1977.

39. Memorandum from Paul Cromwell to director, NIH, "Proposed Final Environmental Impact Statement on Recombinant DNA Research," February 2, 1977. DSF Papers, Folder dna\dna77\finaleis.

40. J. Sambrook, "Adenovirus Amazes at Cold Spring Harbor," *Nature* **268:**101–104 (1977).

41. Richard E. Neustadt and Harvey V. Fineberg, *The Swine Flu Affair*, p. 73, 74 (Washington, D.C.: DHEW, 1978).

42. Memorandum from director of environmental affairs (Custard) to the secretary, HEW, "Proposed Issuance of Final Environmental Impact Statement on Recombinant DNA Research," February 25, 1977. DSF Papers, Folder dna\dna77\finaleis.

43. Memorandum from Steven Ebbin to Paul Cromwell, "Review of the Final EIS, NIH Guidelines for Research Involving Recombinant DNA Molecules," February 23, 1977. Ibid. (Mr. Ebbin is described in the Custard-to-HEW-secretary memorandum as "a Senior Associate of the Environmental Studies Board of the National Academy of Sciences, Staff Director of the Senate Subcommittee on Government Research and Development, and Assistant to the [former] Majority Leader of the Senate.")

44. Memorandum from deputy assistant secretary for management, planning, and technology to the HEW secretary, "Final Environmental Impact Statement—Recombinant DNA Molecules—ACTION," March 10, 1977. Ibid.

45. Memorandum from director, NIH, to James F. Hinchman, Office of the General Counsel, HEW, "NIH Environmental Impact Statement on Recombinant DNA Research," March 24, 1977. Ibid.

46. Memorandum conveying Hinchman's views from Joseph Perpich to Donald S. Fredrickson, "Environmental Impact Analyses," April 8, 1977. Ibid.

47. Memorandum from principal environmental officer to director, NIH, "Publication of a Review of the Environmental Implications of the NIH Recombinant Research Guidelines," May 13, 1977. Accompanied by draft statement for the director, NIH. Ibid.

48. Note from Joseph Perpich to Donald S. Fredrickson, May 9, 1977. Ibid.

49. In sum, this complaint of the Friends of the Earth alleges that (i) We failed to comply with the requirements of NEPA, the department's regulations implementing NEPA, with respect to (a) funding of rDNA research, (b) promulgation of the NIH Guidelines, and (c) proposing national legislation to regulate recombinant DNA activities. (ii) We violated the Administrative Procedure Act in developing and implementing the NIH Guidelines. (iii) We violated the Federal Advisory Committee Act in respect to the creation and operation of the RAC. DSF Papers, Folder dna\aagenl\dsf\notes.

50. *RDR*, vol. 3, Appendix C, p. 3–242.

51. Memorandum from director, NIH, to general counsel, HEW, "Suits Filed Against the Department on the Matter of Recombinant DNA Research," June 9, 1977. DSF Papers, Folder dna\dna77\finaleis.

52. Memorandum from Richard J. Riseberg to Messrs. Fredrickson, Libassi, Cotton, Feiner, and Hinchman, "Memorandum to the Secretary on Recombinant DNA Research," June 20, 1977, and addendum to Fredrickson and Perpich on June 27, 1977. Ibid.

53. Note from Rick Cotton to Jim Hinchman, "Thoughts on the DNA EIS," September 28, 1977. Ibid.

54. Letter from Maxine Singer to Donald Fredrickson, October 5, 1977. Ibid.

55. Note from Donald S. Fredrickson to Maxine Singer, July 28, 1978. Ibid.

56. Letter from director, NIH, to Bernard Talbot, July 28, 1978. Ibid. It was one of gratitude to members of the Kitchen RAC: ". . . No one arrived earlier or stayed longer in the pits, and your prodigious output has been a keystone to our success . . . [for] the link you have provided between the scientists and the rest of us, in interpretation of the nuances of questions considered and disposed of by the Recombinant Committee, in the traffic of communiques between experts of many persuasions, and in the scaling of the issues confronting us. . . . I admire, too, the understated grace with which you assisted the Chairman of the RAC [Hans Stetten] with his duties and in overcoming his periodic and understandable torments by the ghosts of Rene Descartes and Francis Bacon . . . for your stubborn honesty and cheerful willingness (after having aired your doubts) to take one more step toward the precipice along with the rest of our beleaguered band. Many, many thanks, Bernie, for simply being indispensable in this affair."

57. Memorandum from special assistant for intramural affairs (Bernard Talbot) to director, NIH, October 3, 1977. Ibid.

58. Memorandum from director of research safety to director, NIH, "EIS on NIH Guidelines," October 7, 1977. Ibid.

59. Letter from Wallace Rowe to Donald S. Fredrickson, October 7, 1977. Ibid.

60. Letter from Donald S. Fredrickson to Wallace Rowe, October 10, 1978. Ibid.

61. Memorandum from director, NIH, to general counsel, DHEW, October 7, 1977. Ibid.

62. Memorandum from director, NIH, to secretary, HEW, October 11, 1977. Ibid.

> Subject: SCIENTIFIC OBJECTIONS TO THE REDUNDANCY OF AN EIS ON THE RESEARCH
>
> An EIS on research would be a tacit admission that the current EIS on the Guidelines is deficient on its discussion of possible environmental hazards. I believe this admission is unwarranted because the EIS is not so deficient. For those experiments permitted under the Guidelines, there are no *known* impacts on the environment, and this is clearly stated to the maximum possible extent in the current EIS. Also the admission that the current EIS is deficient could present problems in defending the two pending lawsuits against the Department. New scientific information significantly diminishes arguments for the presence of possible environmental hazard from the research permissible under the Guidelines . . . [cited in the EIS] and additional scientific evidence to be published next month suggests that experiments performed in the laboratory and those performed by nature are possibly very similar [Stanley Cohen experiments]. . . .
>
> ADMINISTRATIVE/LEGAL OBJECTIONS
>
> An EIS on basic research per se that has no known environmental impact, as opposed to an EIS on guidelines that set safety standards, establishes a significant precedent for statements on experiments with no known environmental impact. Such a precedent established by the Department would affect all other Federal research agencies. To our knowledge other agencies have not required Environmental Impact Statements to be filed on basic laboratory research. . . . [not in the regulations of the EPA (no EIS for laboratory work there) and similarly not required by the Energy Research and Development Administration.] . . . There are no cases, so far as I am aware, in which an EIS on the early stages of basic research has been required by the courts. An EIS on Guidelines makes an important contribution to the public debate and to the decision-making requirements in the Department; but another EIS on the research would have nothing more to contribute.

63. Memorandum from Rick Curtin to Joseph Perpich, "Environmental Impact Statements for Basic Research Projects at the Department of Energy (DoE)," October 13, 1977. Ibid.

64. Memorandum from director, NIH, to Peter Libassi, general counsel, "Approval of NIH Environmental Impact Statement," October 19, 1977. Ibid.

65. Memorandum from Rick Cotton, deputy executive secretary, to the secretary, HEW, Peter Libassi, Donald Fredrickson, Julius Richmond, and Jim Hinchman, "Subject Meeting of October 18 on Recombinant DNA Environmental Impact Statement," October 26, 1977. DSF Papers, Folder dna\dna77\finaleis.

66. National Institutes of Health, *Final Environmental Impact Statement on NIH Guidelines for Research Involving Recombinant DNA Molecules*, Part I (DHEW publication no. [NIH] 77-1489) and Part II (DHEW publication no. [NIH] 77-1490) (October 1977).

67. *Kleppe v. Sierra Club*, 427 U.S. 390, 410 n.21 (1976), citing *Natural Resources Defense Council v. Morton*, 458 F. 2d 827, 838; 148 U.S. App. D.C. 5, 16 (1972).

68. *RDR*, vol. 3, Appendix C, p. 10–11.

69. Ibid., p. 33.

70. United States District Court for the District of Columbia. In re *Mack v. Califano et al.*, Brief for the American Society for Microbiology. *RDR*, vol. 3, Appendix C, p. 143–154.

71. Richard Goldstein, "Public Health Policy and Recombinant DNA," *N. Engl. J. Med.* **296:**1226–1228 (1977).

72. Green Diary 1, March 18, 1978, p. 74–75.

73. M. A. Israel, H. W. Chan, W. P. Rowe, and M. A. Martin, "Molecular Cloning of Polyoma Virus DNA in *Escherichia coli:* Plasmid Vector System," *Science* **203:** 8833–8837 (1979).

74. Stetten, Gartland, and Talbot, *Bioessays*, p. 185–186. See chapter 2, note 19.

Chapter 7

1. An example of actions by the RAC on a given meeting day. Minutes of the RAC meeting, February 15–16, 1979. *RDR*, vol. 5, p. 47–79.

2. Victoria Harden, *Inventing the NIH: Federal Biomedical Research Policy* (Baltimore: Johns Hopkins University Press, 1986).

3. Thomas Parran (surgeon general, 1936–1954) was the principal architect of the NIH expansion in 1944. Lewis ("Jimmy")Thompson was the fifth director of NIH (1937–1942). He was responsible for the procurement of land, donated by the Luke Wilson family, on which the NIH campus is now located. Both men were visionaries whose efforts on behalf of NIH were invaluable.

4. The Office of Scientific Research and Development, headed by Dr. Vannevar Bush during World War II.

5. National Archives, Records of the OSRD, RG 227, Records of the CMR Minutes, Minutes of December 1, 1944, Box 8, 8-3-44 to 12-21-44.

6. For reference to detailed history of this period, see Stephen P. Strickland, *The Story of the NIH Grants Program* (New York: University Press of America, 1989);

Donald S. Fredrickson, "Biomedical Science and the Culture Warp," in *Health Policy Annual*, vol. III. *Emerging Policies for Biomedical Research* (W. N. Kelley, M. Osterweis, and E. Rubin, ed.) (Washington, D.C.: Association of Academic Health Centers, 1993).

7. Speakers included Drs. Liebe Cavalieri, Robert Pollack, Maxine Singer, and Robert Sinsheimer.

8. *Congressional Record*, House, January 19, 1977.

9. "Petition of Environmental Defense Fund and Natural Resources Defense Council." *RDR*, vol. 2, p. 325–345.

10. Janet L. Hopson, "Recombinant Lab for DNA and My 95 Days in It," *Smithsonian* **8**(3):55–62 (1977). After 3 months' internship in Herb Boyer's UCSF laboratory, the author complained that "half the researchers here follow the guidelines fastidiously, others seem to care little. . . . among the graduate students and postdoctorates it seemed almost chic not to know the NIH rules." In the Stevenson hearings on November 8, 1977, Dr. Boyer professed outrage at her perceptions of how his laboratory operated.

11. Draft of H.R. 5020, sponsored by Mr. Ottinger and Congressmen Edgar, Gephardt, Hannaford, Harkin, Martin, Rodino, Seiberling, Traxler, and Tsongas. DSF Papers, Folder dna\dna77\legisl\fed\March.

12. Federal DNA legislation, 95th Cong., 1st sess. *RDR*, vol. 2, p. 506–508.

13. An article written by Dr. Liebe Cavalieri, a molecular biologist at Sloan-Kettering Institute.

14. On March 1, Congressman Stephen Solarz (D-NY) submitted H.R. 4232, The Commission on Genetic Research and Engineering Act of 1977. The commission would include two senators and two representatives among the 17 members. On March 8, Senator Metzenbaum introduced S. 945, The Recombinant DNA Standards Act of 1977. Its 13-member commission would report its study of recombinant DNA issues 27 months after all members of the commission had taken office.

15. Those attending included Philip Abelson, president, Carnegie Institution; Christian Anfinsen, NIH protein chemist and Nobel Prize winner; Beruj J. Benacerraf, Harvard immunologist; Charles K. Bockelman, Yale provost; Herbert E. Carter, chemist and former chancellor of the University of Illinois; Gerald Edelman, professor, Rockefeller University; John Edsall, professor emeritus, Harvard; Rae Godell, MIT, Richard Goldstein microbiologist, Harvard; Irwin C. Gunsalus, professor of biochemistry, University of Illinois; Harlyn O. Halvorson, basic scientist, Brandeis University, and president of the American Society for Microbiology; David Hamburg, president, Institute of Medicine; Philip Handler, president, National Academy of Sciences; Peter Hutt, Covington & Burling; Donald Kennedy, professor of biology, Stanford; Leon Heppel, professor of biochemistry, Cornell University; Julius Krevans, dean, University of California at San Francisco Medical School; John Littlefield, professor of pediatrics, Johns Hopkins; Sherman Mellinkoff, dean, UCLA Medical School; Thomas E. Morgan, Association of American Medical Col-

leges; Robert Murray, chief of medical genetics, Howard University Medical School; Charles Overberger, vice president for research, University of Michigan; J. E. Rall, director of intramural research, NIAMD; Frank Rauscher, senior vice president for research, American Cancer Society; Walter Rosenblith, provost, MIT; Lewis Thomas, chancellor, Memorial Sloan-Kettering Cancer Center; Daniel C. Tosteson, dean, University of Chicago Medical School; and Helen R. Whiteley, professor of microbiology, University of Washington. Among those invited but unable to attend were Paul Berg, Robert Berliner, Stanley Falkow, David Hogness, James F. Kelly, Philip Siekevitz, Robert L. Sinsheimer, and George Wald.

16. "Licensing of DNA Research Expected," *U.S. Medicine* **13:**1, 22 (1977). DSF Papers, Folder dna\dna77\legisl\fed\March.

17. For a detailed agenda and record of the reactions of participants, see DSF Papers, Folder dna\dna77\special mtg NIH:2-19-77.

18. Section 361 (42 U.S.C. §264) "The Surgeon General, with the approval of the Secretary, is authorized to make and enforce such regulations as in his judgment are necessary to prevent the introduction, transmission, or spread of communicable diseases from foreign countries into the States or possessions, or from one State or possession into any other State or possession. For purposes of carrying out and enforcing such regulations, the Surgeon General may provide for such inspection, fumigation, disinfection, sanitation, pest extermination, destruction of animals or articles found to be so infected or contaminated as to be sources of dangerous infection to human beings, and other measures, as in his judgment may be necessary." *RDR*, vol. 3, Appendix A, p. 258.

19. *RDR*, vol. 1, p. 327. Hutt, former counsel general of the FDA, was an expert in the relevant statutes and modes of regulation.

20. *Interim Report of the Interagency Committee on Recombinant DNA Research and Its Applications*, March 13, 1977. *RDR*, vol. 3, Appendix B, p. 321–323.

21. Memorandum from Joseph Perpich to Donald S. Fredrickson, "Relevant Documents Distributed at Meeting with Gil Omenn," January 11, 1978. DSF Papers, Folder dna\dna78\legisl\fed\Jan-Feb.

22. Letter from Stanley Edelman to Peter Libassi re meeting at OSTP on January 12, 1978. Ibid.

23. Memorandum from director, NIH, to HEW general counsel, January 17, 1978. "Reconsideration of Section 361." Ibid.

24. B. F. Armstrong, *A Profile of the United States Public Health Service 1798–1948.* DHEW publication no. 73-369 (1973).

25. The story of the National Commission begins in 1971 when the Tuskegee Project, in which treatment was withheld from patients with syphilis, was uncovered to justifiable outrage. About this same time, Senator Kennedy became chairman of the Senate Health Subcommittee and began numerous hearings on the ethics of research. He proposed the creation of a free-standing commission, operating out-

side the Public Health Service's grant-awarding agencies, to report on biomedical ethics. An amendment of the Public Health Service Act included Title II of Public Law 93-348 in 1974, with a provision under which the commission was formed.

26. *Interim Report of the Federal Interagency Committee on Recombinant DNA Research: Suggested Elements for Legislation.* Submitted to the secretary of HEW, March 15, 1977. *RDR,* vol. 2, p. 279–345.

27. Press release, *HEW News, RDR,* vol. 2, p. 276–278.

28. Memorandum for the record from Donald S. Fredrickson to staff on OMB hearing on administration bill, April 7, 1977. DSF Papers, Folder dna\dna77\legisl\fed\April.

29. ". . . (a) No State or political subdivision of a State may establish or continue in effect with respect to recombinant DNA activities any requirement which is different from, or in addition to, any requirement applicable under this Act to any such activities. . . . (b) Upon application of a State or political subdivision of a State, the Secretary shall exempt from subsection (a) a requirement of that State or political subdivision applicable to recombinant DNA activities if he determines that the requirement is, and will be administered so as to be, as stringent as, or more stringent than, a requirement under this Act. The Secretary may not withdraw any such exemption for so long as he finds that such requirement remains unchanged and continues to be so administered."

30. "The VA does not believe that the total responsibility for controlling research in any area should be invested in a single individual who can exercise unilateral control over the VA's activity from outside the agency. . . . The VA proposes that a new bill create a Commission. . . ." Memorandum from John D. Chase, chief medical director, VA, to acting deputy assistant general counsel, March 1977. DSF Papers, Folder dna\dna77\legisl\fed\April.

31. H.R. 4759 sponsored by Congressmen Rogers, Maguire, Preyer, Scheuer, Waxman, Florio, Markey, Walgren, Carter, and Madigan. *RDR,* vol. 2, p. 543–552.

32. Legislative Highlights, 95th Cong., 1977, DLA, NIH, "House Holds Hearing on Recombinant DNA Research," March 29, 1977. DSF Papers, Folder dna\dna77\legisl\fed\April.

33. Statement of director, NIH, before the Subcommittee on Health and the Environment of the House Committee on Interstate and Foreign Commerce, March 17, 1977. *RDR,* vol. 2, p. 840.

34. Dr. Ruth Hubbard (Harvard) felt that the use of physical containment methods established in the Guidelines was inadequate, that the use of *E. coli* hosts was very dangerous, and that a moratorium against such use should be instituted. Drs. Maxine Singer, Dan Nathans, Bernard Davis, Wallace Rowe, Melvin Chalfen, Mark Ptashne, Richard Novick, Frank Young, Alex Rich, and Robert Alberty, agricultural scientist C. F. Lewis of the Department of Agriculture, and Emmett Barkley all believed the Guidelines to be adequate safeguards, and most thought that the use of *E. coli* K-12 was not troublesome. Professor George Wald (Harvard) opined that

recombinant DNA technology is dangerous and has the potential to alter millions of years of evolution. Dr. Liebe Cavalieri expressed alarm about the possibility recently discovered by Sanger that pure DNA fragments might contain overlapping information that "could also specify a second protein"; he was also concerned about the rising incidence of *E. coli* infections of the bloodstream. He felt that the Guidelines were weak because they rely on the advice of scientists and not laypersons. Drs. Christine Oliver and Stephen Havas from the Massachusetts General Hospital had decided that the risks outweighed the benefits and that either there should be a moratorium or the research should be limited to only one or two facilities. Ethan Signer (MIT) argued for a total and immediate ban on work with such a highly dangerous "gimmick" and did not believe that it might hold the key to many health problems. Instead, he advocated "reducing environmental pollution and better medical care." Marcia Cleveland, a staff attorney for the Natural Resources Defense Council, favored at least a partial moratorium, questioned the need to protect the freedom of private industry to exploit this new and potentially dangerous technique for profit, and felt strict liability should be imposed on those who conduct the research. Mr. Jeremy Rifkin maintained that recombinant DNA research would lead to genetic engineering, making an analogy to the Nazi eugenics. He suggested a 4-year moratorium during which time a "commission of 1400 randomly chosen citizens would carry the issues to the public in preparation for the holding of a national referendum."

35. Leon Heppel, March 21, 1977. Personal communication.

36. *Mr. Ottinger:* Give us some odds, as roughly as you can . . . on there being developed some strain that would be damaging to either human beings or the plant life on which human beings depend?

Dr. Fredrickson: My opinion is that the odds are very small indeed.

Mr. Ottinger: What range are we talking about? Are we talking about one in 1,000, one in 100,000, one in 1,000,000 or one in 10,000,000?

Dr. Fredrickson: I would have to say, giving you my own personal opinion, derived from the sources I have described to you, that they might be one in 1,000,000.

Mr. Ottinger: What kind of odds do you place on there being major, beneficial breakthroughs derived from this experimentation?

Dr. Fredrickson: Again that is a qualified statement as to what is major or beneficial, but the odds are already one to one because the use of these techniques [has resulted] in obtaining gene material to a degree of purity that cannot be achieved by any other known techniques.

37. Ray Thornton, "Commentary," p. 29–31, in *The Gene-Splicing Wars.* See chapter 2, note 3.

38. Congressmen, usually chairs of powerful committees with sufficient clout, can request the Speaker of the House to refer a particular piece of legislation for examination to that committee, even though it is in the domain of another. The period of such "sequential referral" is limited, usually about a month.

39. Legislative Highlights, DLA, NIH, "Recombinant DNA Research Hearings and Bills," April 20, 1977. DSF Papers, Folder dna\dna77\legisl\fed\April.

40. *Congressional Record*, Senate, April 1, 1977, p. 5335–5337.

41. "Guidelines for Research Involving Recombinant DNA Molecules," *Fed. Regist.* **41:**27911 (July 7, 1976). Also *RDR*, vol. 1, p. 11.

42. *Congressional Record*, Senate, April 1, 1977, p. 5335.

43. Letter from Maxine Singer to Donald S. Fredrickson, April 5, 1977. DSF Papers, Folder dna\dna77\legisl\fed\April.

44. Legislative Highlights, DLA, NIH, April 20, 1977, p. 20. Ibid.

45. Statement of director, NIH. Senate hearings on April 6, 1977. *RDR*, vol. 2, p. 859–876.

46. Green Diary 1, April 20, 1977, p. 47.

47. In March 1989, I attended a meeting in Washington, and I ran into a familiar face who proved to be Allan M. Fox, now a Washington attorney. He had been counsel of the subcommittee in the memorable spring of 1977 when he summoned Libassi and me to his office. He asked me if I knew that he was the author of the revision we had received on April 20. I was curious as to why he had proposed the "abominable regulation of DNA research by a commission." He replied that the staff at that time feared that the scientists could not objectively regulate the research. He also said that their fears were often buttressed by material sent by Stan Cohen to Lawrence Horowitz. I said I understood, for I often was the recipient of much of it myself. As we amicably went our separate ways. I was elated to have uncovered the anonymous adversary, with whom, like Don Quixote, I had been tilting for so long.

48. The legislative analysts at NIH, particularly Joseph Hernandez, Esq., played an invaluable role in both dissecting legislation and relating the interplay of tensions among the legislators and staff as the weeks rolled on. See DSF Papers, Folder dna\dna77\legisl\fed\April.

49. *RDR*, vol. 2, p. 690–739.

50. Harlyn O. Halvorson, "The Impact of the Recombinant DNA Controversy on a Professional Scientific Society," p. 73–91, in *The Gene-Splicing Wars*. See chapter 2, note 3.

51. Legislative Highlights, 95th Cong., 1977, DLA, NIH, "House Holds Markup on DNA Bills." DSF Papers, Folder dna\dna77\legisl\fed\May.

52. "Proposed Penalties for Violation of Guidelines." House markup of H.R. 4759, May 3, 1977. Ibid.

53. H.R. 7418, May 24, 1977. *RDR*, vol. 2, p. 583.

54. Burke K. Zimmerman, "Science and Politics: DNA Comes to Washington," p. 39, in *The Gene-Splicing Wars*. See chapter 2, note 3, and memorandum from Joseph A. Hill to Anthony C. Liotta, acting deputy assistant attorney general, Land and Resources Division, "The Patenting of Recombinant DNA Inventions," May 5,

1977. DSF Papers, Folder dna\dnagenl\patent. The legislative intent was concisely described by the chief of the Patent Section in the Civil Division of the Department of Justice, in a note to the Justice Department representative on the FIC.

55. Note from Joseph A. Califano, Jr., to Libassi, Warden, Fredrickson, and Richmond, May 21, 1977. DSF Papers, Folder dna\dna77\legisl\fed\May, item 8.

56. Memorandum prepared by Peter Libassi for the secretary, HEW, "Position on Kennedy Subcommittee DNA Bill." Ibid., item 5.

57. Memorandum from Donald S. Fredrickson and Peter Libassi to secretary, HEW, "Proposed Position on Kennedy Subcommittee DNA Bill," April 26, 1977. Ibid., item 5b.

58. Memorandum from Henry Aaron to Peter Libassi, "Proposed Position on Senate Health Draft . . . ," April 25, 1977. Ibid., item 5c.

59. Diana D. Dutton, with contributions by Thomas A. Preston and Nancy E. Pfund, *Worse Than the Disease: Pitfalls of Medical Progress*, p. 334 (New York: Cambridge, 1988).

60. S. 1217, Subcommittee Print No. 3, May 19, 1977. DSF Papers, Folder dna\dna77\legisl\fed\May, item 6.

61. Memorandum from DLA, NIH, "Recombinant DNA, the First Amendment and Freedom of Inquiry." DSF Papers, Folder dna\dna77\legisl\fed\May, item 16.

62. Memorandum from Donald Hirsch to secretary, HEW. "Talking Paper for Secretary's Planned Telephone Conversation on Tuesday, May 31 with Senator Kennedy on His Subcommittee's Draft DNA Bill (Print 3 of S. 1217)." DSF Papers, Folder dna\dna77\legisl\fed\May, item 19; see also Fredrickson notes on items in this folder.

63. Letter from secretary, HEW, to chairman, House Committee on Interstate and Foreign Commerce. DSF Papers, Folder dna\dna77\legisl\fed\June, item 1.

64. Memorandum from Joseph Hernandez for the record, "House Recombinant DNA Markups (June 7, 8, 10)," June 14, 1977. Ibid.

65. *RDR*, vol. 2, p. 662.

66. Joseph Hernandez, "DLA Reports: House Recombinant DNA Markups." DSF Papers, Folder dna\dna77\legisl\fed\June.

67. Letter from Secretary Joseph A. Califano, Jr., to Dr. Kenneth Ryan, chairman of the National Commission for the Protection of Human Subjects of Biomedical and Behavioral Research, June 10, 1977. Ibid.

68. Memorandum from director, NIH, to secretary, HEW, "Meeting with Representative Ottinger," June 13, 1977. Ibid.

69. Memorandum from Joseph Hernandez for the record, "House Recombinant DNA Legislation," June 20, 1977. Ibid.

70. The full text of H.R. 7897 is found in *RDR*, vol. 2, p. 618–669.

71. Memorandum from Joseph Hernandez for the record, "Senate DNA Legislation," June 23, 1977. DSF Papers, Folder dna\dna77\legisl\fed\June.

72. Memorandum from Joseph Perpich to Donald S. Fredrickson, June 22, 1977. Ibid.

73. Memorandum from director, NIH, to general counsel, June 28, 1977. Ibid.

74. Memorandum from general counsel and director, NIH, to secretary, HEW. June 28, 1977. Ibid.

75. Letters from Harlyn Halvorson to Senator Edward Kennedy, May 27, 1977, and mailgram, May 9, 1977. DSF Papers, Folder dna\dna77\legisl\fed\May.

76. Memorandum from Joseph Hernandez for the record, "DNA Legislative Efforts of the American Society for Microbiology," June 28, 1977. DSF Papers, Folder dna\dna77\legisl\fed\June.

77. "The American Society for Microbiology Position on H.R. 7418," undated draft. Ibid.

78. The second document supplied by Joseph Hernandez, "Supplemental Views of Mr. Nelson," was a six-page draft bearing evidence of considerable editing. Ibid.

79. Letter from Intersociety Council president Robert F. Acker to Congressman Paul Rogers, June 30, 1977. Ibid.

80. Note from Joseph Perpich to Donald S. Fredrickson, June 30, 1977. Ibid.

81. Memorandum from Joseph Hernandez for the record, July 11, 1977. DSF Papers, Folder dna\dna77\legisl\fed\July.

82. Hon. Olin Teague, "International Interest in Genetic Research," Extension of Remarks, *Congressional Record*, July 11, 1977, p. E-4331. Ibid.

83. Letter from Stansfield Turner to Paul Rogers, July 9, 1977. Ibid.

84. The scientists were Harlyn Halvorson of ASM (spokesperson); Frank Young, American Society of Academic Societies; Peter Day, American Institute of Biological Sciences; Oliver Smithies, American Society of Biological Chemists, representing FASEB; Tracy Sonneborn, Genetics Society of America; Lawrence Bogorad, FASEB; and John Sherman, Association of American Medical Colleges.

85. *FASEB Newsl.*, vol. 10, no. 6, July 1977.

86. Letter from Donald S. Fredrickson to Lawrence G. Horowitz, July 20, 1977. DSF Papers, Folder dna\dna77\legisl\fed\July.

87. Letter from Sherwood L. Gorbach to Donald S. Fredrickson, "Report on June 20–21 Worcester Workshop for Assessment of Potential Risks," July 14, 1977. Ibid.

88. Letter from director, NIH, to secretary, HEW, July 20, 1977. Ibid.

89. Draft letters prepared for Senator Kennedy and Congressman Rogers from secretary, HEW, July 21, 1997. Ibid.

90. *Senate Recombinant DNA Safety Act, July 22, 1977*. S. Rept. 95-359. *RDR*, vol. 2, p. 741–803. (See especially p. 755.)

91. Ibid., p. 796.

92. Views of Senator Nelson on S. 1217. Ibid., p. 797–803.

93. Ibid., p. 801–803.

94. Ibid., p. 741–775.

95. Ibid., p. 748–754. Page numbers in text are of actual report, not *RDR* reprint.

96. For a lecture in 1977, a staff artist drew for me a caricature of the present congressional scene. A knight (with Senator Kennedy's face) surrounded by five pawns stands on one part of a chessboard. Confronting him is a rook (with Congressman Rogers's face) with only two pawns to serve as his staff. See D. S. Fredrickson, "Aesculapian Merry-Go-Round," *Trans. Assoc. Am. Physicians* **xc**:59–73 (1977).

97. Memorandum from director, NIH, for the record, "Meeting with Senator Kennedy on S. 1217, Recombinant DNA Safety Regulation Act," July 26, 1977. DSF Papers, Folder dna\dna77\legisl\fed\July.

98. Letter from Professor Charles Yanofsky to Donald S. Fredrickson, July 21, 1977. Ibid.

99. "Amendment in the Nature of a Substitute to S. 1217," *Congressional Record*, Senate, August 2, 1977, p. S13312–13319. DSF Papers, Folder dna\dna77\legisl\fed\August. Amendment no. 754 to S 1217. *RDR*, vol. 2, p. 804.

100. "Comparison of House and Senate DNA Legislation," DLA, NIH, September 28, 1977. DSF Papers, Folder dna\dna77\legisl\fed\September.

101. Memorandum from Robert Watkins (ASM) to Judith Robinson (care of Senator Nelson), "Revised Definition of Recombinant DNA," September 1, 1977. Ibid.

102. Memorandum from associate director for program planning and evaluation to director, NIH, "Follow-Up to Your Conversation with Larry Horowitz," September 8, 1977. Ibid.

103. Memorandum from director, NIH, for the record, September 12, 1977. Ibid.

104. Memorandum from director, NIH, to secretary, HEW, September 15, 1977. Ibid.

105. Memorandum from Joseph Hernandez for the record, September 9, 1977. Ibid.

106. Memorandum from Joseph Perpich on director's briefing of secretary, September 13, 1977. Ibid.

107. Memorandum from Joseph Perpich to Donald S. Fredrickson, "The Rogers Subcommittee Report," September 16, 1977. Ibid.

108. Memorandum from Donald S. Fredrickson on Halvorson's call concerning amendments discussed by Rogers committee, September 27, 1977. Ibid.

109. Memorandum from Donald S. Fredrickson for the record, September 19, 1977. Ibid.

110. Harlyn O. Halvorson, "Recombinant DNA Legislation—What Next?" *Science* **198:**1 (1977).

111. Memorandum from Joseph Hernandez for the record, "Staggers-Kennedy Discussion on DNA Legislation," October 14, 1977. DSF Papers, Folder dna\dna77\legisl\fed\October.

Chapter 8

1. Letter from Senator Adlai E. Stevenson to Frank Press, September 7, 1977. DSF Papers, Folder dna\dna77\legisl\fed\September.

2. *Congressional Record,* Senate, no. 148, September 22, 1977.

3. Senate, *Report on Regulation of Recombinant DNA Research, November 2, 8, and 10, 1977.* 95th Cong., 1st sess., 1977, serial no. 95-52. The testimony from these hearings in this chapter is excerpted and edited by the author. Description of this hearing by Office of the Director, Division of Legal Affairs, NIH, is located in DSF Papers, Folder dna\dna77\legisl\fed\November.

4. Hearing before House Subcommittee on Investigations and Oversight of the Committee on Science and Technology, March 31, 1981.

5. These witnesses included Jonathan King, Roy Curtiss, Bruce Levin, and Daniel Callahan. Also testifying were Clifford Grobstein, professor of biological science and public policy at UCSD, a persistent advocate for creation of a national commission; Marshall Shapiro, professor of law at the University of Virginia; and Marc Lappé, chief, Office of Health, Law and Values, Department of Health, Sacramento, California, sometimes noted for his apocalyptic views in advocating legislation to declare recombinant DNA research a national resource and requiring a commission to oversee this activity in the United States.

6. Nicholas Wade, "Recombinant DNA: NIH Rules Broken in Insulin Gene Project," *Science* **198:**1342–1345 (1977).

7. Green Diary 1, November 8, 1977, p. 56–59.

8. Stephen S. Hall, *Invisible Frontiers: the Race to Synthesize a Human Gene* (New York: Atlantic Monthly Press, 1987).

9. *University of California v. Eli Lilly and Co.*, MDL Docket No. 912, No. Ip-92-0224-C-D/G. 39 USPQ2d 1225 (1996), and Eliot Marshall, "A Bitter Battle over Insulin Gene," *Science* **277:**1028–1030 (1997).

10. "Guidelines for Research Involving Recombinant DNA Molecules," *Fed. Regist.* **41:**27911–27943 (July 7, 1976). See especially p. 27916–27917. Also *RDR,* vol. 1, p. 16.

11. I remembered that the new data for this particular plasmid were to be supplied by the laboratory of Stanley Falkow, whose conservatism and enviable knowledge of microbiology I had learned to appreciate during his early association with the

RAC. Within a week after the RAC received his information, it was forwarded to me for final approval. I then signed off on the certification with no delay. The stipulation of the requirement of the Director's approval of all certification recommendations was not published in a proposed set of revised guidelines until September ("Proposed Revised Guidelines for Research Involving Recombinant DNA Molecules," *Fed. Regist.* **42:**49596–49609 [September 27, 1977].) Also *RDR*, vol. 3, p. 166–174.

12. Memorandum from Joseph Hernandez for the record, November 9, 1977. DSF Papers, Folder dna\dna77\legisl\fed\November.

13. Memorandum from Joseph Perpich to Donald S. Fredrickson, "The Stevenson Hearings," November 9, 1977. Ibid.

14. Personal communications from Norton Zinder and Robert Berliner as recorded in Green Diary 1, p. 59.

15. Barbara Culliton, "Recombinant DNA Bills Derailed: Congress Still Trying To Pass a Law," *Science* **199:**274–277 (1978).

16. Memorandum from Joseph Hernandez to Joseph Perpich, January 24, 1978. DSF Papers, Folder dna\dna78\legisl\fed\jan 24hern,mem.

17. Press release, U.S. House of Representatives, February 10, 1978. DSF Papers, Folder dna\dna78\legisl\fed\Jan-Feb.

18. Harold M. Schmeck, Jr., "New Gene-Splicing Bill Criticized at Science Meeting," *New York Times,* February 16, 1978.

19. Letter from Roy Curtiss III to Congressman Harley Staggers, March 8, 1978; letter from Frank Young to Donald S. Fredrickson, March 9, 1978. DSF Paper, Folder dna\dna78\legisl\fed\March-April.

20. Memorandum from Joseph Perpich to Donald S. Fredrickson, March 7, 1978. Ibid.

21. Memorandum from Donald S. Fredrickson for the record, March 8, 1978. Ibid.

22. Letter from legal advisor, NIH, to HEW general counsel, March 9, 1978. Ibid.

23. Green Diary 1, March 10, 1978.

24. Those voting "no" were Representatives Richard Ottinger (D-NY), Henry Waxman (D-Calif), Andrew Maguire (D-NJ), Edward Markey (D-Mass), Barbara Mikulski (D-Md), and Marc Marks (R-Pa). Research Highlights, April 10, 1978. DSF Papers, Folder dna\dna78\legisl\fed\March-April.

25. House Committee on Interstate and Foreign Commerce, *Recombinant DNA Act, H.R. 11192,* 95th Cong., 2d sess., 1978, H. Rept. 95-1005, part 1. Also *RDR*, vol. 3, Appendix B, p. 3–23.

26. House Committee on Interstate and Foreign Commerce, *Recombinant DNA Act, To Accompany H.R. 11192, with Dissenting and Separate Views,* 95th Cong., 2d sess., 1978, H. Rept. 95-1005, part 1. (For referral to Committee on Science and Technology.) Also *RDR*, vol. 3, Appendix B, p. 131–174.

27. House Subcommittee on Science, Research, and Technology, Committee on Science and Technology, *Science Policy Implications of DNA Recombinant Molecule Research*, 95th Cong., 2d sess., 1978. Also *RDR*, vol. 3, Appendix B, p. 45–127. For a complete record of the hearings, see House Subcommittee on Science, Research, and Technology, Committee on Science and Technology, *Hearings on Science Policy Implications of DNA Recombinant Molecule Research*, March 29, 30, 31; April 27, 28; May 3, 4, 5, 25, 26; September 7 and 8, 1977, 95th Cong., 1st sess., 1977.

28. House Committee on Science and Technology, *Recombinant DNA Act, To Accompany H.R. 11192*, 95th Cong., 2d sess., 1978, H. Rept. 95-1005, part 2. Also *RDR*, vol. 3, Appendix B, p. 189–220; see especially "Committee Views, General," p. 201.

29. *RDR*, vol. 3, Appendix B, p. 206.

30. Letter from Dick Warden to Joseph A. Califano, Jr., May 19, 1978. DSF Papers, Folder dna\dna78\legisl\fed\May-June.

31. Letter from Senator Adlai E. Stevenson to Donald S. Fredrickson, November 30, 1977. *RDR*, vol. 3, p. 549–551.

32. Letter from Donald S. Fredrickson to Senator Adlai E. Stevenson, December 20, 1977. *RDR*, vol. 3, p. 552–555.

33. Senate Subcommittee on Science, Technology and Space, Committee on Commerce, Science and Transportation, *Recombinant DNA Research and Its Applications. Oversight Report, Together with Minority Views*. 95th Cong., 2d sess., 1978. Also *RDR*, vol. 3, Appendix B, p. 253–315.

34. Letter from Donna Parratt to Senator Harrison Schmitt, May 12, 1978, "Questions Concerning the Applicability of Section 361 of the Public Health Service Act to Recombinant DNA Research." *RDR*, vol. 3, Appendix B, p. 324–327.

35. Letter from Senator Edward M. Kennedy to Donald S. Fredrickson, May 2, 1978. DSF Papers, Folder dna\dna78\legisl\fed\May-June.

36. Letter from Senator Harrison H. Schmitt to Secretary Joseph A. Califano, Jr., May 2, 1978. Ibid.

37. Letter from Undersecretary Hale Champion to Senators Edward M. Kennedy and Harrison A. Williams, May 4, 1978. Ibid. Also *RDR*, vol. 3, Appendix B, p. 332.

38. Memorandum from Joseph Hernandez to Donald S. Fredrickson, May 11, 1978. DSF Papers, Folder dna\dna78\legisl\fed\May-June.

39. Letter from Frank Press to Senator Harrison H. Schmitt, May 18, 1978. *RDR*, vol. 3, Appendix B, p. 338–339.

40. Letter from Senators Edward M. Kennedy, Adlai E. Stevenson, Jacob J. Javits, Harrison A. Williams, Gaylord Nelson, and Richard S. Schweiker to Secretary Califano, June 1, 1978. DSF Papers, Folder dna\dna78\legisl\fed\May-June.

41. Letter from Dick Warden to the secretary, HEW, "DNA Legislation," June 6, 1978, with letter from note 40 attached. Ibid.

42. Draft letter prepared by Donald S. Fredrickson for secretary's reply to Senate (see note 40), as revised by Carole Emmott and Dick Warden, June 8, 1978. Ibid.

43. Letter from Michael Goldberg to Joseph Perpich, August 1, 1978. DSF Papers, Folder dna\dna78\legisl\fed\July-Aug.

44. Senator Adlai Stevenson, Address to the Senate, "The Status of Recombinant DNA Research," *Congressional Record*, vol. 124, October 14, 1978. DSF Papers, Folder dna\dna78\legisl\fed\Sept-Oct.

45. Identical letters from secretary of HEW to Senators Kennedy, Javits, Nelson, Schweiker, Stevenson, and Williams, September 12, 1978. Ibid.

46. Hon. Harley O. Staggers, Extension of Remarks, *Congressional Record*, House, January 31, 1979. DSF Papers, Folder dna\dna79\legisl\fed.

Chapter 9

1. "Decision of the Director, NIH, To Release Guidelines for Research on Recombinant DNA Molecules," *Fed. Regist.* **41:**27902–27911 (July 7, 1976). Also *RDR*, vol. 1, p. 2–11. "Guidelines for Research Involving Recombinant DNA Molecules," *Fed. Regist.* **41:**27911–27943 (July 7, 1976). Also *RDR*, vol. 1, p. 11–43.

2. "Proposed Revised Guidelines for Research Involving Recombinant DNA Molecules," *Fed. Regist.* **42:**49596–49609 (September 27, 1977). Also *RDR*, vol. 3, p. 163–177.

3. "Proposed Revised Guidelines for Research Involving Recombinant DNA Molecules," *Fed. Regist.* **43:**33042–33178 (July 28, 1978). Also *RDR*, vol. 3, p. 3–139.

4. "DHEW Public Hearing on the Proposed Revised Guidelines, September 15, 1978." *RDR*, vol. 4, p. 91–278.

5. "Guidelines for Research Involving Recombinant DNA Molecules," *Fed. Regist.* **43:**60080–60131 (December 22, 1978). Also *RDR*, vol. 4, p. 3–53.

6. "Guidelines for Research Involving Recombinant DNA Molecules," *Fed. Regist.* **45:**6724–6749 (January 29, 1980). Also *RDR*, vol. 5, p. 16–41.

7. "Guidelines for Research Involving Recombinant DNA Molecules," *Fed. Regist.* **45:**77384–77409 (November 21, 1980). Also *RDR*, vol. 6, p. 1–28.

8. "Guidelines for Research Involving Recombinant DNA Molecules," *Fed. Regist.* **47:**17166–17198 (April 21, 1982). Also *RDR*, vol. 7, p. 400–431.

9. "Proposed Guidelines for Research Involving Recombinant DNA Molecules, January 1976." *RDR*, vol. 1, p. 71–120.

10. Memorandum from director of NIH to secretary of HEW, "On Efforts To Revise the NIH Guidelines on Recombinant DNA Research," September, 14, 1977. DSF Papers, Folder dna\dna77\GLRev.

11. The report of the NIH\EMBO Workshop on Parameters of Physical Containment, held at the Ariel Hotel, Heathrow, London, March 21–23, 1977, was

forwarded to NIH with a letter from John Tooze, March 24, 1977. DSF Papers, Folder dna\dnagenl\intl\EMBO\76-77.

12. Among the participants at Falmouth were E. S. Anderson (London); David Botstein, Donald Brenner, Sydney Brenner (Cambridge); Allan Campbell, Charles Carpenter, Roy Curtiss III, Ronald Davis, Stanley Falkow, Harry Feldman, Bernard Fields, Samuel Formal, Rolf Freter, Eugene Gangarosa, Peter Gemski, Walter Gilbert, Richard Goldstein, Sherwood Gorbach, Richard Hornick, Lansing Hoskins, Gerald Keusch, Jonathan King, Bruce Levin, Arnold Levine, Stuart Levy, Werner Maas, Malcolm Martin, John Montgomerie, Richard Novick, Frits Orskov (Copenhagen); Nathaniel J. Pierce, John T. Potts, Wallace P. Rowe, R. Bradley Sack, H. Williams Smith (Huntington, England); and Grace Thorne. *RDR*, vol. 3, p. 151–152.

13. Letter from Sherwood L. Gorbach to Donald S. Fredrickson, July 14, 1977. DSF Papers, Folder dna\dna77\Falmouth.

14. "Risk Assessment of Recombinant DNA Experimentation with *Escherichia coli* K-12. Proceedings of a Workshop Held at Falmouth, Mass., June 20 and 21, 1977 (Sherwood L. Gorbach, ed.), *J. Infect. Dis.* **137:**615–714 (1978).

15. Letter from Bruce R. Levin, associate professor of zoology, University of Massachusetts, to Donald S. Fredrickson, July 29, 1977. DSF Papers, Folder dna\dna77\Falmouth.

16. Telegram from Jonathan King, associate professor of biology, MIT, to NIH, July 20, 1977. Ibid.

17. Letter from Wallace P. Rowe, chief, Laboratory of Viral Diseases, NIAID, to Jonathan King, July 27, 1977. Ibid.

18. *Research with Recombinant DNA: an Academy Forum, March 7–9, 1977* (Washington, D.C.: National Academy of Sciences, 1977). A 295-page document containing remarks from the following list of invited speakers: David Hamburg, Daniel Koshland, John Abelson, Maxine Singer, Daniel Callahan, Erwin Chargaff, David Nathans, Alexander Rich, Paul Berg, Robert Sinsheimer, Francisco Ayala, Sir John Kendrew, Stephen Toulmin, Bruce Dull, Bernard Davis, Delbert Barth, Anthony Mazzocchi, Irving S. Johnson, Ruth Hubbard, Raymond C. Valentine, Ethan R. Signer, David Baltimore, Jonathan Beckwith, Roger Noll, Paul Thomas, Donald S. Fredrickson, a summation by Tracy Sonneborn, and final remarks by FDA commissioner Donald Kennedy and other members of the audience.

19. Personal communication.

20. *Research with Recombinant DNA: an Academy Forum.* See note 18 above.

21. June Goodfield, *Research with Recombinant DNA: an Academy Forum*, p. 279. See also J. Goodfield, *Playing God: Genetic Engineering and the Manipulation of Life* (New York: Random House, 1977).

22. Letters from director, NIH, to Secretary Joseph A. Califano, Jr., and Assistant Secretary Julius Richmond, September 14, 1977. DSF Papers, Folder dna\dna77\GLRev.

23. "Comparison of NIH Guidelines for Recombinant DNA Research of June 23, 1976, and the Proposed Revised Guidelines of September 27, 1977," prepared by the Office of the Director, NIH, October 28, 1977. Also, "Justification for Proposed Revisions of NIH Guidelines for Research involving Recombinant DNA Molecules," prepared by the Recombinant DNA Advisory Committee, October 31–November 1, 1977. DSF Papers, Folder dna\dna77\GLRev\Jan-Sept.

24. Memorandum from Florence Hassell to Joseph Perpich, "Comments on Analysis of Comments Pertaining to the NIH Guidelines and the Environmental Impact Statement," March 28, 1977. DSF Papers, Folder dna\dna77\GLRev\Jan-Sept.

25. "Proposed Revised Guidelines," September 27, 1977. *RDR*, vol. 3, p. 164–177.

26. "Proposed Revised Guidelines," July 28, 1978. *RDR*, vol. 3, p. 8. A fine-print footnote in this dense 136-page document reads as follows: "(11) Prof. James Watson, in testimony at the December 1977 DAC meeting and in print, has sought repentance for his earlier activities in support of special precautions for recombinant DNA research."

27. All letters in this series are in *RDR*, vol. 3, Appendix A, p. 1–329.

28. Letter from Leroy Walters, ethicist and RAC member. Ibid., p. 7.

29. Letter from Paul Berg. Ibid. p. 12.

30. Letter from David Baltimore, molecular biologist. Ibid., p. 21.

31. Letter from Barbara E. Echols, secretary, Duke University Biohazards Committee. Ibid., p. 26.

32. Letter from Ronald W. Davis, microbiologist. Ibid., p. 31.

33. Letter from Waclaw Szybalski, oncologist and member of the original RAC. Ibid., p. 46.

34. Letter from Stanley Cohen. Ibid., p. 49.

35. These "veterans" included Attorney Peter Hutt; MIT provost Walter Rosenblith; the molecular biologist Robert Sinsheimer, now chancellor of the University of California at Santa Cruz; and physician-attorney Marjorie Shaw.

36. New members were A. Karim Ahmed, scientist with the Natural Resources Defense Council; Jon Beatty, student member of an institutional biohazards committee from Oregon State University; Roger deRoos, environmental safety expert, University of Minnesota; Harold Ginsberg, chairman of microbiology at Columbia University; James Gustafson, theologian-ethicist from the University of Chicago; Dennis Helms, special assistant to the attorney-general from New Jersey; Sir John Kendrew, EMBO; Patricia King, professor of law, Georgetown University Law Center, and member of the National Commission for the Protection of Human Subjects of Biomedical and Behavioral Research; Rosemary Menard, laboratory technician and IRB member from the University of Washington at Seattle; Mario Molina, chemist, University of California, Irvine, and codiscoverer of the effect of fluorohydrocarbons on the ozone layer; James Neel, geneticist, University of Mich-

igan; Jeanne Sinkford, dean of dentistry at Howard University; Katherine Sturgis, physician and environmental health specialist from Pennsylvania; and Ann Vidaver, plant biologist from the University of Nebraska. Participants are also listed in *RDR*, vol. 3, p. 200–203.

37. John Adams, Pharmaceutical Manufacturers Association; Robert Bock, dean, University of Wisconsin; Ronald Cape, Cetus Corporation; Dennis Chamot, AFL-CIO; Mary-Dell Chilton, plant biologist, University of Washington; Leslie Dach, Environmental Defense Fund; Bernard Davis, Harvard; Donald Duvick, industrial plant breeder; Jonathan King, MIT; Nancy Pfund, Sierra Club; Sheldon Samuels, director of health, safety and environment, AFL-CIO; and David T. Suzuki, Canadian geneticist and television commentator. Suzuki came from Vancouver as part of our conscious effort to hear talented "antiestablishment" scientists who might bring views outside the conventional ones.

38. Among the "public witnesses" were Naum Bers, safety engineer; David Gelfand, scientist, Cetus Corp.; D. Liberman, MIT biohazard assessment officer; Pamela Lippe, assistant legislative director, Friends of the Earth; Arthur Schwartz, University of Michigan mathematician; Francine Simring, executive director, Coalition for Responsible Genetic Research/Friends of the Earth; Anne Skalka, molecular biologist, Roche Institute; Waclaw Szybalski, molecular biologist and former RAC member; Scott Thatcher, graduate student; John Tooze, executive secretary, EMBO; James Watson, director, Cold Spring Harbor Laboratory; and Susan Wright, historian.

39. A. M. Skalka, in *First Report to COGENE from the Working Group on Risk Assessment* (A. M. Skalka, G. Benard, and V. Sgaramella, ed.), July 1978. DSF Papers, Folder dna\dna78\risk assessment.

40. Transcription of the proceedings of the Director's Advisory Committee meeting, December 15–16, 1977. *RDR*, vol. 3, p. 205–498. Individual comments are identified by page numbers in the text.

41. Nicholas Wade, "Gene-Splicing Rules: Another Round of Debate," *Science* **199:** 30–32 (1978).

42. Green Diary Supplement. Impressions of DNA meeting, December, 21, 1977. Written in our quarters in Brattas, above St. Moritz, as the moon shown down on the mountain.

43. I had consumed two Stoppard plays, *Dirty Linen* and *Jumpers*, on the plane as we crossed over to Europe from New York.

44. James Watson, Director's Advisory Committee meeting, December 15–16, 1977. *RDR*, vol. 3, p. 438–439.

45. Letters to Donald S. Fredrickson. *RDR*, vol. 3, Appendix A. Excerpts from letters are identified by page numbers in the text.

46. Letter from Donald S. Fredrickson to Recombinant DNA Molecule Program Advisory Committee, "Review of Proposed Revisions in NIH Guidelines," April 12, 1978. *RDR*, vol. 3, p. 497–516.

47. Meeting of the RAC, April 27–28, 1978. "Summary of Recommendations for Changes in the Proposed Guidelines—Selected Issues for Committee Review." *RDR*, vol. 3, p. 521–529.

48. Stetten, Gartland, and Talbot, *Bioessays*, p. 231–222. See chapter 2, note 19.

49. DeWitt Stetten, Jr., "Valedictory by the Chairman of the NIH Recombinant DNA Molecule Program Advisory Committee," *Gene* **3**:265–268 (1978).

50. This story of my trip to Beijing is based on Green Diary 1, p. 95–98.

51. United States Science and Technology Delegation to the People's Republic of China, July, 1978. The delegation leader was Frank Press, the president's science advisor. At our briefing in the Roosevelt Room of the White House by Mr. Brzezinski on June 28, I learned that my other compatriots would include Richard Atkinson, director of NSF; Jordan Baruch, assistant secretary of commerce; Rupert Cutler, assistant secretary of agriculture; John Deutsch, director of research, Department of Energy; Robert Frosch, director of NASA; William Menard, head of the Geological Survey; Roger Sullivan, deputy assistant secretary for East Asia and Pacific affairs, and Scott Halford, country officer for East Asia, State Department; Ann Keatley, OSTP; Ben Huberman and Mike Ochenburg, staff members of the National Security Council; and Flo Broussard, secretary to Frank Press.

52. "Leading Government Science Officials Will Visit Beijing," *Washington Post*, June 28, 1978.

53. Richard E. Neustadt and Harvey V. Fineberg, *The Swine Flu Affair* (Washington, D.C.: U.S. Government Printing Office, 1978).

54. Green Diary 1, p. 95–103.

55. Witnesses at the Libassi hearing: Representative Richard L. Ottinger; Robert M. Bock, University of Wisconsin, Federation of American Societies for Experimental Biology; Leslie Dach, Environmental Defense Fund; June E. Osborn, University of Wisconsin; Harlyn Halvorson, American Society for Microbiology; W. J. Whelan, Committee on Genetic Experimentation, International Council of Scientific Unions; Dhun Patel, New Jersey Department of Health; Robert M. Faust, U.S. Department of Agriculture; Philip L. Bereano, University of Washington; Deborah Feinberg, New York Attorney General's Office; Norton Zinder, Rockefeller University; Pamela T. Lippe, Friends of the Earth; Jonathan King, MIT and Steering Committee of Coalition for Responsible Genetic Research; Thomas Blessing, Washtenaw County (Mich.) drain commissioner; L. Christine Oliver, Oil, Chemical, Atomic Worker's International Union; Carl E. Kline, Western Maryland Clergy and Laity Concerned; Daniel F. Liberman, biohazard assessment officer, MIT; Grace Ungers, Sloan-Kettering Institute; Daniel Nathans, Johns Hopkins Medical School; Carl Anderson, Brookhaven National Laboratory; Nancy Pfund, Stanford University; Donald Brown, Carnegie Institution of Washington; Frank Ruddle, Yale University; Myron Levine, University of Maryland School of Medicine; Marshall Edgell, University of North Carolina; Oscar L. Miller, University of Virginia; Charles Emerson, University of Virginia; Richard Hartzman, Friends of the Earth; and David Coplin, Ohio State University Agricultural Center.

56. "Comments on NIH-Proposed Revised Guidelines Published in July 1978." *RDR*, vol. 4, Appendix A, p. 1–451.

57. "Issues Arising in Testimony at Public Hearing, September 1978, on NIH-Proposed Revised Guidelines." DSF Papers, Folder dna\dna78\Libassi.

58. Green Diary 1, p. 111.

Chapter 10

1. "Telegram" carried to Secretary Joseph A. Califano, Jr., on December 14, 1978, a few hours after the revision of the Guidelines was released.

2. Letter from Richard C. Atkinson to Donald S. Fredrickson, "Comments on NIH-Proposed Revised Guidelines," September 28, 1978. *RDR*, vol. 4, Appendix A, p. 394–396.

3. Meeting of the Federal Interagency Committee, October 12, 1978. *RDR*, vol. 4, p. 426–439.

4. Following text is edited from Green Diary 1, October to December, 1978, p. 113–146.

5. Workshop for Chairmen of Institutional Biosafety Committees, November 19–21, 1978. Agenda. *RDR*, vol. 4, p. 442–445.

6. Green Diary 1, p. 105–107.

7. Maxine F. Singer, "Editorial: Spectacular Science and Ponderous Process," *Science* **203:**9 (1979).

8. Donald S. Fredrickson, "Note on an Editorial," February 8, 1979. DSF Papers, Folder dna\dna78\forming neorac.

9. "Guidelines," December 22, 1978. *RDR*, vol. 4, p. 49.

10. Letter from Philip Handler, president, NAS, to Secretary Califano, January 4, 1979. DSF Papers, Folder dna\dna79\Guidelines.

11. Green Diary 1, January 29, 1979, p. 150.

12. James D. Watson, "DNA Folly Continues; a Biologist's Plea for Science," *New Republic*, January 13, 1979.

13. "The Public Interest in Gene Splitting," *Washington Post*, January 13, 1979, Editorial.

14. Press release, HEW, announcing appointment of 14 new members of the Department of Health, Education and Welfare's Recombinant DNA Advisory Committee. *RDR*, vol. 4, p. 446–449. (Includes lists of current and new members.)

15. "Proposed Revised Guidelines," July 28, 1978. *RDR*, vol. 3, p. 1–55.

16. "Guidelines," December 22, 1978. *RDR*, vol. 4, p. 48–49.

> IV-E-1. *Director*: The Director, NIH is responsible for (i) establishing the NIH Guidelines . . . (ii) overseeing their implementation, and (iii) their final interpretation. . . .

IV-E-1-a. *General Responsibilities of the Director, NIH.* . . . shall include the following.

IV-E-1-a-(1). Promulgating requirements . . . to implement the Guidelines

IV-E-1-a-(2). Establishing and maintaining the RAC. . . .

IV-E-1-a-(3). Establishing and maintaining ORDA. . .

IV-E-1-a-(4). Maintaining the Federal Interagency Advisory Committee on Recombinant DNA Research. . . .

IV-E-1-b. *Specific Responsibilities of the Director, NIH.* . . .

The Director shall weigh each proposed action, through appropriate analysis and consultation, . . . to determine that it complies with the Guidelines and presents no significant risk to health or the environment.

IV-E-1-b-(1). The Director is responsible for the following major actions. For these, the Director must seek the advice of the RAC and provide opportunity for public and Federal agency comments. Specifically, the agenda of the RAC meeting citing the major actions will be published in the Federal Register at least 30 days before the meeting and the Director will also publish the proposed actions in the Federal Register for comment at least 30 days before the meeting. In addition, the Director's proposed decision, at his discretion, may be published in the Federal Register for 30 days of comment before final action is taken. The Director's final decision, along with a response to the comments, will be published in the Federal Register and the Recombinant DNA Technical Bulletin. The RAC and IBC chairpersons will be notified of this decision.

IV-E-1-b-(1)-(a). Changing containment levels for types of experiments that are specified in the Guidelines when a major action is involved.

IV-E-1-b-(1)-(b). Assigning containment levels for types of experiments that are not explicitly considered in the Guidelines when a major action is involved.

IV-E-1-b-(1)-(c). Certifying new host-vector systems. . . .

IV-E-1-b-(1)-(d). Promulgating and amending a list of classes of recombinant DNA molecules to be exempt from the Guidelines because they consist entirely of DNA segments from species that exchange DNA by known physiological processes, or otherwise do not present risk to health or the environment. . . .

IV-E-1-b-(1)-(e). Permitting exceptions to the prohibited experiments in the Guidelines, for risk assessment studies.

IV-e-1-b-(1)-(f). Adopting other changes in the Guidelines.

IV-E-1-b-(2). The Director is also responsible for the following lesser actions. (For these the Director must seek the advice of the RAC, [and transmit his decisions] . . . to the RAC and IBC chairpersons and published in the Recombinant DNA Technical Bulletin.

IV-E-1-b-(2)-(a) to -(d). Interpreting and determining containment levels . . . Changing containment levels . . . Assigning containment levels . . . Designating certain class 2 agents as class 1 for the purpose of these Guidelines.

IV-E-1-b-(3). The Director is also responsible for the following actions: (The Director's decisions will be transmitted to the RAC and the IBCs and published for comment): (3)-(a) to -(i): Interpreting the Guidelines . . . Determining appropriate containment conditions . . . Case-by-case analysis of experiments explicitly considered in the Guidelines but for which no containment levels have been set.

IV-E-1-b-(3)-(e) to -(i). Authorizing, under procedures set by the RAC, large-scale experiments for recombinant DNAs that are rigorously characterized and free of harmful sequences.

IV-E-1-b-(3)-(f). Lowering containment levels one step for characterized clones involving primate DNA . . . or other characterized clones or purified DNA . . . Approving minor modifications . . . Decertifying already certified host-vector systems . . .

IV-E-1-b-(4). The Director shall conduct, support and assist training programs in laboratory safety. . . .

IV-E-1-b-(5). The Director, at the end of 36 months . . . will report on the Guidelines, their administration, and the potential risks and benefits of this research . . . [Consulting with the RAC and the Interagency Committee] . . . Public comment will be solicited on the draft report. . . .

17. Memorandum from Donald S. Fredrickson for the record, February 15, 1979. DSF Papers, Folder dna\dna78\forming neorac.

18. Green Diary 1, p. 162.

19. Keith Gibson, "European Aspects of the Recombinant DNA Debate," p. 55–71, in *The Gene-Splicing Wars*. See chapter 2, note 3.

20. Mark F. Cantley, "The Regulation of Modern Technology: a Historical and European Perspective: a Case Study on How Societies Cope with New Knowledge in the Last Quarter of the Twentieth Century," p. 518–519, in *Biotechnology*, vol. 12. *Legal, Economic and Ethical Dimensions* (D. Brauer, ed.) (Weinheim: VCH, 1995). I am indebted to Mr. Cantley for providing information used in this account.

21. European Commission, "Proposal for a Council Directive Establishing Safety Measures against the Conjectural Risks Associated with Recombinant DNA Work," *Official Journal of the European Communities* **C301:**5–7 (1978).

22. I first met Princess Liliane in 1962 when she toured the National Heart Institute, of which I was then clinical director. She astounded us all with her knowledge of heart surgery, much of it learned from the late Robert Gross, who had operated successfully on her child. She was also a great friend of Michael E. DeBakey, who had urged me to take an active interest in her Medical Research Foundation. In the ensuing years I occasionally saw the princess again prior to the last illness of King Leopold. Visits to Argenteuil were rare, breathtaking samplings of elite European taste and manners.

23. Cantley, "The Regulation of Modern Biotechnology," p. 519. See note 20 above. See also DSF Papers, Folder dna\dna79\OECD\Schuster&Cantley.

24. European Commission, "Draft Council Recommendation, Concerning the Registration of Recombinant DNA (Desoxynucleic Acid) Work," *COM* **80:**467 (1980).

Chapter 11

1. "FDA Notice of Intent to Propose Regulations," *Fed. Regist.* **343:**60134–60135 (1978). Also *RDR*, vol. 4, p. 55–56.

2. Meeting of the Federal Interagency Committee on Recombinant DNA, July 17, 1979. *RDR*, vol. 5, p. 132–143.

3. "Guidelines," January 29, 1980. *RDR*, vol. 5, p. 38–39.

4. "Straddling the Boundaries of Theory and Practice: Recombinant DNA Research as a Case of Action in the Process of Inquiry," p. 234, in *Recombinant DNA: Science, Ethics, and Politics* (John Richards, ed.) (New York: Academic Press, 1977).

5. Ibid., p. 257.

6. Minutes of the RAC meeting, September 6–7, 1979. *RDR*, vol. 5, p. 152.

7. Minutes of the RAC meeting, May 21–23, 1979. Ibid., p. 104.

8. Minutes of the RAC meeting, December 6–7, 1979. Ibid., p. 416.

9. Ibid., p. 420.

10. "Physical Containment Recommendations for Large-Scale Uses of Organism Containing Recombinant DNA Molecules," *Fed. Regist.* **45:**24968–24971 (1980). Also *RDR*, vol. 6, p. 86–88.

11. "Proposed Actions Under Guidelines: Recombinant DNA Advisory Committee; Meeting\254\," *Fed. Regist.* **45:**7182–7184 (1980). Also *RDR*, vol. 6, p. 29–32.

12. Minutes of the April 1, 1980 meeting of the Industrial Practices Subcommittee of the Federal Interagency Advisory Committee on Recombinant DNA Research. *RDR*, vol. 6, p. 347–354.

13. Minutes of the RAC meeting, May 21–23, 1979. *RDR*, vol. 5, p. 96–126.

14. Minutes of the RAC meeting, June 5–6, 1980. *RDR*, vol. 6, p. 97–139.

15. "Recombinant DNA Research; Proposed Actions Under Guidelines," *Fed. Regist.* **44:**22315–22316 (1979). Also *RDR*, vol. 5, p. 94.

16. "Actions Under the Guidelines," *Fed. Regist.* **44:**45088–45089 (1979). Also *RDR*, vol. 5, p. 145.

17. *Research on Health Effects of Radiation*, vol. 1, part A. NIH publication no. 81-2195 (1980); Interagency Radiation Research Committee, Federal Strategy, *Responding to the Congressional Mandate under the Biological Research Extension Act of 1978 (P.L.95-622)*. NIH publication no. 81-2402 (1981).

18. Green Diary 2, p. 65–69.

19. Ibid., p. 86.

20. Ibid., p. 126.

21. Ibid.

22. Letter from Donald S. Fredrickson to Joseph A. Califano, Jr., July 20, 1979. Ibid., p. 134.

23. Ibid., p. 132–134.

24. Ibid., p. 147.

25. Ibid., p 134.

26. Ibid., p. 149.

27. Minutes of the RAC meeting, September 6–7, 1979. *RDR*, vol. 5, p. 150–194.

28. Ibid., p. 164.

29. *RDR*, vol. 5, p. 239–384 and 435–648. Pagination for cited letters: Bereano, 346, 378, 532; Bross, 343, 373; Cavalieri, 325; Curtiss, 260, 339, 612; Dach, 302, 352; DeNike, 345, 370; Garb, 341, 371; Hartzman, 326; Simring, 252, 335, 369; Senator Stevenson, 374; Congressman Waxman, 353; Wright, 306, 625.

30. "The E. Coli K-12/P1 Recommendation, Made by the RAC at the September 6–7, 1979, Meeting," *Fed. Regist.* **44:**69235–69251 (1979). Also *RDR*, vol. 5, p. 221–237.

31. Letter from Susan Gottesman to Donald S. Fredrickson, December 20, 1979. *RDR*, vol. 5, p. 563.

32. Minutes of the RAC meeting, December 6–7, 1979. *RDR*, vol. 5, p. 396–428.

33. "Recombinant DNA Research; Actions Under Guidelines. Notices," *Fed. Regist.* **45:**3552 (1980). Also *RDR*, vol. 5, p. 430.

34. Letters commenting on guidelines as proposed on November 30, 1979. *RDR*, vol. 5, p. 435–648.

35. Green Diary 2, p. 141.

36. Ibid., p. 153.

37. Ibid., p. 159.

38. *New York Times*, September 21, 1979.

39. Diary note two days later. "Forty-odd staff members welcome Joe Califano back to town for dinner. Joe's offerings included gag that 'I see that 48 hours after we've gone, Fredrickson has relaxed the guidelines.' " Green Diary 2, p. 163.

40. Ibid., p. 179.

41. Ibid., p. 183.

42. Ibid., p. 188.

43. Ibid., p. 192.

44. Ibid., p. 194.

45. (i) "Secretary To Keep NIH Director on Short Leash and in Permanent Heel Position in Waxman Bill H.R. 7036," *FASEB Newsl.*, May 1980; (ii) Daniel S. Greenberg, "Washington Report: NIH, With Friends Like That . . ." *N. Engl. J. Med.*, June 26, 1980; (iii) Victor Cohn, "House Votes for More Control Over NIH," *Washington Post*, August 29, 1980; (iv) "NIH: If It Works, Don't Fix It," *Med. World News*, September 29, 1980; (v) "NIH Bills. Legislators May Or May Not Compromise," *Science*, October 3, 1980. A section of the DSF Papers, entitled "Authorization," contains a large record of the reaction to this proposed legislation, which was eventually defeated.

46. Green Diary 3/4, p. 180–184.

47. On January 20, 1981, during the transition period, I received a memorandum from President Reagan: "In accordance with the provisions of section 3347 of Title 5, U.S. Code you are directed to perform the duties of the Office of Secretary Of Health and Human Services." I had seniority for a day. In the afternoon, there came a phone call from the White House, informing me that two resignations from the NIH were missing. I asked if one of those was Dr. DeVita. "Why, yes," said the voice. "Then the other must be mine," I responded. "Secretary Harris has instructed us not to submit our resignations." No response to this was ever received.

48. "NIH Director's Decision Document and Revised NIH Guidelines Involving Recombinant DNA Molecules," *Fed. Regist.* **45:**6718–6749 (1980). Also *RDR*, vol. 5, p. 9–41.

49. *RDR*, vol. 5, p. 13.

50. "DNA, Risks and Guidelines," *Washington Post*, February 4, 1980, Editorial.

51. From arguments presented by Maxine Singer, discussed at the RAC meeting, September 25–26, 1980, for her proposal of a series of changes in the administrative requirements specified by the Guidelines. *Fed. Regist.* **45:**55926–55927 (1980). Also *RDR*, vol. 6, p. 168–170.

52. "Guidelines for Research Involving Recombinant DNA Molecules," *Fed. Regist.* **45:**77376–77409 (1980). Also *RDR*, vol. 6, p. 211–213.

53. *RDR*, vol. 6, p. 291, 296–308, 311–313.

54. Letter from Paul Berg to Maxine Singer. July 29, 1980. Ibid., p. 291.

55. Summation of the "Fredrickson formula for Guidelines" by Chairman Thornton at the RAC meeting, February 8–9, 1982. *RDR*, vol. 7, p. 334.

56. "Recombinant DNA Research. Proposed Actions Under Guidelines," *Fed. Regist.* **46:**17995 (1981). Also *RDR*, vol. 7, p. 75.

57. Minutes of RAC meeting, April 23–24, 1981. *RDR*, vol. 7, p. 87–132.

58. Letter from Norton Zinder to ORDA. *RDR*, vol. 7, p. 99.

59. The members are listed in *RDR*, vol. 7, p. 237.

60. Ibid., p. 232.

61. "Guidelines for Research Involving Recombinant DNA Molecules," *Fed. Regist.* **47**:17166–17198 (April 21, 1982). Also *RDR*, vol. 7, p. 400–431.

62. Comments of Robert Fildes, vice president of the Industrial Biotechnology Association and president of Biogen, Inc., at a meeting of the Large Scale Review Working Group of the Recombinant DNA Advisory Committee, June 29, 1982. *RDR*, vol. 7, p. 500–505.

63. Larry Thompson, *Correcting the Code: Inventing the Genetic Cure for the Human Body* (New York: Simon and Schuster, 1994). This is a detailed story of this unfortunate incident, including as well the description of the first approved gene therapy experiments in 1990. The author quotes some passages in my Green Diaries relating to the Cline case.

64. Robert Neville, "Philosophical Freedom of Inquiry," p. 1115–1128, in *Recombinant DNA: Science, Ethics, and Politics* (John Richards, ed.) (New York: Academic Press, 1977).

65. Hans Jonas. Ibid., p. 357.

66. Green Diary 7, p. 3–4.

67. Debarment and Suspension from Eligibility for Financial Assistance. October 9, 1980 . . . On March 19, 1979, the Secretary of Health, Education and Welfare published in the *Federal Register* (44 FR 16444) a notice proposing "regulations for debarment and suspension of individuals and institutions for specified causes and process safeguards." *Fed. Regist.* **45**:67262–67269 (1980).

68. Jon A. Wolff and Joshua Lederberg, "An Early History of Gene Transfer and Therapy," *Hum. Gene Ther.* **5**:469–480 (1994).

69. Robert M. Cook-Deegan, "Human Gene Therapy and Congress," *Hum. Gene Ther.* **1**:163–170 (1990).

70. President's Commission for the Study of Ethical Problems in Medicine and Biomedical and Behavioral Research, *Splicing Life. The Social and Ethical Issues of Genetic Engineering with Human Beings* (Washington, D.C.: U.S. Government Printing Office, 1982).

71. House Subcommittee on Investigation and Oversight, Committee on Science and Technology, *Hearing on Human Genetic Engineering*. 97th Cong, 2d sess., November 16–18, 1982.

72. Donald S. Fredrickson, "Comments to the Annual Meeting of the American Association for the Advancement of Science, January 7, 1982." Revised and expanded as "The Recombinant DNA Controversy: the NIH Viewpoint," p. 13–26, in *The Gene-Splicing Wars*. See chapter 2, note 3.

Chapter 12

1. Donald S. Fredrickson, "A History of the Recombinant DNA Guidelines in the United States," *Recombinant DNA Tech. Bull.* **2**:87–89 (1979). Also in *Recombinant DNA and Genetic Experimentation* (J. Morgan and W. J. Whelan, ed.) (Oxford: Pergamon Press, 1979).

2. From "IV-E-1-b. Specific Responsibilities of the Director, NIH Guidelines for Recombinant DNA Research," *Fed. Regist.* **43**:60126 (1978). Also *RDR*, vol. 4, p. 48.

Chapter 13

1. Bernard John and George L. Gabor Miklos, *The Eukaryotic Genome in Development and Evolution* (London: Allen and Unwin, 1988).

2. "Guidelines," April 21, 1982. *RDR*, vol. 7, p. 400–431.

3. An early summary of the introduction of recombinant DNA technology into the discipline of molecular genetics can be found in R. A. Weisberg and P. Leder, "Fundamentals of Molecular Genetics," in *The Metabolic Basis of Inherited Disease*, 5th ed. (J. B Stanbury, J. B. Wyngaarden, D. S. Fredrickson, J. L. Goldstein, and M. S. Brown, ed.) (New York: McGraw Hill, 1983).

4. "Government R & D Funding Spawns a New Industry," *Washington Post*, December 16, 1984.

5. *Washington Post*, August 26,1991. French Anderson and Michael Blaese talk about the arrangements between Gene Therapy Inc. and NIH.

6. J. H. Comroe, Jr., and R. D. Dripps, "Scientific Basis for the Support of Biomedical Science," *Science* **192**:105–111 (1976); J. H. Comroe, Jr., "Roast Pig and Scientific Study," *Am. Rev. Respir. Dis.* **115**:853–860 (1977).

7. David H. Guston, "Retiring the Social Contract for Science," *Issues Sci. Technol.*, Summer 2000, p. 32–36.

8. Robert Frodeman and Carl Mitcham, "Beyond the Social Contract Myth." Ibid. p. 37–41.

9. Gerald Holton and Gerhard Sonnert, "A Vision of Jeffersonian Science," *Issues Sci. Technol.*, Fall 1999, p. 61–65.

10. Bryan Appleyard, *Brave New Worlds: Genetics and the Human Experience*, p. 126–127 (London: Harper Collins, 1999).

11. Minutes of the RAC meeting, April 11, 1983. *RDR*, vol. 8, p. 85–125.

12. "Points to Consider" was published for public comment in *Fed. Regist.* **50**:2940–2945 (1985). Also *RDR*, vol. 9, p. 290–295. See also Eric T. Juengst, "The NIH 'Points to Consider' and the Limits of Human Gene Therapy," *Hum. Gene Ther.* **1**:425–433 (1990).

13. Leroy Walters, "Human Gene Therapy: Ethics and Public Policy," *Hum. Gene Ther.* **2**:115–122 (1991).

14. Nelson A. Wivel and W. French Anderson, "Human Gene Therapy: Public Policy and Regulatory Issues," in *The Development of Human Gene Therapy* (Cold Spring Harbor, N.Y.: Cold Spring Harbor Press, 1999).

15. From the archives on the website of ORDA (now OBA, Office of Biotechnology Activities): http://www.nih.gov/od/oba.

16. "Recombinant DNA Research: Notice of Intent to Propose Amendments to the NIH Guidelines for Research Involving Recombinant DNA Molecules (NIH Guidelines) Regarding Enhanced Mechanisms for NIH Oversight of Recombinant DNA Activities," *Fed. Regist.* **61**:35774–35777 (1996). Also available at http://www4.od.nih.gov/oba/fr-small.htm.

17. Statement from American Society for Microbiology to NIH Office of Science Policy, August 1, 1996. See http://www.asmusa.org/pasrc/rac.htm. This brief contains a valuable contemporary view on the merits and achievements of the RAC.

> . . . In view of the pivotal role that the NIH Recombinant DNA Advisory Committee has played in the public and scientific debate over the safety of recombinant DNA and gene therapy, it is not surprising that the proposal to discontinue RAC is controversial. The RAC has earned a reputation for thoughtfulness, integrity, independence and scientific validity. For over twenty years it has been constituted to have the breadth of knowledge, experience and prestige necessary to deal effectively with issues raised by molecular genetics and its applications.
>
> . . . Certainly, there is no need for duplication with gene therapy and approval processes by the Food and Drug Administration (FDA) and other agencies. On the other hand, however, the elimination of the RAC removes an historically important and time tested means of ensuring future progress and public scrutiny of recombinant DNA research. RAC activities have taken a careful approach to public participation and understanding, and RAC's sensitivity to public as well as scientific opinion has contributed to the confidence scientists who conduct genetic research and the lay public have in the NIH Guidelines, and in the RAC's capacity to modify the Guidelines as dictated by advances in knowledge. The RAC has greatly increased the visibility of how science contributes to the development of recombinant DNA technologies. In so doing, it has improved public confidence in ways that the normal regulatory review process does not. Since it was established to provide guidance rather than regulatory oversight, the RAC has been able to involve the most able and best qualified scientists and members of the laity in the continued review and development of the NIH Guidelines for Recombinant DNA Research. This is especially important because the Guidelines serve as a model for the rest of the world and, in some cases, the legal framework for biosafety related to recombinant DNA activities. . . .

18. "Recombinant DNA Research: Proposed Actions Under the Guidelines," *Fed. Regist.* **61**:59725 (1996). Also available at http://www4.od.nih.gov/oba/fedreg.htm.

19. Recombinant DNA Advisory Committee meeting, December 8–10, 1999. http://www.nih.gov/od/oba.

20. Edward H. Ahrens, Jr., *The Crisis in Clinical Research: Overcoming Institutional Obstacles* (New York: Oxford, 1992).

APPENDIX 1.1

Participants at Asilomar

International Conference on Recombinant DNA Molecules

February 24–27, 1975, Asilomar Conference Center, Pacific Grove, Calif.

ORGANIZING COMMITTEE

Paul Berg, *Chair;* Professor, Department of Biochemistry, Stanford University Medical Center, Stanford, Calif.

David Baltimore, American Cancer Society Professor of Microbiology, Center for Cancer Research, Massachusetts Institute of Technology, Cambridge, Mass.

Sydney Brenner, Member, Scientific Staff of the Medical Research Council, United Kingdom, Cambridge, England

Richard O. Roblin III, Professor of Microbiology and Molecular Genetics, Harvard Medical School, and Assistant Bacteriologist, Infectious Disease Unit, Massachusetts General Hospital, Boston

Maxine F. Singer, Biochemist, National Institutes of Health, Bethesda, Md.

Sherman Weissman, Professor, Department of Medicine, Biology and Molecular Biophysics, Yale University, New Haven, Conn.

Norton D. Zinder, Professor, The Rockefeller University, New York, N.Y.

PARTICIPANTS—UNITED STATES

Edward A. Adelberg, Department of Microbiology, Yale University, New Haven, Conn.

W. Emmett Barkley, Head, Environmental Control Section, National Cancer Institute, Bethesda, Md.

Louis S. Baron, Chief, Department of Bacterial Immunology, Walter Reed Army Institute of Research, Washington, D.C.

Michael Beer, Department of Biophysics, Johns Hopkins University, Baltimore, Md.

Jerome Birnbaum, Basic Microbiology, Merck Institute, Rahway, N.J.

J. Michael Bishop, Professor of Microbiology, University of California Medical Center, San Francisco

David Botstein, Cold Spring Harbor Laboratory, Cold Spring Harbor, N.Y.

Herbert Boyer, Department of Microbiology, University of California, San Francisco

Donald D. Brown, Staff Member, Department of Embryology, Carnegie Institution of Washington, Baltimore, Md.

Robert H. Burris, Professor of Biochemistry, University of Wisconsin, Madison

Allan M. Campbell, Department of Biology, Stanford University, Stanford, Calif.

Alexander Capron, University of Pennsylvania School of Law, Philadelphia

John A. Carbon, Professor of Biochemistry, Department of Biological Science, University of California, Santa Barbara

Dana Carroll, Department of Embryology, Carnegie Institution of Washington, Baltimore, Md.

A. M. Chakrabarty, Physical Chemistry Laboratory, General Electric Company, Schenectady, N.Y.

Ernest Chu, Department of Human Genetics, University of Michigan Medical School, Ann Arbor

Alfred J. Clark, Department of Molecular Biology, University of California, Berkeley

Eloise E. Clark, Division Director, Division of Biological and Medical Sciences, National Science Foundation, Arlington, Va.

Royston C. Clowes, Professor of Biology, Institute for Molecular Biology, University of Texas, Dallas

Stanley Cohen, Associate Professor, Department of Medicine, Stanford University Medical School, Stanford, Calif.

Roy Curtiss III, Department of Microbiology, University of Alabama Medical Center, Birmingham

Eric H. Davidson, Department of Developmental Biology, California Institute of Technology, Pasadena

Ronald W. Davis, Assistant Professor, Department of Biochemistry, Stanford University Medical Center, Stanford, Calif.

Peter Day, Connecticut Agricultural Experiment Station, New Haven, Conn.

Vittorio Defendi, Chairman, Department of Pathology, New York University Medical Center, New York, N.Y.

Roger Dworkin, Department of Biomedical History, University of Washington Medical School, Seattle

Marshall Edgell, Department of Bacteriology, University of North Carolina, Chapel Hill

Stanley Falkow, Department of Microbiology, University of Washington School of Medicine, Seattle

W. Edmund Farrar, Jr., Department of Medicine, Medical University of South Carolina, Columbia

Maurice S. Fox, Department of Biology, Massachusetts Institute of Technology, Cambridge, Mass.

Theodore Friedman, Department of Medicine, University of California at San Diego, La Jolla

William Gartland, National Institute of General Medical Sciences, National Institutes of Health, Bethesda, Md.

Harold Green, Fried, Frank, Harris, Schriver, and Kampelman, Washington, D.C.

Irwin C. Gunsalus, Professor of Biochemistry, University of Illinois, Urbana

Donald R. Helinski, Professor, Department of Biology, University of California at San Diego, La Jolla

Robert B. Helling, Department of Botany, University of Michigan, Ann Arbor

Alfred Hellman, Head, Biohazards and Environmental Control, National Cancer Institute, Bethesda, Md.

David S. Hogness, Professor, Department of Biochemistry, Stanford University Medical Center, Stanford, Calif.

David A. Jackson, Department of Microbiology, University of Michigan Medical School, Ann Arbor

Leon Jacobs, Associate Director for Collaborative Research, National Institutes of Health, Bethesda, Md.

Henry Kaplan, Department of Radiology, Stanford University Medical Center, Stanford, Calif.

Joshua Lederberg, Professor, Department of Genetics, Stanford University Medical Center, Stanford, Calif.

Arthur S. Levine, Head, Section on Infectious Diseases, National Cancer Institute, Bethesda, Md.

Andrew M. Lewis, Laboratory of Viral Diseases, National Institute of Allergy and Infectious Diseases, Bethesda, Md.

Herman Lewis, Head, Cellular Biology Section, Division of Biological and Medical Sciences, National Science Foundation, Arlington, Va.

Paul Lovett, Department of Biological Sciences, University of Maryland, Baltimore, Md.

Morton Mandel, Department of Biochemistry and Biophysics, University of Hawaii School of Medicine, Honolulu

Paul Marks, Vice President, Medical Affairs, College of Physicians and Surgeons, Columbia University, New York, N.Y.

Malcolm A. Martin, Head, Physical Biochemistry Section, National Institute of Allergy and Infectious Diseases, National Institutes of Health, Bethesda, Md.

Robert G. Martin, Biochemist, National Institute of Arthritis, Metabolism, and Digestive Diseases, National Institutes of Health, Bethesda, Md.

Carl R. Merril, Laboratory of General and Comparative Biochemistry, National Institute of Mental Health, National Institutes of Health, Bethesda, Md.

John Morrow, Department of Embryology, Carnegie Institution of Washington, Baltimore, Md.

Daniel Nathans, Boury Professor and Director, Department of Microbiology, Johns Hopkins University School of Medicine, Baltimore, Md.

Elena O. Nightingale, Resident Fellow, Division of Medical Sciences, National Academy of Sciences, Washington, D.C.

Richard P. Novick, Department of Microbiology, Public Health Research Institute, New York, N.Y.

Ronald Olsen, Department of Microbiology, University of Michigan, Ann Arbor

Richard J. Roberts, Cold Spring Harbor Laboratory, Cold Spring Harbor, N.Y.

William Robinson, Department of Infectious Diseases, Stanford University Medical Center, Stanford, Calif.

Stanfield Rogers, Department of Biochemistry, University of Tennessee Medical Units, Memphis, Tenn.

Bernard Roizman, Professor of Microbiology and Biophysics, University of Chicago, Chicago, Ill.

Joe Sambrook, Cold Spring Harbor Laboratory, Cold Spring Harbor, N.Y.

Jane Setlow, Brookhaven National Laboratory, Upton, Long Island, N.Y.

Philip Sharp, Center for Cancer Research, Massachusetts Institute of Technology, Cambridge, Mass.

Aaron J. Shatkin, Member, Roche Institute of Molecular Biology, Nutley, N.J.

George R. Shepherd, Los Alamos Scientific Laboratory, Los Alamos, N.M.

Artemis P. Simopoulous, Staff Officer, Division of Medical Sciences, National Research Council, National Academy of Sciences, Washington, D.C.

Daniel Singer, Vice President, Institute of Society, Ethics and Life Sciences, Hastings, N.Y.

Robert L. Sinsheimer, Chairman, Division of Biology, California Institute of Technology, Pasadena

Anna Marie Skalka, Associate Member, Department of Cell Biology, Roche Institute of Molecular Biology, Nutley, N.J.

Mortimer P. Starr, Department of Bacteriology, University of California, Davis

DeWitt Stetten, Jr., Deputy Director for Science, National Institutes of Health, Bethesda, Md.

Waclaw Szybalski, McArdle Laboratory, University of Wisconsin, Madison

Charles A. Thomas, Jr., Department of Biological Chemistry, Harvard Medical School, Boston, Mass.

Gordon M. Tompkins, Professor of Biochemistry, Department of Biochemistry and Biophysics, University of California, San Francisco

Jonathan W. Uhr, Professor and Chairman, Department of Microbiology, University of Texas Southwestern Medical School, Dallas

Raymond C. Valentine, Assistant Professor in Residence, Department of Chemistry, University of California, San Diego

Jerome Vinograd, Professor of Chemistry and Biology, California Institute of Technology, Pasadena

Duard Walker, Department of Medical Microbiology, University of Wisconsin, Madison

Rudolf G. Wanner, Associate Director for Environmental Health and Safety, Division of Research Services, National Institutes of Health, Bethesda, Md.

James Watson, Professor, Department of Biology, Harvard University, Cambridge, Mass.

Peter Weglinski, Department of Biology, Massachusetts Institute of Technology, Cambridge, Mass.

Bernard Weisblum, Professor, Department of Pharmacology, University of Wisconsin Medical School, Madison

Pieter Wensink, Brandeis University, Waltham, Mass.

Frank Young, Department of Microbiology, University of Rochester, Rochester, N.Y.

PARTICIPANTS—INTERNATIONAL

Ephraim S. Anderson, Director, Enteric Reference Laboratory, Public Health Laboratory Service, London, England

Toshihko Arai, Department of Microbiology, Keio University, Shinjuku, Tokyo, Japan

Werner Arber, Department of Microbiology, University of Basel, Basel, Switzerland

A. A. Bayev, Academician, Institute of Molecular Biology, Moscow, USSR

Douglas Berg, Department of Molecular Biology, University of Geneva, Geneva, Switzerland

Yuriy A. Berlin, Professor, M. M. Shemyakin Institute of Bioorganic Chemistry, Academy of Sciences of the USSR, Moscow

G. Bernardi, Institut de Biologie Moleculaire, Faculte des Sciences, Paris, France

Max Birnstiel, Institute of Molecular Biology II, University of Zurich, Zurich, Switzerland

Walter F. Bodmer, Genetics Laboratory, Department of Biochemistry, Oxford, England

N. H. Carey, G. D. Searle and Company, Ltd., Research Division, England

Y. A. Chabbert, Professor, Bacteriology Department, Institut Pasteur, Paris, France

Francois Cuzin, Institut Pasteur, Paris, France

Julian E. Davies, Professor, Department of Molecular Biology Sciences, University of Geneva, Geneva, Switzerland

Ray Dixon, ARC Unit of Nitrogen Fixation, University of Sussex, Brighton, England

W. A. Englehardt, Professor, Institute of Molecular Biology, Academy of Sciences of the USSR, Moscow

Walter Fiers, Laboratorium voor Moleculaire Biologie, Ghent, Belgium

Murray J. Fraser, Professor, Department of Biochemistry, McGill University, Montreal, Quebec, Canada

W. Gayewski, Professor, Department of Genetics, Warsaw University, Ujazdowskie, Poland

Stuart W. Glover, Department of Psychology, University of Newcastle-Upon-Tyne, England

Walter Goebel, Professor Gesellschaft fur Molekularbiologische Forschung, Braunschweig, West Germany

Carleton Gyles, Department of Veterinary Microbiology and Immunology, The Ontario Veterinary College, University of Guelph, Guelph, Ontario, Canada

Gerd Hobom, Institut fur Biologie II der Universitat Freiburg, West Germany

Peter H. Hofschneider, Professor, Max-Planck-Institut fur Biochemie, Munchen, West Germany

Bruce W. Holloway, Department of Genetics, Monash University, Victoria, Australia

H. S. Jansz, Professor, Nederlandse Dereniging voor Biochemie, Amsterdam, The Netherlands

Mikhail N. Kolosov, Academician, M. M. Shemyakin Institute of Bioorganic Chemistry, Academy of Sciences of the USSR, Moscow

Philippe Kourilsky, Institut Pasteur, Paris, France

Ole Maaloe, Professor, Department of Microbiology, University of Copenhagen, Copenhagen, Denmark

Alastair T. Matheson, Senior Research Officer, Division of Biological Sciences, National Research Council, Ottawa, Ontario, Canada

Kenichi Matsubara, Department of Biochemistry, Kyushu University, Fukuoka, Japan

Andrey D. Mirzabekov, Professor, Institute of Molecular Biology, Academy of Sciences of the USSR, Moscow

Kenneth Murray, Senior Lecturer, Department of Molecular Biology, University of Edinburgh, Edinburgh, Scotland

Haruo Ozeki, Department of Biophysics, Faculty of Sciences, University of Kyoto, Kyoto, Japan

James Peacock, Division of Plant Industries, CSIRO, Canberra City, Australia

Lennart Philipson, Department of Microbiology, The Wallenberg Laboratory, Uppsala University, Uppsala, Sweden

James Pitard, Department of Microbiology, University of Melbourne, Parkville, Victoria, Australia

Mark H. Richmond, Head, Department of Bacteriology, University of Bristol, Bristol, England

A. Rorsch, Department of Biochemistry, Leiden State University, Leiden, The Netherlands

Vittorio Sgaramella, Instituto di Genetica, Pavia, Italy

Luigi G. Silvestri, Gruppo Lepetit, Milano, Italy

Lou Siminovitch, University of Toronto, Department of Medical Genetics, Toronto, Ontario, Canada

H. Williams Smith, Houghton Poultry Research Station, Huntingdon, England

Peter Starlinger, Institut fur Genetik der Universitat Koln, Germany

Pierre Tiollais, Institut Pasteur, Paris, France

Alfred Tissieres, Professor, Department of Molecular Biology Sciences, University of Geneva, Geneva, Switzerland

John Tooze, EMBO, Heidelberg, West Germany

Alex J. Van der Eb, Laboratory of Physiological Chemistry, Leiden, The Netherlands

Robin Weiss, Imperial Cancer Research Fund Laboratories, London, England

Charles Weissmann, Professor, Institut fur Molekularbiologie, University of Zurich, Zurich, Switzerland

Robert Williamson, Beatson Hospital, Glasgow, Scotland

Ernest Winocour, Professor, Department of Genetics, Weizmann Institute of Science, Rehovot, Israel

E. L. Wollman, Institut Pasteur, Paris, France

Hans G. Zachau, Professor, Institut fur Physiologische Chemie und Physikalische Biochemie, University of Munchen, Munchen, West Germany

PARTICIPANTS—PRESS

George Alexander, *Los Angeles Times*

Stuart Auerbach, *Washington Post*

Jerry Bishop, *Wall Street Journal*

Graham Chedd, *New Scientist* and *Nova*

Robert Cooke, *Boston Globe*

Rainer Flohl, *Frankfurter Allgemeine*

Angela Fritz, Canadian Broadcasting Corporation

Gail McBride, *Journal of the American Medical Association*

Victor McElheny, *New York Times*

Colin Norman, *Nature*

Dave Perlman, *San Francisco Chronicle*

Judy Randal, *Washington Star-News*

Michael Rogers, *Rolling Stone*

Cristine Russell, *Bioscience*

Nicholas Wade, *Science*

Janet Weinberg, *Science News*

APPENDIX 9.1
Decision Document

Excerpts from the decision document prepared by the DHEW Task Force after the "Libassi hearing"† (an illustration of the application of legal analysis to the process of revision of recombinant DNA guidelines).

Naked DNA

Issue: Witness 10 opposes exemptions from Guidelines of experiments involving naked DNA. Cites Rowe-Martin Polyoma experiment and says, "this experiment showed that naked polyoma DNA, when injected into the bloodstream or tube-fed into the colon of mice caused low level infection."

Analysis: Witness is incorrect that feeding of naked polyoma DNA to mice caused infections. It did not. Injected, it caused low grade infection but with much reduced ability as compared with the whole virus. Were naked DNA molecules to be ingested by humans, we can be quite certain they would be destroyed in the intestinal tract because of the variety of enzymes within the tract that degrade DNA. . . .

Recommendation: No change in Guidelines.

On NEPA

Issue: Witness 13 says exemptions and exceptions from prohibitions . . . as well as . . . all exceptions from the guidelines should be undertaken only after a full environmental impact statement to allow for assessment as well as better public input and awareness. . . .

Analysis: NIH will comply with NEPA.

Recommendation: No change in Guidelines.

---

† Issues arising in testimony at a public hearing, September 1978, on NIH-proposed revised Guidelines. DSF Papers, Folder dna\dna78\Libassi.

Who Approves?

Issue: Witness 2 notes that the Guidelines state that new host-vector systems may not be used unless they have been certified by NIH. But there is no indication of what office within NIH certifies . . .

Analysis: Paragraph II-D-2-a of the Guidelines does explain the certification process and says, "When new host-vector systems are certified, notice of certification will be sent . . . to the applicant. . . ." Paragraph IV-B-1-c specifies that the Director, NIH has responsibility for certifying new host-vector systems. . . . Paragraph IV-B-2-c specifies that the RAC only recommends approval to the NIH Director.

Recommendation: To make this clearer, add text to Paragraph II-D-2-a stating again that the RAC recommends . . . The Director certifies.

Containment Alterations

General: Witnesses 1, 3, 4, 22, 24, 26, 27 and 33 concurred with the scientific arguments in the Director's Decision Document. Many cited the reports of the Ascot and Falmouth meetings as confirmatory evidence. . . . Witness 27 concurred with the lowered containment levels for viruses. Witness 26 stated that relaxation would permit major studies to be made of human genes. Witness 33 stated that the proposed revisions would permit important research with plants and plant-associated microorganisms.

Other witnesses criticized the scientific arguments upon which the proposed reductions in containment levels are based. Witnesses 10, 14, and 20, particularly, state that the revisions . . . are a consequence of decisions arrived at by scientific methods as a result of "semi-authorized reports held at closed scientific meetings." They state that the basis for downgrading containment for cloning foreign DNA does not take into account enough experiments designed to test the escape from the laboratory of EK1 and EK2 strains of *E. coli*. . . . More general concerns are expressed by Witness 10, who notes . . . Because "we have not yet witnessed an instance of pathogenicity resulting from recombinant DNA activities is not proof of safety. . . ." Witness 17 suggests a prospective study of the ability of *E. coli* K-12 to colonize the urinary tract of the exposed laboratory worker should be done. Witness 14 notes that in "numerous, well-studied cases, the pathogenicity of *E. coli* strains for a particular host is associated with plasmid genes. . . ." The witness continues by reviewing the characteristics of bacterial pathogenicity including its association with immunologically compromised individuals. . . .

Analysis: Both the Director's Decision document and the Environmental Impact Assessment describe in considerable detail the scientific rationales for the revised containment levels. Considerable support for these revisions was given by a series of conferences, the result of which were available for public consideration . . . Falmouth, Ascot, Virus Working Group . . . Additional pages of citations . . . scientific evidence from Environmental Impact Assessment.

Recommendation: No change in the Guidelines.

Roles and Responsibilities: Who's Responsible?

Issue: Witness 17 says that delegation in the proposed revised guidelines to IBCs of "Key technical and procedural decisions . . . is analogous to the allocation to the states of responsibility for monitoring and enforcing industrial compliance with OSHA standards. Such a scheme has proved to be disastrous. . . ." Witness 10 recommends that "grievance procedures for workers be established under the Guidelines." Witness 8 says, "The guidelines should require that practices within institutional laboratories comply with OSHA standards, and the standards of any state agencies concerning occupational health and safety."

Analysis: The Occupational Safety & Health Act does not cover State and local institution employees, unless that State operates under an OSHA approved State plan. Twenty-three States have approved plans. It is not felt that specific grievance procedures should be detailed other than existing OSHA procedures and Section IV-D-1-c of the proposed Guidelines.

Recommendation: No changes at this time.

Membership of RAC

Issue: Witnesses 2, 8, 10, 13, and 20 recommend broader public participation on the RAC. Witness 2 recommends that "at least one-third of the RAC be composed of individuals who are not engaged in biomedical research and who can reasonably be expected to represent the interests of the general public. Such should include representatives of labor, public interest groups and elected or appointed public health groups. A subcommittee . . . or a majority of RAC members who represent the interests of the general public should be given authority to make recommendations to the Director of NIH concerning exceptions to prohibited experiments and exemptions from the guidelines." Witness 8 says, "additional public members need to be appointed." Witness 10 urges . . . "A greater portion of representatives of the public interest and individuals with knowledge of public policy . . ." Witness 13 says, "EPA, OSHA, FDA, and CEQ should have full voting membership. Majority of open spaces on the RAC should be filled with individuals nominated directly by the public. . . ."

Analysis: These issues are discussed intensively by the Director in his Decision (p. 33067, Col 3 and p. 33068, Col 1). There the following points are made:

- The RAC responsibility has been primarily a scientific and technical one with recommendations for revisions of the Guidelines reviewed by the Director's Advisory Committee, a public advisory group.
- In order to ensure fairness and sensitivity to public commentators, solicitations or nominations for openings on the RAC will be in accord with the recommendations of the NIH Grants Peer Review Study Team.

. . .

- By means of the nomination process, the Director will be able to consider carefully a wide spectrum of nominations and assure appropriate representation suited to the needs of the committee.
- Several Federal agencies have liaison representatives who come regularly to the RAC, and the Federal Interagency Committee is kept fully informed of the RAC's activities.

Recommendation: DHEW Committee to review these issues as *part of the procedural review of the Guidelines* [italics mine].

IBC Membership

Issue: Several witnesses (1, 2, 4, 8, 10, and 13) would mandate categories of membership on the IBCs, each giving a somewhat different plan:

- Mandate a local health official and a nondoctoral person from a laboratory technical staff.
- One third of the members "represent scientific disciplines related to risk assessment."
- "Community leaders who are in touch with grassroot attitudes."
- Environmental groups which have been diligent in public interest protection of the environment.
- According to demographic variables.
- Publicized meetings and open to the public.

Analysis: The Director's Decision (page 33065) urges that IBC meetings be open and announced. (Also mandated in section IV-A-2-f of the proposed revised guidelines) . . . NIH position is that IBCs are not subject to FOIA or FACA, because they are advisory to the institution, not NIH.

Recommendation: Except for FOIA and FACA issues, make part of the procedural review of the Guidelines.†

NIH as Police

Issue: Witness 10 suggests that the responsibility for inspection and oversight be removed from the IBCs to NIH, which would monitor and inspect all recombinant DNA research.

Analysis: As noted in the Decision Document (pages 33064-5), it is not considered feasible to have the Guidelines standards enforced externally, including Federal (NIH) surveillance, As cited on page 33064 of the Decision Document, this view is endorsed by a report of the House Committee on Interstate and Foreign Commerce.

Recommendation: DHEW Committee to consider as part of procedure review of the Guidelines.†

† The chairman of the DHEW Task Force intends to review these issues as a supplementary exercise to spell out in greater and more precise detail the duties and responsibilities of NIH and the Director.

Appendix 9.2
Comments in Letters

Excerpts from letters are coded by page number in the collection in *Recombinant DNA Research* (*RDR*), vol. 4, Appendix A, p. A1–A451.

Richard Atkinson (Director NSF) (A394)

> . . . We believe that proper separation has not been made between the NIH responsibility for the development of the guidelines and the designation of certain aspects of implementation and interpretation. . . . By removing the flexibility in the current guidelines that allows other agencies to use discretionary powers, the proposed revised guidelines in effect subject the research of other agencies to the additional requirement for NIH approval procedures. This approaches establishing a regulatory role for NIH. Just as the proposed revised guidelines recognize that the responsibilities of the IBCs must be enhanced to make them partners in implementing the guidelines, so too, should the discretionary powers of other agencies be recognized. . . .
>
> We believe each agency can exercise its independent responsibility without diminishing the central role of NIH in coordinating and communicating all decisions. Specifically, the RAC should be recognized as the *de facto* national advisory body for all agencies. Accordingly, RAC recommendations could be considered directly by each agency. You pointed out in your Decision Document that, in addition to recommendations from RAC, you may seek additional advice for NIH purposes. In the same vein, we believe that other agencies should have this option. No agency should unilaterally act on a RAC recommendation; but within 30 days after each RAC meeting, representatives from concerned agencies should meet to review the basis for accepting, rejecting or postponing action on RAC recommendations. . . .
>
> . . . NSF accepts Sections I, II, and III of the current guidelines, but has developed its own Section IV, substituting NSF roles and responsibilities for those of NIH. . . . Requiring institutions to register with NIH all

projects approved by the IBC should be limited to NIH supported research. For Federally supported research supported by agencies other than NIH, each agency should maintain its own registry, which can routinely be added to the NIH registry.

David Baltimore (MIT) (A7)

. . . Most important . . . is the increased responsibility given to the institutional committees.

James C. Beardon (U. Hawaii) (A50)

My first recommendation for remedying . . . this situation [bureaucratic regulation and red tape] is to return to the original ideas of the conferees at Asilomar and elsewhere, and delegate the responsibility of all enforcement of the . . . guidelines to the local IBCs, what Paul Berg and most other scientists at Asilomar had in mind.

Paul Berg (Stanford University) (A212)

. . . I support enthusiastically the change that permits local IBCs to approve proposals and the investigator to initiate the experiments after obtaining such approval. The previous procedures were cumbersome, time-consuming and unnecessary. In my view the greatest risk to non-compliance with the Guidelines is for investigators to be confronted with unresponsive, time-wasting bureaucracy. . . .

Kenneth Berns (U. Florida) (A53)

. . . Local committees which are acceptable to NIH will be qualified in every way to make these decisions. . . .

Eula Bingham (Director, OSHA) (A366)

. . . I feel that too much responsibility for interpretation and enforcement has been placed at the local level. I anticipate many problems with this structure. . . .

Marcia Cleveland and Lovis Slesin (Natural Resources Defense Council) (A200)

. . . The same philosophy which has prompted the proposed reduction in containment levels is also the basis of the proposed increased reliance on IBCs and increased discretion for them and the Director of NIH. According to the Director, the lack of information about risks warrants looser administration of the guidelines. . . .

. . . Instead of "traditional regulation," NIH proposes a form of enforcement which institutionalizes conflict of interest and provides even less accountability for the Director of NIH and the IBCs than the previous guidelines.

Roy Curtiss III (A308)

. . . NIH has considered all the evidence, opinions . . . in drafting a well organized, comprehensive and realistic set of Guidelines. . . . Two of my significant criticisms deal with . . . microorganisms that exchange genetic information, and containment for some experiments involving cloning of eukaryotic viral DNA sequences in EK1. . . .

Rene Dubos (Rockefeller University) (A448)

One of my colleagues . . . showed me yesterday his copy of "Comments of the Natural Resources Defense Council on the Proposed Revision of the . . . Guidelines." . . . I had no idea that NRDC was involved in the Recombinant DNA problem, for which it has no competence. I find the text of the "Comments" misleading because it is grossly unbalanced and displays a complete lack of understanding of laboratory practices. . . . Failure on the part of NRDC to communicate with me, a member of its Board of Trustees familiar with the technical aspects of recombinant DNA research reveals either an . . . irresponsible lack of familiarity with the literature . . . or intellectual dishonesty. . . . For these reasons, I am resigning today my membership in the Board of Trustees of NRDC.

Barbara Mazur (University of Chicago) (A39)

. . . I am especially pleased that more responsibility has been delegated to the IBCs.

Cornelius Pettinga (Eli Lilly) (A272)

. . . Individual laboratories should have the option of securing one or more non-affiliated (IBC) members, but the inclusion of such individuals should not be mandated. . . .

Herbert Semmel and Mark Kleiman (Consumer Coalition for Health) (A377)

. . . The proposed revisions concentrate in the IBC a great deal of power and great responsibility to make both technical and value judgments affecting all members of the community. Nevertheless, the Guidelines substitute meaningless tokenism for meaningful public composition and vague lip service for substantial guarantees of open proceedings. . . . The IBC should have a minimum of seven members. . . .

David Saxon (President, Systemwide Administration, University of California) (A440)

. . . I fully agree that the principal responsibility for ensuring compliance should be borne by the institutions where the research is done. . . .

Francine Simring (Coalition for Responsible Genetic Research) (A285)

. . . Thus far, controls . . . have been based largely on principles of self-regulation and peer review. Experience during the past year, at two prestigious schools, has shown that irresponsible or thoughtless actions do occur under this form of control. Yet the response of the NIH has been to propose to *increase*, rather than to balance the responsibilities of local institutions by transferring primary responsibility for control to these institutions.

Adlai Stevenson (U.S. Senate) (A414 to A427)

[Accompanied by a letter sent to Secretary Califano on October 13. It acknowledged a recent letter from the secretary and called it an "unconscionably late" reply to the six senators who since June had unsuccessfully sought the secretary's terms for using Section 361. The senator now gently brandishes the threat of new legislation next year, if for no other reason than to sweep all or as much of recombinant research he can under the same set of rules.]

The Senator has some very specific views about the Guideline revisions. . . . the proposed revised guidelines represent . . . a . . . considerable advance in more informed assessment of risk comprehensiveness and specificity over the present guidelines that will permit further progress in recombinant DNA research under appropriate safeguards. . . . I commend Dr. Fredrickson, his colleagues and advisors for their efforts. . . . On the other hand, there are ambiguities and weaknesses, particularly with respect to administrative procedures. . . . [An eight-page supplement of suggested changes was included, with attention especially to the following.]

- Applicability. Guidelines should apply to all recombinant DNA research where NIH provides *any* support to the institution.
- Exceptions. Proposed Guidelines permit exceptions to prohibited experiments or the 10 liter rule. NIH should rethink conditions for granting exceptions.
- Clarify the standards and procedures for future changes in the Guidelines. Precise statement of who does what.
- Compliance.
- Clarify expectations of IBCs with periodic monitoring by NIH.
- Representation on the IBCs and the RAC should be broader. . . . The number of lay members should be increased in view of the social implications of the research.
- Training.

Scott Thatcher (Science for the People) (A370)

. . . All IBC meetings should be open to the public. . . . Representatives of non-doctoral, technical staff should be required as members. . . . Local community representatives should have their nominations . . . ratified by local community governments. . . .

Lewis Thomas (Sloan-Kettering Institute) (A410)

. . . I have been told that the Friends of the Earth are pressing their case against modification of the NIH Guidelines. I am in flat disagreement on straightforward scientific grounds, with the rigid position taken by their organization. . . . I resigned from [their Advisory] Council on December 14, 1977 because of the civil action suit. . . .

David Unger (Department of Agriculture) (A391)

. . . Agriculture is pleased to have contributed to revisions . . . through the USDA-NSF workshop on Recombinant DNA Research in Relation to Plant and Animal Sciences and . . . workshop on Risk Assessment of Agricultural Pathogens . . . Review of research by a local Institutional Safety Committee is endorsed. . . . USDA is assisting colleges and universities in establishing IBCs.

Murray Welch (Upjohn Co.) (A257)

. . . The Upjohn Company has voluntarily established an Institutional Biosafety Committee with 12 members, four of whom are from the local community and are not employed by the company. . . . If the minutes of the meeting are made public . . . the data and information in the minutes would be available to competitors and premature disclosure might prevent the company from . . . adequate protection. . . .

APPENDIX 10.1
Members of RAC

NIH Recombinant DNA Advisory Committee (RAC), December 29, 1978

CURRENT MEMBERS

Jane Setlow, Ph.D., *Chair;* Biologist, Brookhaven National Laboratory, Upton, N.Y.

Allan Campbell, Ph.D., Professor, Department of Biology, Stanford University, Stanford, Calif.

Peter Day, Ph.D., Chief, Division of Genetics, Connecticut Agricultural Experiment Section, New Haven

Susan Gottesman, Ph.D., Senior Investigator, Laboratory of Molecular Biology, National Cancer Institute, Bethesda, Md.

Richard Hornick, M.D., Director, Division of Infectious Diseases, University of Maryland School of Medicine, Baltimore

Elizabeth Kutter, Ph.D., Member of the Faculty in Biophysics, Evergreen State College, Olympia, Wash.

Emmette Redford, Ph.D., LL.D., Ashbel Smith Professor of Government and Public Affairs, University of Texas, Austin

Wallace Rowe, M.D., Chief, Laboratory of Viral Diseases, National Institute of Allergy and Infectious Diseases, Bethesda, Md.

John Spizizen, Ph.D., Chairman, Department of Microbiology, Scripps Clinic and Research Foundation, La Jolla, Calif.

LeRoy Walters, Ph.D., Director, Center for Bioethics, Kennedy Institute, Georgetown University, Washington, D.C.

Milton Zaitlin, Ph.D., Professor, Department of Plant Pathology, Cornell University, Ithaca, N.Y.

NEW MEMBERS

A. Karim Ahmed, Ph.D., Senior Staff Scientist, Natural Resources Defense Council, New York, N.Y.

David Baltimore, Ph.D., American Cancer Society Professor of Biology, Massachusetts Institute of Technology, Cambridge, Mass.

Francis Broadbent, Ph.D., Professor of Soil Microbiology, Department of Land, Air and Water Resources, University of California, Davis

Zelma Cason, Chief of Cytotechnology, Department of Cytology, University of Mississippi Medical Center, Jackson

Richard Goldstein, Ph.D., Assistant Professor of Microbiology and Molecular Genetics, Harvard Medical School, Boston, Mass.

Patricia King, J.D., Professor of Law, Georgetown University Law Center, Washington, D.C.

Sheldon Krimsky, Ph.D., Acting Director, Program in Urban Social and Environmental Policy, Tufts University, Medford, Mass.

Richard Novick, M.D., Chairman of Plasmid Biology, Public Health Research Institute, New York, N.Y.

David Parkinson, B.M., B.Ch., Associate Professor of Occupational Health, University of Pittsburgh, Pittsburgh, Pa.

Ramon Piñon, Ph.D., Assistant Professor of Biology, University of California, San Diego

Samuel Proctor, Ph.D., Professor of Education, Rutgers University, New Brunswick, N.J.

Ray Thornton, J.D., retiring member, U.S. House of Representatives (D-Ark)

Luther Williams, Ph.D., Assistant Provost, Department of Biological Sciences, Purdue University, West Layfayette, Ind.

Frank Young, M.D., Ph.D., Professor and Chairman, Department of Microbiology, University of Rochester, Rochester, N.Y.

Index

C

D

F

I

J

K

L

M

N

P

Q

R

Y

Z